When Money Grew on Trees

A. B. Hammond and the Age of the Timber Baron

GREG GORDON

University of Oklahoma Press : Norman

Also by Greg Gordon

Landscape of Desire: Identity and Nature in Utah's Canyon Country (Logan, Utah, 2003)

This book is published with the generous assistance of the Frances K. and Charles D. Field Foundation

Library of Congress Cataloging-in-Publication Data

Gordon, Greg, 1963–
When money grew on trees : A.B. Hammond and the age of the timber baron / Greg Gordon.

pages cm
Includes bibliographical references and index.

ISBN 978-0-8061-4447-4 (hardcover) ISBN 978-0-8061-9200-0 (paper)
1. Hammond, Andrew B. (Andrew Benoni), 1848–1934. 2. Lumber trade—United States—History.
3. Lumbermen—United States—Biography. 4. Businessmen—United States—Biography. I. Title.

HD9760.H36G67 2014
338.1'7498092—dc23
[B]

2013039703

The paper in this book meets the guidelines for permanence and durability of the Committee on Production Guidelines for Book Longevity of the Council on Library Resources, Inc. ∞

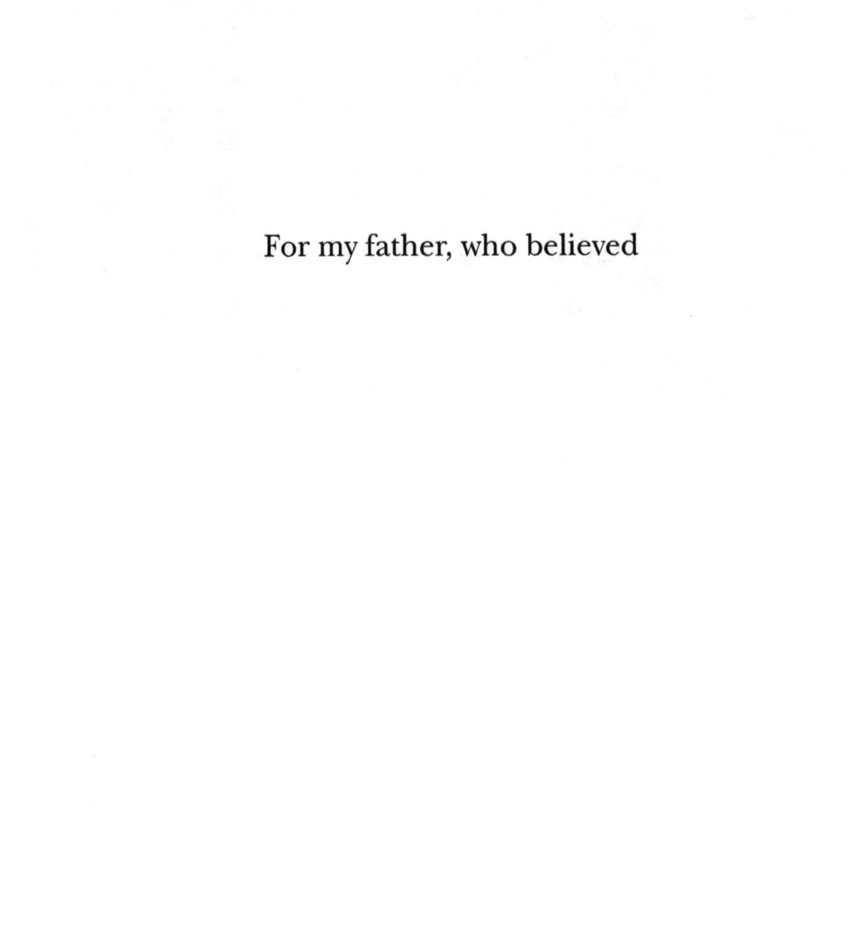

For my father, who believed

Contents

Illustrations

MAPS

FIGURES

Acknowledgments

Only nominally is writing a solitary endeavor. So many people have shaped this book—some in profound ways, in everything from providing obscure source material to support, editing, and intellectual development—that it feels like a collective enterprise. Only the mistakes are mine alone.

This book began as my Ph.D. dissertation at the University of Montana at the suggestion of my advisor, Dan Flores, who steered the project through uncharted waters and provided unwavering endorsement. I am also indebted to the rest of my committee, all of whom pushed, pulled, and prodded the manuscript in the direction it needed to go. Forest ecologist Paul Alaback ensured scientific accuracy, and sociologist Jill Belsky furnished conceptual development on land tenure and political ecology. Historians Richard Drake, David Emmons, and Jeff Wiltse aided in historical analysis and methodology. Although they might not agree with the conclusions I've drawn, this work is, in large part, a product of the influence these six individuals have had on my thinking; I can't thank them enough.

I also owe an immeasurable debt to Dale Johnson, former archivist at the Mansfield Library, University of Montana. Indeed, without him, I would still be scrolling through microfilm. By generously supplying access to his personal files, Johnson greatly expedited my research, and his local anecdotes provided insight into Hammond's character. Future researchers will certainly appreciate Johnson's many years of dedication in organizing and cataloging the UM archives. Archivists Donna McCrea, Mark Fritch, and Teresa Hamann also provided eager assistance and made the Mansfield Library the most pleasant place to conduct historical research.

Likewise, Joan Berman and Edith Butler at Humboldt State University went out of their way to help me locate materials and make local contacts. I am grateful to Brian Shovers and the staff the Montana Historical Society for their assistance. I also thank the Oregon Historical Society, the Huntington Library, the Bancroft Library, the Provincial Archives of New Brunswick, the Missoula Public Library, the Astoria Public Library, the Clatsop County Historical Society, the Tillamook Historical Museum, the Fort Missoula Historical Museum, the Van Buren (Maine) District Library, the Grand Falls (New Brunswick) Museum, the Humboldt County Historical Society, and the Forest History Society. I especially appreciate the willingness of the Timber Heritage Association to permit access to its recently uncovered files of the Little River Redwood Company.

I was most fortunate to inherit the collections of three previous researchers of the Hammond Lumber Company courtesy of Jack Blanchard. Blanchard's vast assortment of photographs and data included work by Dan Strite, an HLC employee who served as the company's unofficial historian, and Bill Stoddard, who intended to write a company history and produced a rough manuscript before he died. All of their work is now complied into a single collection, which is in my possession.

I would like to thank Daniel Cornford, Michael Kazin, Lowell Mengel II, and William Parenteau for helping to clarify many of my thoughts and local historians Guy Dubay of Van Buren, Maine, and Susie Van Kirk of Eureka, California, for providing valuable specifics.

I am especially appreciative of my fellow graduate students at the University of Montana who read (and reread) many portions. Offering suggestions, critiques, and insight, they substantially improved the manuscript. Many thanks to the Rattlesnake Writing Group: Happy Avery, Shawn Bailey, David Brooks, Jeff Gailus, Monika Bilka, Ros Lapier, John Robinson, Ian Stacey, and Rachel Toor. David's "big picture" observations and Happy's editorial brilliance proved a powerful one-two punch. Jay Dew at the University of Oklahoma Press took over from there and provided unhesitating editorial leadership, guiding the manuscript to finished product.

William G. Robbins and Thomas Cox brought over a century of collective wisdom and insights on the West Coast timber industry and provided a through reading and peer review of the manuscript. I could not have asked for more knowledgeable or diligent readers. Any omissions or errors are a result of my own obstinacy in failing to heed their valuable suggestions.

From beginning to end, the entire project would not have been possible without the generous support of Charles Field, Jr., and the Frances K. and Charles D. Field Foundation. On a personal level, I am deeply grateful for the unflagging encouragement and moral support of Al Gordon and Happy Avery.

Note on Language

As much as possible I have tried to preserve the flavor of the time period by using terms that were in common parlance in the late nineteenth and early twentieth centuries. While some terms, such as "crimps" and "swampers," might be unfamiliar to modern readers, the context should provide the meaning; others, like "lumberjacks," have fallen out of favor in recent years. I have consciously maintained the use of gendered terms, such as "lumbermen" and "entrymen." Although women did file entry claims under the Homestead and Timber and Stone Acts, they were still considered "entrymen." Also, in keeping with the editorial convention of the University of Oklahoma Press, I use the term "Indian" rather than "Native American." My intent in selecting the terms I use is not to perpetuate racist or sexist language but to preserve historical accuracy and context.

WHEN MONEY GREW ON TREES

Introduction

> The social structure and the state are continually evolving out of the life-process of definite individuals.
>
> —Karl Marx, *The German Ideology*

In 1934, when eighty-five-year-old timber baron A. B. Hammond died, a story quickly circulated among the lumbermen of San Francisco. According to legend, Hammond—impeccably dressed, with a shock of white hair and neatly trimmed goatee, just as he was in life—sat up in his coffin at the approach of the pallbearers. "Six pallbearers?" he thundered. "Fire two, and cut the wages of the others by ten percent."[1]

The anecdote reflected Hammond's hardnosed business reputation and the attitude of his employees and fellow lumbermen toward him. Notoriously anti-labor, Hammond vigorously opposed collective bargaining, minimum wages, and the eight-hour day. He waged a thirty-year war against the unions of lumber workers, sailors, and longshoremen, crushing every strike against his enterprises. For nearly forty years, Hammond attempted to consolidate the redwood lumber industry by buying up smaller mills, acquiring timberlands through nefarious means, browbeating his competitors into forming associations to control prices, and then underselling everyone. Finally, in 1931, taking advantage of the Depression, he negotiated the largest merger of his career, cementing the Hammond Lumber Company as the nation's leading redwood lumber company. With Hammond's death, however, the age of the timber baron came to an end.

Born in the timber colony of New Brunswick in 1848, Andrew Benoni Hammond was the scion of generations of lumbermen. After an inauspicious start as a teenage lumberjack in Maine, Hammond moved to

Montana, where he combined his intelligence and ruthless determination with Gilded Age opportunity to reach the forefront of the region's nascent timber industry. Not content to be the economic and political powerhouse of western Montana, Hammond sought increased wealth and opportunity and relocated to California. Stitching together timberlands, sawmills, railroads, shipping lines, and retail yards, Hammond eventually amassed an empire of wood stretching from the Puget Sound to Arizona. With sixty-five lumberyards in California alone—including the world's largest, in Los Angeles—the Hammond Lumber Company name was ubiquitous across the Golden State for the first half of the twentieth century.

The history of the American West is, in large part, a story of the unfolding conflict between different groups over access to and control over natural resources. Many Indian tribes battled each other for access to bison while fighting with white settlers and the federal government over territory. Settlers, for their part, clashed with big business over minerals and timber. And perhaps most prominent during the last half of the nineteenth century was the dispute between such companies and the federal government over the nation's forests. Hammond's story illustrates how this struggle over natural resources gave rise to the two most pervasive forces in modern American life: the federal government and the modern corporation. But rather than a triumph of one over the other, the conflict resulted in a huge increase in the power and ubiquity of both and defined the increasingly close relationship between them.

Hammond also provides a rare glimpse into the motivations and character of a turn-of-the-century timber baron. These lumbermen formed an extraordinarily homogenous group—white Anglo-Saxon Protestants who were fiercely individualistic, usually lacking formal education, and proudly regarded themselves as "self-made men." As historian Norman Clark wrote, "They were men to whom privacy was a proof of character. They regarded their various business ventures and industrial operations as ultimate expressions of their very private lives."[2] As such, they either kept no records or methodically had them destroyed. A. B. Hammond was no exception. Nonetheless, letters, newspaper articles, recently unearthed company records, and lumber industry scholarship allow one to sketch a portrait of this archetypal lumberman.

To develop a thorough understanding of the past, historian Jacob Burckhardt implored us to unearth the zeitgeist, or spirit, of the times. He argued that this is properly revealed in certain individuals who best mirror their age. In seeking to understand the U.S. industrial age (approximately 1877–1920), historians have traditionally looked toward the top echelon of American businessmen, such as John D. Rockefeller, Jay Gould, and J. P. Morgan, largely ignoring scores of second-tier and regional capitalists. The era of laissez-faire capitalism was not the work of only a few individuals but, rather, of hundreds of men of a type forged in the Protestant tradition and who embodied principles of thrift and acquisition—men such as A. B. Hammond.

While the Rockefellers, Goulds and Morgans provided the capital, it was the regional industrialists who were primarily responsible for the transformation of the American West. In Colorado, it was Otto Mears; in Oregon, Simon Benson; in Montana, it was Marcus Daly and A. B. Hammond. In just a few decades these men dammed the rivers, logged the forests, and leveled the mountains. They also built cities, towns, and a vast transportation network of steamships and railroads to export natural resources and import manufactured goods. In doing so, they established the modern American nation. Westerners were not colonists under the yoke of eastern capital; instead, they created the institutions, conditions, and opportunities that converted the nation's natural resources into private wealth.

Beginning with the fur trade, regional entrepreneurs established the linkage between the resource-rich West and the financial East. Mining and railroads cemented the connection between the two regions in what appeared to be a colonial relationship.[3] Capital, however, being extremely liquid, flows in many directions and is hardly confined by geography. As Hammond would demonstrate, home-grown capital from the Montana hinterland could be used to exploit resources in a more developed region. While the forests and people of western Montana and the opportunities therein had made him a millionaire, he longed for more. In 1894 Hammond began selling off his Montana investments and moved to the West Coast. Nevertheless, Hammond's Missoula Mercantile Company—and, indirectly, the citizens of western Montana—continued to provide much-needed capital and credit for Hammond's California enterprises.

There were, of course, hundreds of men who captured the zeitgeist of the industrialization of the American West, as numerous studies on western railroad builders, mining magnates, and cattle kings attest. Nearly all of these works, however, focus on the relatively narrow time frame of the frontier in which laws, institutions, and cultural restraints were still in flux. A. B. Hammond, because of his long life, spanned the era from the fur trade to the Great Depression, by which time the American West was largely industrialized. Furthermore, Hammond engaged in nearly every possible western industry, including banking, railroads, shipping lines, fishing and canning, livestock, mining, real estate, merchandising, land speculation, and, most of all, lumbering.

Although he began his career as a pioneer entrepreneur, Hammond, unlike many of his associates, successfully negotiated the transition to corporate businessman. As a product of the frontier, however, he retained many characteristics of a nineteenth-century proprietary capitalist. Seeking control over his workforce, he built exclusive company towns for mill workers and was one of the first to establish year-round, company-owned logging camps. His opposition to organized labor led him to employ his own stevedores, longshoremen, sailing and railroad crews, and retail merchants. Hammond also maintained a firm control over his board of directors, which he appointed. In a modern corporation, a board of directors separates owner-stockholders from managerial decisions; instead, Hammond held majority stock in each of his enterprises and acted as president, CEO, and general manager of each of his companies, approving every major purchase and sale. Yet, from the extraction of raw materials to the retail sales of finished products, the Hammond Lumber Company illustrates how vertical integration gave rise to the modern corporation.

The first step in vertical integration was to ensure a continuous supply of raw materials by gaining access to forests of the public domain. In following Hammond, we can trace the trajectory of how this played out in the lumber industry as a whole. In Montana, Hammond, like many frontier lumbermen, simply took the trees he wanted, regardless of ownership. He even hired agents to patrol the forests against individual woodcutters, assuming de facto ownership over public resources. Before long, Grover Cleveland's administration singled out Hammond for his egregious and unrepentant depredations of federal timber. Timber theft, though illegal, was so prevalent that was it was considered a customary practice dating back to colonial times; most lumbermen, when caught, simply paid the fine and went about their business of felling more trees. Not so with A. B. Hammond. Having imbibed the backwoods mentality of his native New Brunswick—where his father and grandfather nearly provoked a war between the United States and Great Britain over timber—Hammond believed he was in the right when he took federal timber to assist in the economic development of the American West.

When Hammond shifted his operations to Oregon, he, along with scores of other lumbermen, engaged in land fraud to acquire timberlands and drew the attention of the first Roosevelt administration. His role in timber poaching and the Oregon land frauds helped catalyze the creation of the U.S. Forest Service by supplying Progressive Era reformers with a primary reason to end the privatization of forests. The conflict over natural resources framed Hammond's entire life—he did indeed mirror the spirit of the age.[4]

By the 1890s, the nation's rapid industrialization increased demand for lumber, prompting the modernization of sawmills. Such upgrades required capital, which in turn demanded consistent and reliable production. To achieve dependable production, lumbermen believed they needed to secure control over raw materials to ensure future supply. Once they acquired vast acreages of the public domain, often illegally, debt required rapid liquidation of those timberlands, while new technologies made it possible. The resulting environmental consequences dramatically reshaped the landscape, with repercussions that continue to the present.

A. B. Hammond embodied the mythological promise of the West: independence, prosperity, and optimism. Both his contemporary and historical reputations, however, were generally negative, largely because of his anti-labor practices and his role in timber poaching on federal lands. Was Hammond a "robber baron" who stifled competition, exploited workers, monopolized industry, and appropriated public resources for his own wealth? Or was he one of the "crucial catalysts in the most profound transformation of civilization" who built railroads, developed the

West Coast timber industry, helped rebuild San Francisco after the 1906 earthquake, and aided in the growth of Southern California?[5]

Whether men like Hammond were robber barons or captains of industry has spawned endless debate from the dawn of the Gilded Age to the present. Historians generally agree that industrialization was one of the most profound changes in American history, but they dispute exactly *how* it occurred and to what end. Business historians focus on changes in organizational structure stimulated by technological advances that increased production. Others point to legal and social institutions of power. An analysis of A. B. Hammond's rise to prominence suggests a dynamic interplay between these forces, illustrating that industrialization proceeded through alternative pathways of historical contingency—individual actions and past events shaped the course of history in unpredictable ways. While individuals operated within the structures of capitalism, they also molded the institutions that guided this process, tailoring it to fit their own needs, desires, and specific circumstances. In doing so, they also shaped the world around them.

Two highly contrasting views inform our perception of America's Industrial Age. The first contends that America's industrial and economic might depended on creative capitalists freely pursuing individual wealth. The opposing position insists that the capitalist free-market system engendered a great disparity of wealth and resulted in social and environmental costs while accruing benefits to the few. This conflict of values not only informs our interpretation of history but also shapes our current political and economic debate over issues such as globalization and free trade. This yin/yang aspect of the Industrial Revolution underscores a fundamental divide in American culture and politics, one that had its genesis in the Industrial Age with the transformation from an agrarian to an industrial economy.

In the Pacific Northwest—stretching from western Montana to Northern California—the lumber industry was the single largest source of environmental change and, until recently, provided the region's economic, and consequently political and social, foundation. Just as it had in the East, Midwest, and South, the lumber industry tied local economies to a boom-and-bust cycle with long-term social and environmental consequences. The owners of Hammond's first sawmill, at Bonner,

Montana, announced in 2008 that it would close permanently, after 120 years of more or less continuous operations, due to diminishing supply and plummeting lumber prices. This action signaled the symbolic end to the timber era.[6] In an otherwise prosperous region, many timber-dependent communities of the Pacific Northwest are now among the most impoverished in the United States. Throughout Oregon, Montana, and California, the mills of the former Hammond Lumber Company are shuttered. Their roofs are collapsing, machinery is rusting, and weeds are reclaiming the tarmac. The houses of the company towns have fallen into disrepair and most of their residents are retired, are unemployed, or commute long distances to fading timber jobs.

The forests have fared no better. In Montana, a century of logging the pine has led to an invasion of the less-fire-resistant Douglas fir, contributing to an increase in catastrophic forest fires. In Oregon, the logged-over lands of the Hammond Lumber Company went up in flames in the infamous Tillamook fires of 1933–51, while clear-cutting has so restructured West Coast forests that they verge on ecological collapse. In California, a century of heavy logging has all but eliminated the coastal redwoods, and the cutover lands of Hammond's vast redwood empire are now tree farms featuring fast-growing Douglas fir. In all three areas, the activities of Hammond and other timber barons transformed the forests from functioning natural ecosystems into industrial landscapes. Was it worth it?

CHAPTER 1

Standing at the Crossroads

For as long as any of them could remember, Indian tribes living west of the northern Rocky Mountains had traveled the "road to buffalo" by following the Clark Fork River to the Continental Divide and over to the vast hunting grounds of the upper Missouri River. For an equally long time, Great Plains Indians had resented this intrusion, and the battle over resources seesawed back and forth. At the eastern end of a broad valley, in what would become western Montana, twin promontories pinched the Clark Fork into a narrow canyon. On more than one occasion, the Blackfeet, who inhabited the buffalo country, ambushed the mountain-dwelling Salish at this point to prevent their passage on the road to buffalo. The bones and skulls littering the site caused French trappers to call the canyon *La Port d'Enfer* (the gate of hell). English-speaking fur traders then applied the name Hell's Gate Ronde to the wide valley that emerged from the canyon's mouth.[1]

Twelve thousand years earlier, Hell's Gate Ronde lay beneath two thousand feet of water as the immense Glacial Lake Missoula spread two hundred miles to the west, making it larger than Lake Erie and Ontario combined. When the ice dam holding back the water broke, the lake emptied in forty-eight hours, and the ensuing flood, at ten times the combined flow of all the world's rivers, scoured the Columbia River Basin. The lake left behind a wide, flat, fertile valley surrounded by rolling hills. The hills yielded to forested mountains and ridges, behind which rose sharp, treeless peaks. Entering through Hellgate Canyon, the Clark Fork River meandered across the valley floor to merge with the Bitterroot River flowing from the south. The Bitterroot, in turn, followed a long, narrow valley nestled between two parallel mountain ranges. Sheltered from the snow and cold, these low-elevation bottomlands provided early spring grass for horses and abundant bitterroots, a dietary staple, and formed an ideal wintering ground for the Salish.[2]

White travelers also found the valley amenable. On their westward journey in 1805, Lewis and Clark rested their men and horses near the banks of the Bitterroot River before tackling the perilous crossing of the Bitterroot Mountains. The following year, they greatly simplified their return by following the road to the buffalo back to the Missouri River country. Half a century later Lt. John Mullan attempted to link the Missouri and Columbia watersheds by building a military road from Walla Walla, Washington, to Fort Benton on the Missouri River. For much of the route, Mullan simply followed the incised travois ruts of well-established Indian roads leading from the Columbia Plateau, along the Clark Fork, through Hell's Gate, and on to the Great Plains. By the 1860s traders recognized the geographical and economic advantages of establishing posts on these Indian thoroughfares, and the town of Missoula, Montana, slowly began to grow at the crossroads of the Mullan Road and the Salish trail up the Bitterroot Valley.[3]

One day in mid-October 1891, the trajectories of two men's lives intersected at that crossroads, which had become Higgins Avenue and Front Street, the heart of a now booming city. The first man, Andrew B. Hammond, was dressed in well-fitted suit with wide lapels, a four-in-hand tie, and a stylish bowler. The chill air exacerbated his rheumatism, and he walked with a slight limp. At forty-two, he was a thin, handsome man with sharp features and a close-cropped beard that was just beginning to grey. As a frontier capitalist, Hammond's star was ascending; in another ten years he would be one of the premier lumbermen on the West Coast. Riding the wave of American industrialization, Hammond accumulated undreamed-of wealth and power.

Industrialization, however, did not result in universal prosperity. In contrast to Hammond, Charlo, hereditary chief of the Flathead Salish, was sliding deeper into poverty and alienation. Nevertheless, on this day, proudly astride his horse, Charlo sported buckskin leggings, a beaded sash over one shoulder with bracelets and anklets to match, a wool jacket over a red vest, and his signature plug hat. In contrast to the wealthy

Hammond, Chief Charlo was destitute, despondent over the loss of territory, and desperately trying to keep his people from starvation.

Not only did Charlo's and Hammond's individual paths cross on this day—so did the cultures, economies, and eras that each inhabited. The Missoula and Bitterroot Valleys were a microcosm of the profound and cataclysmic changes that occurred throughout western North America at the close of the nineteenth century. In a single decade the arrival of the railroad, the conversion of forests into lumber, the rise of agricultural production, and pervasive mining activity transformed the landscape and marginalized the people whose ancestors inhabited it for thousands of years. In most of North America, this transformation required decades, even centuries. Montana's geographic and economic isolation until late in the nineteenth century compressed this into just a few years. The transcontinental railroad, for example, did not arrive until 1883, the same year as the last bison hunt. Emblematic of this condensed process, the events of October 16, 1891, provide a vivid snapshot. Differing only in particulars, the stories of Hammond and Charlo played out throughout the American West.

Standing at the intersection of Higgins and Front, Hammond could survey his domain, which included the four largest buildings in town. Each anchored a corner of the bustling intersection and formed the nucleus of his empire. Hammond, no doubt, felt pride and a sense of accomplishment. While the unique geography of Hell's Gate Ronde—a low-elevation valley bisected by two rivers and blessed with rich topsoil and timber—was the major factor in the establishment of Missoula, the city's economic growth as a mercantile, railroad, lumbering center, and regional hub was largely due to the business activities of A. B. Hammond.

When Hammond arrived in 1870, Missoula was just another frontier trading post consisting of a smattering of log cabins strung out along the muddy Mullan Road. Now, two decades later, it was a thriving city of five thousand and the focal point of the region's economic and political life. As the town's most successful businessman and one of Montana's wealthiest individuals, Hammond was responsible for much of the change. His political and economic influence extended throughout northern Idaho and western Montana and even to the nation's capital. Although

Hammond's name would eventually be purged from Missoula's history, in 1891 it was familiar in every corner of the valley, though rarely uttered with reverence. While many acknowledged his business acumen, his ruthless practices and brusque manner won him few friends. In the 1889 election for town council he received one vote.[4]

Securing the southeast corner of Higgins and Front, the First National Bank was the city's architectural masterpiece, and as bank president, Hammond had overseen its construction the previous year. Built in a classical Queen Anne style, rough-hewn granite framed the entryway of the imposing four-story edifice. Heavy wooden doors opened into a circular foyer that rose into a turret, ninety-six feet high, that towered above all else in town.[5]

Across the street, even newer and nearly as ornate, the Hammond Building also stood four stories high and had a castle-like rampart running the perimeter of the roofline. Minarets sprouted from each end of the rampart, while a turret rose from the corner. West on Front Street, past the Hammond Building, were the Mascot Theater, wood-frame saloons, gambling houses, and Chinese laundries; farther on were the euphemistically termed "female boarding houses."[6]

On the northwest corner sat Missoula's largest and most luxurious hotel, which boasted steam heat and electric lights powered by its own power plant in the basement. Part of Missoula's recent building boom, the Florence Hotel (named after Hammond's wife) had opened three years earlier. Even at three to four dollars per night, the Florence's two hundred rooms were nearly always full with land speculators, railroad executives, shoppers, merchants, masons, and investors from Helena, Butte, New York, and London.[7]

Although dwarfed by the other buildings in height, Hammond's Missoula Mercantile Company (MMC) spread over an entire block on the northeast corner. Scores of windows lined the recently added second story of the largest department store between St. Paul, Minnesota and Portland, Oregon. Shaded by red-and-white awnings over the wooden sidewalks, a passerby could examine the merchandise displayed behind the large, plate glass windows. Inside, shoppers could purchase clothing, hats, cloth, canned goods, fresh fruit and vegetables, furniture, carpets, saddles, drugs, farm implements, hardware, guns, saddles, shoes, and even carriages and wagons. While less impressive than Hammond's other buildings, the Merc, as it was known, served as the hub of the valley's activity. With branch stores across western Montana, the MMC dominated the region's commerce in both retail and wholesale markets. The grocery department alone accounted for one million dollars in annual sales. For the previous ten years, the company had been Missoula County's largest employer.[8]

From this one corner, Hammond's power and influence radiated into the region's five valleys. Branch stores dotted the Flathead Valley to the north and the Bitterroot Valley to the south. His timber operations extended far up the Blackfoot to the northeast, as well as east and west along the Clark Fork River. In addition to the bank, the Florence Hotel, the MMC, and real estate development, Hammond's massive sawmill on the Blackfoot River dominated the region's lumber industry. He also controlled the Missoula Street Railway and the power and water company. Little wonder that Harrison Spaulding, editor of the *Weekly Missoulian,* portrayed Hammond as the "Missoula Octopus, that is undertaking to reach its slimy arms over the county and strangle the life out of it."[9] Nevertheless, in large part due to Hammond's commercial activities, the population of Missoula County, including the Bitterroot Valley, exploded from 2,554 in 1880 to 14,427 ten years later, excluding Indians. And none felt this pressure more than those who were excluded from the census.[10]

Indians felt the settlement pressure most intently in the Bitterroot Valley, where Hammond had punched in a spur line in 1888 to connect with the Northern Pacific Railroad and integrate the valley into the national economy. Cattle, wool, wheat, ore, and lumber could now be shipped profitably to markets in the Midwest, and Hammond began developing the valley, buying, subdividing, and platting town sites. Not only did the railroad encourage more settlement, but it profoundly changed the character of the valley. Cattle and sheep replaced elk and deer. The clear, flowing river was now choked with sediment and sawdust from Hammond's logging and milling operations. Loggers and miners swarmed over the forests and mountains, felling trees and evacuating earth. Where hundreds of teepees once stood, houses, stores, and barns dotted the valley. The incessant chugging and rattling of the Bitterroot Valley

Railroad was never far from earshot, and telegraph poles punctuated the landscape. Only a short train ride from Missoula, farmers could now send their wheat to Hammond's mills near the city, and ore could be shipped from Hammond's Curlew copper mine. But the primary cargo was timber.

Blocked from the winter snow by the mountains and blessed with good soil and abundant water, the valley had become a paradise for settlers, supporting vegetable gardens, orchards, and wheat and dairy farms. One farm in the Bitterroot boasted fifty thousand fruit trees with fifty varieties of apples, as well as plums, cherries, strawberries, and raspberries. Nearly all this produce funneled into Missoula to feed the rapidly growing city and was shipped to Butte and other mining centers too high and cold to support agriculture. The settlers built houses, stables, and stockyards. They fenced, irrigated, and grazed thousands of sheep, all on lands that the U.S. government had granted to the Salish Indians in the Hell Gate Treaty of 1855.[11]

For twenty years Chief Charlo had clung to this treaty, even as white settlers pouring into the fertile valley made it increasingly anachronistic. Feeling constricted by Indian tenure, in 1871 settlers petitioned President Ulysses S. Grant to have the Salish removed to the Flathead Reservation in the Jocko Valley. Opposing the move, Charlo produced a copy of the original 1855 treaty—owned by his father, Victor, one of the signatories—as evidence of his claim that the Bitterroot Valley was his birthright. Nonetheless, Grant authorized the settlers' request, and the next year Congress appointed James Garfield to carry out the removal.[12]

After visiting both the Jocko and Bitterroot Valleys, Garfield submitted an agreement to Charlo, Arlee, and Adolph, the three Salish chiefs. Arlee and Adolph consented to the arrangement, and they moved their bands to the Jocko, sixty-five miles north, in exchange for $50,000. Charlo, however, repeatedly stated that he would kill himself rather than sign. Yet when Garfield published the contract, he put Charlo's name first as primary chief and signed it with an X. The future president would later confess the forgery, saying he thought he was acting in the Indians' best interest.[13]

Incensed at the government's treachery, Charlo became intransigent, repeatedly insisting that he never signed the document. But believing

Charlo's band would soon vacate, white immigrants flooded into the valley following publication of the agreement. In 1883 the secretary of interior opened unoccupied land in the Bitterroot to homesteading, but still Charlo refused to budge. While Charlo held out, his people suffered. "The Indians became listless, indolent, shiftless. They traded their grain and few furs for groceries, calicoes and blankets, but their dependent conditions and doubt as to the future was driving their young men to the saloons, demoralization and ruin," wrote General Henry Carrington, the aging Indian fighter who had been called up from retirement to oversee the ejection of Charlo's band. "Removal of the Indians from the Valley had become equally necessary for all its inhabitants," he concluded.[14]

By 1889, the lure of money for their individual allotments and the notion that they would be reunited with the rest of the tribe induced several of Charlo's people to agree to the removal. Charlo, however, reiterated his position that he had never signed Garfield's agreement, at which point Carrington, in a brilliant stoke of diplomacy, produced the original document, which indeed indicated a blank space next to Charlo's name. Publicly vindicated, Charlo finally acquiesced, but the material condition of his people also factored greatly in his decision to relocate. Carrington noted that "many of the Indians were aged or lame, and supplies from the [Flathead] Reservation were absolutely insufficient to prevent great suffering and extreme want." Charlo acknowledged that the young Salish men often sold their game for alcohol and "followed the words of bad white men and stole what they wanted to eat without working for it." Anticipating the move to the Jocko, Charlo's band neglected to plant crops, and they slipped into poverty so dismal that they were reduced to hanging around butcher shops begging for entrails and scraps. Faced with the dilemma of whether to wait until their lands were sold or leave before winter set in, Charlo set aside his pride and agreed to move to the Flathead Reservation, north of Missoula.[15]

Thus, on October 15, 1891, Charlo clutched an eagle feather in his right hand and bid farewell to his ancestral home with a large feast accompanied by singing, drumming, and dancing. Refusing to regard the relocation as a defeat, his followers saw it as cause for celebration since it would mean an end to starvation and hardship. Dressed in their finest, the Salish wore colorful blankets, feathers, necklaces, leggings, bone breastplates, and freshly painted faces. Early the next morning, they loaded all their belongings onto wagons and horses, rolled up their teepees, and lashed them to the long lodge poles the horses would drag. In high spirits, the Indians traveled down the valley past the orchards, the apple and plum trees turning crimson. "The great mass of horses, wagons and people melted into a column, more than a mile in length, but in breadth varying according to the eccentric activities of the hundreds of loose ponies that shared the exodus," wrote Carrington.[16]

Larch trees, tinged with gold, highlighted the green forests of ponderosa pine and Douglas fir that marched up the flanks of the Bitterroot Mountains. Higher up, the previous night's snow dusted the jagged peaks that for eighty miles formed a nearly impenetrable barrier on the west side of the valley. Along the riparian bottomlands, cottonwoods and willows dropped their yellow leaves into the river and the musky fall smell of wet bark permeated the air. Dried cattails, their seed heads bursting, lined the marsh. Scattered along the draws that meandered across the valley, chokecherries and sumac blazed orange and red. Wild plums and rose hips hung heavy over the stream banks. Yet for the elders, the Bitterroot Valley was scarcely recognizable from bygone days.

As the Salish entered the broad Missoula Valley, they beheld the largely vacant grassy flats south of the Clark Fork River. Off to the east, however, standing alone was a brand new mansion built for Hammond's business partner, E. L. Bonner. All around the mansion (which had yet to be occupied), Hammond's South Missoula Land Company was busy laying out streets and building sites. Already, Hammond had sold one of the blocks to copper magnate Marcus Daly, and workers were digging the foundation for his mansion. Hammond had surveyed yet another block for his own manor on what would soon become "Millionaire's Row."[17]

The 250 members of Charlo's band coalesced at the south end of the bridge over the Clark Fork River. In preparation for entering Missoula, the Salish paused and applied fresh face paint and cinched their loads. Mothers lifted infants from the wagons and strapped cradleboards to their saddle horns so they could see the city. Children scrambled down from the travois they had been riding upon, while older boys on ponies herded the vast menagerie of dogs. Charlo wished to move through town as quickly as possible. He had eschewed a military escort, as he wished it to appear that they were leaving their home of their own accord rather than being forced onto the reservation. He would remain bitter over the relocation for the rest of his life and would never forget the duplicity of the U.S. government that pushed him out of his beloved Bitterroot Valley.[18]

Crossing the bridge brought the Salish into a different world. While some of the young men often worked as laborers and construction workers and others frequented the saloons, the women and elders had avoided the city. No doubt, stern-faced, intractable Charlo regarded it as a blot upon the landscape, for Missoula sat at the mouth of the canyon plugging the road to the buffalo. Not quite ten years had passed since the Salish last passed this way on their annual bison hunt.[19]

On the other side of the bridge, the horse-drawn trolley that ran from the Florence Hotel to the Northern Pacific depot at the north edge of town carved deep furrows into the streets, muddy from the previous day's rain. Men wearing dark suits and stovepipe hats picked their way through the mud, while women in long dresses, carrying umbrellas and with their hair drawn into tight buns, strolled the wooden sidewalks. Horse-drawn buggies trotted up and down the wide avenue. Older boys carried lard buckets sloshing with beer from the saloons to the workers up and down Higgins Avenue to enjoy on their lunch break. They hurried past Chinese men dressed in baggy pants, loose-fitting silk shirts, and wide-brimmed hats. At the sound of the roundhouse whistle, Central School let out and children filled the street on their way home for lunch.[20]

The children were the first to pause as they glanced south to the end of the Higgins Avenue Bridge. Gradually the trolley drew to a halt, and the carriages pulled off onto the side streets. The stores and businesses emptied. The men and women abandoned the street and crowded onto the sidewalks to watch the procession.[21]

As the Salish passed under the bridge's steel girders, their mood changed dramatically. Instead of the celebratory tone punctuated with whoops and shouts, the group became quiet as they began the slow march across the entryway into the city. Even the dogs and horses fell into line. Below them, the river split into two channels, contouring around a large island. The island, owned by Hammond of course, housed a shantytown of indigent workers and ne'er-do-wells. It also provided the city with a newly constructed wood-burning electrical plant. The wood smoke hung in the air, casting a gray pall over the valley.

Mounted on their finest horses, the young men of the tribe closed ranks and kept order as the Salish proudly paraded through the city. The warriors stood guard at each street corner, as much for protection against unruly whites as for a perfunctory show of force. Charlo led the procession triumphantly carrying an American flag, symbolizing his sense of ownership of place, that he was an American, more American, in fact, than the recent immigrants inhabiting Missoula.[22]

After several blocks, the Salish crossed the railroad tracks and left the city. A few decades earlier, they had gathered here to collect bitterroots, a food staple for generations. For many years thereafter, the Salish continued to set up tepees on the south side of the river to gather the roots, but Charlo, bitter and heartbroken, would never return. Within four years, Hammond, too, would move permanently from the valley, but in a very different direction and with very different prospects.

CHAPTER 2

War against the Pines

> But the pine is no more lumber than man is, and to be made into boards and house is no more its truest and highest use than the truest use of a man is to be cut down and made into manure.
>
> —Henry David Thoreau, *The Maine Woods*

On a winter night in early 1865, sixteen-year-old Andrew Hammond lay on his bunk. Outside the crudely constructed lean-to, a Maine winter storm howled and blew through the logs where the moss chinking had fallen away. The wind entered the open hole in the roof that served as a chimney and blew the smoke from the open fire around the shelter. The other men in their bunks, arranged like spokes in a wheel around the fire, seemed not to mind as their heavy snoring nearly drowned out the storm. Stale tobacco smoke and the stink of wet wool from socks and shirts hanging above the fire permeated the dank air. Exhausted from the day's work, Andrew tried not to think about the bedbugs and lice that infested his straw tick mattress. At five o'clock he, along with everyone else, awoke to the cook's announcement of breakfast, wolfed down a pile of buckwheat pancakes laden with maple syrup, and washed it down with tea heavily sweetened with molasses. The crew then trudged some three miles through the deep snow to the logging operation to begin their fourteen-hour day.[1]

Following the death of his grandfather a few months earlier, Andrew had begun his life in the Maine logging camp. Although this was the first time he had left home, he felt psychologically, if not physically, comfortable in his new surroundings. Not only did his older brother, George, sleep nearby, but the chiding and exhortations of his fellow workers in both French and English provided a recognizable medley. The cultural milieu of New England Protestants and Acadian French Catholics was as familiar as the logging camp itself.

Until now Andrew had lived in the tiny hamlet of St. Leonard's, New Brunswick. Just across the St. John River from Maine, the village was named after Hammond's maternal grandfather, Leonard Coombes, the political and economic pillar of the Madawaska region. Coombes and the Hammonds, however, were virtually the only Anglo-Protestant families in a region settled by Acadian descendants of French colonists. A product of two cultures living side by side but separated by language and religion, Andrew, while thoroughly immersed in the Protestant tradition, would maintain affinities toward the French language, French history, and Catholicism throughout his life. Although Andrew remained fairly secular, the legacy of religion, primarily the Puritan work ethic, infused his being and shaped his political, cultural, and economic views.

In the 1630s, his ancestor Elizabeth Penn Hammond had fled England along with ten thousand other Puritans to pursue their conservative brand of Protestantism in the Massachusetts Bay Colony. With her exodus, Elizabeth Hammond began her prolific family's process of westward peregrinations that would last four centuries and fan across a continent. Orthodox in religious outlook, conservative in social interactions, financially prudent, and largely temperate, the Hammonds and other Puritans would radically transform both the cultural and physical landscape of North America. Although many eventually intermarried with French Catholics, Scottish Episcopalians, and others, the descendants of Elizabeth Hammond maintained a remarkable degree of homogeneity. Instead of integrating into American society, as did the French and Spanish, they would circumscribe the melting pot into which others should assimilate.[2]

For the Puritans, religion and economics became inexorably intertwined. Sociologist Max Weber argued that the Puritan notion of a calling, or vocation, combined with asceticism to foster the development of capitalism. Material wealth became a sign of salvation, while the repudiation of pleasure and luxuries channeled profits back into one's calling, creating the prerequisite of capitalism. Weber maintained that the religious foundations would eventually fall away, but this "ethic of Protestantism" would become embodied in the "spirit of capitalism" in a secular America. He suggested that this is best portrayed in Benjamin Franklin, who famously extolled the virtues of thrift, work, and profit. Where Catholicism found piety in poverty, Protestantism found vice. Franklin declared, "Poverty often deprives a man of all spirit and virtue; it is hard for an empty bag to stand upright."[3] Although, he was hardly pious, Andrew Hammond retained many of his ancestors' Puritan values and embodied both Franklin's aphorisms and Weber's thesis.

Although religious conflict brought the Hammonds to New England, the American Revolution prompted their removal to Canada. The Revolution divided families and polarized communities, and the Hammond clan was one of many that split into Loyalists and Patriots. In a society torn apart by revolution and war, Archelaus Hammond, Andrew's great-grandfather, relocated to Nova Scotia, preceding the flood of Loyalists following George Washington's victory at Yorktown in 1781.[4]

In a strategic move, Nova Scotia's Lieutenant Governor Thomas Carleton opted to resettle the Loyalists along the St. John River to secure the newly established boundary with the United States in case hostilities might resume. Carleton hit upon the idea of dividing the valley into blocks allocated to each regiment. This would preserve the social relations that the men had formed over the years within their own battalions and permit quick activation if a military need should arise. In this manner, Carleton granted prime agricultural land to the second battalion of New Jersey Volunteers, one of whom was Lieutenant John Coombes, Andrew Hammond's maternal great-grandfather. All told, the New Jersey Volunteers received 38,450 acres, with 8,340 acres reserved as a common woodland. Transferred from New England, this notion of a commons would greatly assist in the Hammond family's prosperity.[5]

The arrival of the Loyalists had a cascading effect along the St. John River. Besides the Maliseet Indians, who still roamed the forests and plied the river in birch bark canoes, some five hundred settlers lived in the valley prior to the arrival of the Loyalists. Despite the Acadian diaspora during the Seven Years' War (1756–63)—in which the British conducted a systematic campaign against French Catholic settlers, burning villages, destroying livestock, and confiscating lands—many drifted back to their former villages. None of these groups was especially pleased with the large influx of newcomers. The Indians witnessed the depletion of fish and game, while the Acadians saw their land expropriated as the government clearly privileged the Protestant Loyalists above the Catholic Acadians. In such a manner, John Coombes received 350 acres sandwiched between Archelaus Hammond's two lots.[6]

Following multiple protests of the injustice done to the Acadians, in 1788 Carleton issued land grants to the displaced families 150 miles upriver, in the remote region of Madawaska. This appeared to be a satisfactory solution; it removed the Acadians from the lower St. John Valley, while the Acadians believed they would be secure from Protestant interference in Madawaska. This region had better soil and growing conditions than the lower St. John, but quality farmland remained confined to the river corridor. Thus, this musical-chair approach to land settlement stumbled.[7]

The Loyalist immigration increased the population of Nova Scotia tenfold, and the British decided to split it in two as part of a plan to have a new colony, New Brunswick, replace New England as the center for North Atlantic trade. Although New Brunswick rejected both the Acadian settlers and their religion, it adopted the French system of settlement that divided land into long, thin rectangles that stretched back from the river. This allowed everyone substantial property with river access, flat floodplain farmland, and woodlots toward the rear while permitting people to live in close proximity to one another. New Brunswick historian W. S. MacNutt maintained that these large land grants, ranging from two hundred to one thousand acres, preempted community planning by allowing "the individual to dwell on his land almost wherever fancy pleased. Instead of living in communities, the majority of rural dwellers lived in the bush." MacNutt wrote that the development of New England–type villages "that thrust responsibility as well as privileges upon individuals, was completely absent in New Brunswick." He suggested that this "bred a spirit of individualism that ignored the rights of property . . . [and led to] a general tendency to consider the country as a vast common." According to MacNutt, New Brunswick's system of land distribution shaped cultural attitudes that regarded timber as free for the taking regardless of legal ownership.[8]

MacNutt's argument can also apply to the American West, where the system of land distribution was based upon a grid that scattered settlers into 160- to 640-acre blocks instead of towns. Although it is difficult to trace a direct causal relationship between settlement patterns and social attitudes, both New Brunswick and the American West developed a tradition of putting individuals above the community. Certainly, both regions displayed a flagrant disregard for government control over natural resources.

Timber poaching was widespread throughout colonial America. During the eighteenth century it was common practice to cut timber illegally off commons as well as Crown and private lands. Despite ineffectual laws "prohibiting the common practice of stealing timber," by 1800 timber trespasses had become so established throughout the East Coast that many no longer considered it a crime.[9] The vast majority of western lumbermen originated in New Brunswick and Maine, A. B. Hammond among them, and they viewed the western forests in the same way they did the forests of their homeland.

Since three-quarters of its export revenue came from wood products, New Brunswick was heavily dependent upon timber, as was Maine. In moving to assert property rights, both governments sought to regulate access to the forests but were largely ineffectual. To curb timber poaching on Crown lands, New Brunswick instituted timber licensing in 1817, but unauthorized logging continued unabated. In Maine, timber poachers became so emboldened that they fought off the sheriff who attempted to seize illegally cut logs; prosecution was nearly impossible. Half a century later, this same scenario would repeat itself on the other side of the continent when Hammond's logging crew bullied authorities trying to impound timber harvested from federal lands. Growing up in the hinterland of New Brunswick, where timber poaching was an established tradition with incursions against both government and private lands, Hammond gave little thought to transferring the practice westward.[10]

Wood was a vital resource in colonial North America, necessary for fuel, shelter, transportation, and heat. It also became a valuable global commodity. In Europe, the rise of mercantile capitalism and imperial expansion in the seventeenth century stimulated shipbuilding and iron smelting, which quickly depleted its forests. England and France then turned their attention to the untapped resources of the American colonies. As these imperial powers vied for primacy on the North Atlantic, the towering white pines of Maine and New Brunswick became especially desirable as masts for sailing ships.[11]

While New Brunswick had long provided Great Britain's shipbuilding industry with masts and spars, the Baltic region contributed the bulk of England's timber. But early in the nineteenth century, Napoleon closed the Baltic ports to English exports, and demand shifted to the colonies. Prices doubled, tripled, and quadrupled, making the cost of transatlantic shipment negligible. By 1809 two-thirds of England's timber came from British North America. The Napoleonic blockade increased New Brunswick's lumber production twentyfold, transforming "an undeveloped backwater of 25,000 people to a bustling colony of 190,000."[12] Despite Napoleon's defeat in 1815, colonial timber merchants convinced the English government to maintain duties on colonial timber, providing a protected market for New Brunswick timber. For the next forty years New Brunswick would be Great Britain's primary lumber supplier, and colonial lumbermen quickly depleted the most accessible pines along the St. John and other major rivers. But the remote region of Madawaska on the border between the United States and New Brunswick remained untapped.[13]

Sparsely settled but rich in valuable white pine timber, Madawaska was a political no-man's-land between Maine, New Brunswick, and Quebec. As they filtered into the wide valley, Acadians and Quebecois created a unique regional and cultural identity, and living in a disputed region with minimal government authority suited them just fine. But before long, Madawaskans found themselves in the center of an international confrontation over diminishing natural resources.[14]

Out of political expediency, the 1783 Treaty of Paris was deliberately vague in demarcating the northeastern-most boundary between the U.S. and British possessions. While New Brunswick authorities wished to rid the lower St. John of Acadians to make room for Loyalists, they welcomed their settlement in the Madawaska region to keep the Americans at bay and establish de facto British possession. Encouraging Acadian relocation to Madawaska proved a prescient move on the part of Carleton, for shortly after the War of 1812, settlers from Maine arrived and began cutting timber. Although Lieutenant Governor Carleton authorized Acadian settlement in Madawaska, the uncertain boundary compelled him to suspend conveying any more deeds. Acadians simply squatted on the land, cleared farms, built houses, and planted crops, all without formal land title.[15]

From 1824 to 1830, Madawaska's population jumped from 1,600 to 2,500, leading to increased friction between the United States and New Brunswick. On the north shore of the St. John, Maine settler John Baker celebrated the Fourth of July 1827 by raising the American flag and proclaiming the Republic of Madawaska. Although the Acadians and Quebecois had shown no desire to become part of Maine, Baker's confused symbolism provoked the New Brunswick authorities. His subsequent arrest and imprisonment set off an international crisis.[16]

Following the Baker incident, the provincial government in Fredericton realized it needed to exert control in the backwoods of Madawaska. Concerned over possible Yankee annexation, New Brunswick encouraged English-speaking Protestants in the militia to resettle in Madawaska, but few responded. The government then designated two magistrates—one of whom was Leonard Coombes, A. B. Hammond's grandfather—to watch over the region, report on American activity, and arrest agitators.[17]

At thirty-seven, Coombes was already a well-established gentleman farmer. An Anglican, fluent in French, and a "zealous" supporter of British authority over the region, Coombes was well suited as an emissary. Although he had just received a three-hundred-acre land grant in Kingsclear, just upstream from Fredericton, Coombes packed up his family—except for his oldest daughter, who had just married Andrew Hammond, Sr.—and moved upriver to his new post in Grand-Rivière in the upper St. John Valley.[18]

While Carleton had encouraged Acadian settlement in the disputed region, government policy deliberately excluded French Catholics from government office. Loyalist descendants thus dominated the government, commerce, and most of the resources of the province. Furthermore, the Acadian memory of deportation "reinforced their reluctance to get involved with public life."[19] While many Acadians in Madawaska were willing to have Coombes act as their intermediary, some in the community—the very families his father had displaced in Kingsclear—feared the arrival of Leonard Coombes forecast another diaspora.[20]

When Coombes arrived in 1829, the region was almost entirely French Catholic and nearly all the land on both sides of the St. John River was already occupied. Unlike many of the French settlers, Coombes possessed a modest degree of capital and bought land from the original Acadian grants, becoming the largest landowner in Grand- Rivière.[21]

Although farming was a tenuous proposition given the short growing season, within four years of his arrival Coombes had also become the region's top farmer. Like many men in such positions, Coombes channeled his official government role into one of social and financial status. As magistrate, he received payments for his services from recipients rather than the government. Lacking cash, locals often paid in farm produce or in labor. Thus, not only was Coombes able to purchase the best agricultural land, he could also hire or barter farm labor to clear land and plant crops. In addition, as one of the few in the region who was both bilingual and literate, Coombes could exploit commercial opportunities, such as lumbering, as they arose. Like many men of his day, however, Coombes regarded farming as the highest moral calling and saw lumbering only as an additional source of capital that could be reinvested in the farm.[22] His future sons-in-law—William and Andrew Hammond, Sr.—would soon reverse the equation.

At the beginning of the nineteenth century, small, independent producers characterized New Brunswick's lumber industry. These men worked in the timber trade during the long winter to supplement their farm income. Like many families, the men of the Hammond clan cut trees on back lots and parish commons using their own farm animals and equipment and returned home each evening. Come spring, they floated the logs to their mill. Powered by a waterwheel, the mill ran only when there was enough runoff to rotate the wheel, which turned massive wooden gears driving a single vertical saw blade up and down through the log. Operated by two to four men, the mill could crank out only about two thousand board feet per day.[23] When spring arrived, some sons worked the mill, while others began the planting process. By summer, when the farm demanded full attention and the water level had dropped, the mill no longer functioned. Lumbering was a seasonal, part-time occupation that enabled settlers to establish farms and provided cash for farm improvements or to hire summer labor, thus boosting a family's material condition.[24]

While lumbering provided economic returns for individuals, it also tied their communities to the fluctuation in international markets. In the late 1820s a financial panic in England threw New Brunswick into economic depression. But then, as Great Britain began to develop its railway system with its requisite lumber demands, the timber industry rebounded. As timber cutting reduced the accessible trees near settlements, the scale of operations increased, requiring larger logging crews to travel to more remote regions for six months or longer. Thus began the shift from independent owners toward heavily capitalized producers, with men working for wages rather than for a share in the family business.[25]

Changes in government policy accelerated this transition. For most of his life, Andrew Hammond, Sr., had known Crown forests to be readily accessible either through purchase for a minimal fee or by simply cutting timber and paying the low license fee if caught, which was rare. But beginning in 1827, the government began raising both the price of lands and timber license fees. Three years later, New Brunswick began offering private reservations and granted five-year licenses to those who could guarantee an annual cut. All of these policies favored large capitalists over small producers, so that by 1836 a dozen individuals held half of the timber licenses in the province.[26]

As timber cutting increased, so did the demand for faster, more efficient sawmills, and ownership required substantial capital to build and operate the mill, as well as to ride the economic roller coaster that so characterized the industry. The 1830s marked the boom years in New Brunswick, until the panic of 1837 sent Great Britain into a depression. Again the effects reverberated throughout the colony. But by then, the era of the independent lumber producer was drawing to a close.[27]

The year also marked the low point in Andrew Hammond, Sr.'s, life. After giving birth to her second child, his young wife, Sarah, died. So too had his grandfather and namesake, Andrew Joslin. Joslin bequeathed Andrew, Sr., his sawmill and back lot to be held jointly with his two brothers and two sisters, but this proved an untenable situation. Hammond owned a parcel of land only partly cleared, with marginal soil that occasionally yielded a surplus crop, and partial interest in two crumbling sawmills. Whether compelled by economics, an emotional need for a fresh start, or wanderlust, Hammond left Kingsclear. By the following year, he had relocated to Grand-Rivière, married his deceased wife's sister, Glorianna, and settled in next door to his father-in-law, Leonard Coombes. Three years later, Andrew's younger brother, William, married another of Coombes's six daughters, Caroline.[28]

With Leonard Coombes as the patriarch, the three families merged and established themselves as the regional elite. Coombes built his house on a steep bank overlooking the river, with a wide floodplain below that was suitable for crops. Behind the house the forested land angled gently upward; it would become pasture once cleared. Andrew and Glorianna moved onto the neighboring lot, while William and Caroline settled directly opposite on the south bank of the St. John. The families also set up business arrangements with each other. As bilingual Anglophones, the Coombeses and Hammonds could act as intermediaries between the French Madawaskans and the New Brunswick authorities and businessmen. They quickly put to use their abilities to bridge two languages and cultures, operating as merchants, lumbermen, and farmers on both sides of the river.

As magistrate, Coombes had nearly unconstrained authority. Coombes and Francis Rice, the other magistrate, constituted the entire government presence in the region. They were permitted to make arrests, try cases, impose sentences, notarize documents, and legalize land titles and marriages. Additionally, Coombes was in charge of the region's militia. Much of this control, however, was illusory, for Acadians were highly independent and looked toward the Catholic Church to fill the role of the state. The parish priests sanctified marriages, received tithes, settled disputes, and rendered judgments. Furthermore, the refugee origins of the Madawaskans compounded with the jurisdictional quarrel to create a population that many considered lawless. Coombes, it seemed, would have his work cut out for him.[29]

Rather than impose administrative control over Acadians, Coombes's primary concern became the establishment of British jurisdiction. In 1832, Maine attempted to annex four thousand square miles of

Madawaska by organizing an election on the south side of the St. John River to incorporate a township. Catching wind of the meeting, Rice and Coombes quickly broke it up and took credit for the residents rejecting the overture. When Maine advocates held a subsequent meeting the next month, Coombes and Rice arrested the four organizers but excused all of the French in attendance, demonstrating both territorial jurisdiction and community diplomacy. By 1837 relations had deteriorated to the point where Maine was preparing to send troops into the disputed region. While the United States and Great Britain jockeyed for territory, the real issue was who would control the forests.[30]

Eager to avoid armed conflict, in 1836 New Brunswick had recalled timber licenses and suspended logging in Madawaska. Maine soon followed suit. But with different entities claiming jurisdiction and none having the power to enforce it, illegal cutting continued unabated, with each side accusing the other of infringing upon its territory.[31] New Brunswick, however, held a geographic advantage, as the rivers of northern Maine all drained into the St. John. Transporting timber from the region required floating the logs down the St. John, which, below the Aroostook River, flowed through New Brunswick. Once lumbermen dumped their logs in the river, no one could tell where the timber had been cut, and it could be passed off as having come from legitimate sources. Americans built booms (logs chained together and anchored on shore) to collect the timber coming down the Aroostook before it met the St. John. They then collected the logs into massive rafts and smuggled these past the border guards under cover of darkness. Madawaska historian Béatrice Craig noted that lumbermen kept their activities secret, "playing hide-and seek with state and provincial authorities."[32] Yet many of these same authorities, including Leonard Coombes, ignored or were actively involved in timber trespass themselves.

Despite the ban on timber licenses, in 1839 Coombes received permission to cut one hundred pine logs near the disputed border. With two of Coombes's crews cutting timber, the United States objected to what it believed was an infringement on its territory. When Maine sent officials to investigate the timber depredations, Coombes, once again, attempting to establish official jurisdiction where no boundary existed, arrested them, thus providing the spark that finally ignited the bloodless "Aroostook War."[33]

In response to the timber cutting by Coombes and others, Maine called out some ten thousand militia and sent a force down the Aroostook River, where they practiced their "musketry on effigies of Queen Victoria."[34] New Brunswick responded by assembling a similar number of regular troops and militia of its own. With armed forces amassing on either side of the St. John River, it looked as if the United States and Great Britain were headed for another war.[35]

In a preemptive move, the British government authorized Alexander Baring, Lord Ashburton, to negotiate a treaty with the United States. Baring was an intriguing choice, given that his banking firm held bonds in six U.S. states and he had married into a Philadelphia family that owned six million acres of timberlands in Maine. Baring protested his instructions from the British government to hold out for more territory. Instead, he resolved the crisis by negotiating a treaty that was acceptable to both the United States and his own financial interests.

Ultimately, the 1842 treaty designated the St. John River as the boundary, splitting both the region and the Acadian communities. The vast majority of the land went to the United States, with New Brunswick losing seven thousand square miles and two thousand inhabitants. The primary concern of the province, however, was control of the market: ensuring that the timber would be produced and exported through New Brunswick. Thus, a provision in the treaty allowed Maine timber to be floated down the river and sold duty free in the port of St. John. Ultimately, transnational financial interests took precedence over territorial sovereignty. Despite the official border, however, Madawaska residents continued to move back and forth across the river, cutting timber on either side regardless of ownership, be it state, provincial, or private.[36]

Perched on the cusp between colonial mercantilism and industrial capitalism, New Brunswick developed an intricate system of bringing timber from a cash-poor hinterland to the rapidly industrializing core of Great Britain. In *Timber Colony,* Graeme Wynn detailed how this system worked. Early on, farmers who engaged in seasonal lumbering simply exchanged their timber for consumer goods at local stores. Before long, a tributary system developed that mimicked the physical tributaries of the St. John. Local merchants advanced supplies on credit in the fall to lumbermen, who assembled logging crews. Upon receipt of the logs in the spring, merchants subtracted what the lumbermen owed them. Merchants, who had received supplies from downstream wholesalers, then forwarded the logs on to the wholesaler to settle their accounts. In lieu of cash, which was scarce, timber served as a medium of exchange.[37]

Wholesalers, in turn, had negotiated forward contracts containing set prices for future delivery with English shipping agents and sawmill owners the previous fall. Sawmills, too, had already worked out agreements to supply the agents with lumber. The entire system was based upon credit, and when financial panic hit England, it sent reverberations back upstream as the agents demanded payment, causing the wholesalers to pressure the local merchants, who had to deny credit and force payments from the farmer-loggers. These small-time lumbermen without capital wound up in debt. Having bought their supplies on credit, a decline in the market, difficulty in getting logs out, or poor weather could send them into bankruptcy.[38]

As timber became less accessible, more capital was necessary, and merchants applied for their own timber licenses, which were now beyond the reach of local farmers. Merchants then contracted "master lumberers," who hired crews, supplied them with provisions, paid the crew's wages in the spring, and then delivered the timber. In lower Madawaska, Coombes filled the role of merchant. He sold his excess farm produce on credit to locals, who repaid in kind or in labor. He also loaned small sums of money to locals to start up their own timber operations; he, in turn, borrowed from larger capitalists. With his political connections, Coombes secured timber licenses, and with his ability to hire additional labor, he could plant additional fields to supply timber camps with produce and hay for the animal teams. When Andrew Hammond, Sr., arrived, Coombes folded him into the operations; Hammond likely acted as "master lumber" or Coombes's agent in purchasing timberlands, outfitting the crews, and hauling supplies to the camps. Coombes then turned the timber over to the next tier of lumbermen to transport down the St. John River.[39]

The resolution of the Aroostook War lifted the ban on timber licenses, and what had been the surreptitious activity of timber poaching bloomed into the open. Anticipating a renewed timber boom, Hammond and Coombes borrowed $2,500 to buy timberland and set up logging camps. The rest of the province, however, was cutting indiscriminately, and by 1846 timber exports reached their highest point, resulting in a timber glut. Additionally, the British had substantially reduced the colonial tariff preference. Then, the worldwide depression of 1848 hit. The combination of these factors plunged the timber-dependent colony of New Brunswick into an economic tailspin.[40]

With credit stacked upon credit, the depression sent shock waves throughout the system. Pressed to collect payments, wholesalers in England squeezed New Brunswick merchants, who in turn pressured country storekeepers, who then denied credit and demanded payment from lumbermen. This created a chain reaction of bankruptcies. The Coombes-Hammond network collapsed, and they dissolved their partnership. Overextended financially and in debt to Bangor and Fredericton financiers, Andrew Hammond, Sr., declared bankruptcy in August 1848. In contrast, those who were well capitalized and well connected were able to carry their debts and ride out the fluctuations, reinforcing the advantages of large enterprises over smaller ones.[41]

In addition to his financial woes, Andrew Hammond, Sr., had a rapidly growing family to support. Andrew B. Hammond, the fourth child of Andrew and Glorianna, was born in 1848, a most inauspicious year for a future timber baron. In addition to the global depression, population pressures, disease, and famine fueled wave after wave of revolutions across Europe, as the legacy of feudalism and vast socioeconomic inequities spawned unrest. In England, the working-class Chartist Movement was gaining momentum, and Karl Marx and Frederich Engels published the *Communist Manifesto.* At the end of the year, however, optimism surged in North America when President James Polk announced the discovery of gold in California. While each of these events was to have a significant impact upon A. B. Hammond, none was as great as the long shadow of his father's bankruptcy due to the volatility of New Brunswick's timber industry.

Despite bankruptcy, Andrew Hammond, Sr., still owned his farm and recovered quickly, compared to the French Madawaskans. By 1851 the Hammond household included seven children, an eighteen-year-old niece, three servants, a carpenter servant, and a French teacher, indicating the family's socioeconomic status and the high premium the Hammonds placed upon education.[42]

In 1854, when Andrew, Jr., was only six years old, his father died. Fortunately, his mother proved to be an exceptionally strong-willed and capable woman. In addition, Andrew's extended family provided emotional and financial support. Glorianna's parents lived next door, with grandfather Coombes, an active sixty-two-year-old, serving as a surrogate father to young Andrew. Although his grandfather was the founder and most prominent citizen of the town, Andrew still had to plow the fields, clear the land, pull up stumps, plant potatoes, and cut and haul endless loads of firewood. With the death of their father, even more of the farm work fell to Andrew and his brothers, George, Fred, and Henry. The girls, Sarah and Mary, in addition to cooking, cleaning, and doing farm chores, also spun wool and flax and wove the yarn into clothes and rugs. While they were among the elite of Madawaska, the legacy of the Protestant work ethic precluded idleness.

Despite the hard work and loss of their father, for the boys, at least, life in Madawaska could be considered an idyllic childhood. They spent their summers fishing and swimming in the river or gathering wild berries. In the fall, they collected hazelnuts and hunted deer and moose. During winters, they could sled down the hills and skate on the frozen river. After their chores, they doubtless found time in the late summer evenings to fish off the dock at Coombes Landing. With the sun dropping below the clouds and briefly lighting up the river valley, the Hammond boys would have discussed the world beyond the St. John River Valley. Gold had recently been discovered in Colorado; prospectors were picking gold nuggets right out of the streams, enough to make you rich beyond your wildest dreams, far richer than anyone in New Brunswick. News of the Civil War raging to the south also reached their ears. Thousands fled north to escape the draft, while others crossed the border to join the fight. A military adventure held little appeal for the brothers, but the lure of gold tempted them to leave the quiet valley where they had spent their entire lives.

Much more than his brothers, young Andrew Hammond excelled in his school studies. By the time he was eleven years old, his county boasted the most schools, most students, and highest-paid teachers in Madawaska, in no small part due to Leonard Coombes. Late nights, after finishing his chores, Andrew settled himself on the floor in front of the kitchen fireplace, absorbing all the books he could. His older brothers, George and Fred, teased him for his devotion to reading, calling him "old gravy eyes" because his eyes watered from the smoke while trying to read in the dim light of the fire. Even when Andrew was sent out to plow, Glorianna would find him in the field with his head in a book when she went out to check on him. By the time he was sixteen, Andrew had exhausted the family library that consisted of the Bible, the works of Shakespeare, *Pilgrim's Progress,* and Alexander Kinglake's *Invasion of the Crimea.* While scholarship for its own sake was not a valued pursuit in lower Madawaska, Hammond, nonetheless, would retain a lifelong interest in literature and history, developing a particular affinity for Napoleon.[43]

Although Hammond absorbed Coombes's library, he formed a very different opinion of the role of agriculture than his grandfather had. The educated elites of New Brunswick, like Coombes, assumed that agriculture was the foundation of civilization and believed lumbering to be immoral, as it encouraged drunkenness, vice, and the neglect of farming. They attributed farm failure to participation in the timber trade. They also considered it economically unwise, which, given the wild fluctuations in the lumber market, certainly seemed true. Graeme Wynn, however, maintained that without lumber employment "many settlers might have found it exceedingly difficult to develop successful farms."[44]

Coombes saw lumbering as a temporary expedient to aid the establishment of agriculture but complained that "people cling to old ways" of combining lumbering with subsistence farming. Instead, Coombes promoted export agriculture and petitioned the government for money to buy seeds and ploughs to distribute to area farmers. Recognizing the long-term interests of the community over the short-term profits of the timber industry, Coombes petitioned the government for a law that would protect bridges from damage caused by the spring drive, when lumbermen flooded the rivers with logs.[45]

Coombes's favorable view of agriculture contrasted with the economic reality in Madawaska, where the timber industry expanded the economic base and accelerated immigration, causing a shortage of farmland. Lumberjack work made it possible for families to settle on smaller farms or marginal lands, but it did not raise the standard of living. While economic power resulted from access to good farmland, which led to a position of social and political prominence, the wealthiest were the timber dealers and merchants, as they controlled wages and were the sources of credit. This was certainly true of Andrew's uncle William.[46]

While his older brother struggled financially, William Hammond became the second wealthiest individual in Aroostook County, Maine. William had moved to the south side of the river in 1842 and become a storekeeper and merchant, providing supplies for timber camps, receiving logs in payment, and then forwarding them downriver. Rather than farming, William expanded upon Coombes's commercial network, borrowing money to buy timberlands and set himself up as a merchant. In 1860 his real estate was valued at $4,000, making him a relative millionaire by today's standards. The region's primary capitalists acquired their wealth through business, commerce, and timber exports rather than through agriculture. This lesson was not lost on young Andrew Hammond.[47]

In the spring of 1864, when Andrew Hammond was fifteen years old, Leonard Coombes died and William assumed the role of financial patriarch. Although the family was well positioned, for Andrew and George the excitement of the timber camps overpowered the drudgery of farm work, and they headed south into the Maine woods to seek their own fortune.

By the 1860s the lumber industry had changed dramatically. Unlike earlier operations, in which farmers worked as loggers during the depth of winter, newer and larger lumbering operations kept men in the forest from the fall until late spring. The lengthy absence not only was a hardship for many families but also precluded farming and ensured an economic dependence upon the timber industry with its wildly fluctuating boom-and-bust cycles. As W. S. MacNutt succinctly stated, "Farmers make good lumbermen, but lumbermen do not make good farmers."[48]

The largest lumber crews contained twenty men organized by the master lumberers, who contracted with their suppliers to receive the timber at the end of the drive. In the fall, George, Andrew, and the rest of the crew would have roughed out the logging operation, clearing a road to the river, where they would dump logs to await the spring drive. From the river, they removed downed trees, dynamited boulders and rapids, and tore out root mats, turning creeks into sluices to ease the passage of logs downstream.[49]

To overcome the seasonality of river flows, lumbermen built splash dams on nearly every creek and river in the region to contain the logs until the spring drive. Splash dams could be anything from temporary contraptions of logs and rubble to more permanent constructions equipped with draw gates to control the flow of water. Such dams backed up the river and created temporary log ponds. The substantial labor costs of building splash dams and clearing the river for log drives reinforced the need for capital investment.[50]

With the onset of winter, the crews began chopping trees. Andrew and the other men traded off the various jobs. Unlike later operations, there was little division of labor. With logs exceeding forty feet, hauling the timber out was arduous. The men piled the limbed logs onto bobsleds hitched to horses or oxen, which then towed the sleds over the snow to the rivers. More than one man died from overloaded sleds skidding out of control and toppling over, but the real danger came in the spring drive.[51]

Once the water had backed up behind the splash dams during the spring runoff, lumberjacks threw open the gates or simply blew up the dams with dynamite. The ensuing flood flushed the logs downstream as men rolled the timbers into the water. River drives could range from ten to one hundred miles long and last for two months. Up to their hips in icy water, the men guided the logs, attempting to keep them free of obstructions or getting jammed. Between swigs of rum, men rode the bobbing logs downstream, jumping from one to another. As they picked up speed the logs began hurtling down the river, spinning, leaping like dolphins, and popping high up into the air. Between drowning in the cold water and being crushed by these logs, death was not uncommon for loggers.[52]

Horse-drawn "wangan" (or "wanegan") boats brought oats, beans, flour, salted cod, and molasses to the hungry men. As a mobile forerunner of the company store, the wangan also carried tobacco, socks, boots, mittens, and everything a lumberjack might want during the winter logging or spring drive. The wangan operators sold items at outlandish prices and charged them against the logger's account; many emerged from the woods to discover their paycheck was but a fragment of what they expected.[53]

While the Aroostook War was between two nations battling for control over access to the white pine forests, on his trip to Maine Henry David Thoreau observed that the real Aroostook War was the "war against the pines."[54] Indeed, by 1864, when Andrew Hammond went to work in the woods, lumbermen had practically eliminated white pine from the St. John and Aroostook watersheds. In previous decades, logging operations cut only trees larger than twenty inches in diameter and primarily confined their activities along rivers into which they could drop the trees for downstream transport. But by the 1860s such trees were nearly gone, and lumbermen ranged deeper into the woods, where white pine grew in scattered clumps interspersed with spruce, hemlock, and hardwoods. Fanning out across the landscape, searching for the specific trees they wished to cut, the crews might encounter nearly pure pine stands in the sandy soil of dry upland areas where forest fires or windfall provided gaps in the overstory. Although white pine occasionally occurred in such high, but localized, concentrations, it constituted less than 2 percent of the presettlement forests of northern Maine and New Brunswick. Seeking sunlight above the shade-tolerant beeches and sugar maples, white pines grew extraordinarily tall and straight, making them highly desirable for ship's masts and timbers.[55]

The ecology of the white pine, the export market, existing technology, and labor interacted in ways that at first had a relatively benign ecological impact upon the northern forest. Selective logging could mimic windfalls by opening up gaps in the canopy and allowing for pine regeneration. Nature still constrained the industry. Weather and climate dictated how and when logs would be transported, and human and animal energy limited the entire operation.[56] But continued market pressure from Great Britain's insatiable demand caused loggers to return to previously logged areas every decade or so and cut smaller and smaller trees. This eventually resulted in the loss of local seed sources and, ultimately, the near extirpation of white pine from New Brunswick. With the demise of white pine, the industry shifted toward the more abundant spruce. Simply using axes, horses, and rivers, by the mid-nineteenth century, lumbermen had dramatically changed the composition and structure of New Brunswick's forests.[57]

Perhaps the largest long-term effect of a half century of unrestrained logging was the damage to river ecosystems, many of which still have not recovered. Riparian areas are among the most biologically diverse terrestrial ecosystems and provide essential corridors for wildlife movements. Logging along streams quickly damaged these ecosystems, eroded stream banks, and adversely affected salmon and other migrating fish species. Oxen and horses towed the logs to the riverbank, where logging crews stacked the logs up to twenty feet high, destroying all riparian vegetation in the process. The removal of rocks, trees, and debris for the log drives eliminated aquatic and riparian habitats and altered the hydrology of rivers. Now wider and shallower, rivers became consequently warmer in summer, colder in winter, and thus less habitable for fish populations. The log drives scoured the creek beds, removing the fine gravel that fish needed for spawning as well as invertebrate habitat. Ecologists now estimate that the replacement time for fine gravels in New Brunswick will be at least several hundred years.[58]

The splash dams, too, had significant impacts upon fish populations. Although temporary, the dams, by drying up the water downstream and preventing any movement above the dam, created impediments to migrating fish and especially impacted salmon. By dramatically altering the river's hydrology, dams caused cataclysmic changes in sediment, nutrient, and energy flows. New Brunswick's native fish populations still have not recovered from nineteenth-century logging practices.[59]

Furthermore, the lumber mills on the St. John indiscriminately dumped sawdust and scraps into the river, endangering both fish populations and navigation. New Brunswick authorities, however, deemed the timber industry so important that they ignored the consequences, despite the mounting evidence. As early as 1815, observers noticed the decline in the salmon population. Seventy-four years later, a government report noted that floating sawdust had infiltrated even the smaller streams and killed food sources for fish and marine mammals. The report concluded that sawdust was "a fixed imperishable foreign matter, and adheres to the beds of streams and other waters, and forms a long, continuous mantle of death."[60]

Tree species also vanished. By the 1840s, Madawaska contained the last remnants of New Brunswick's white pine. Twenty years later, the timber industry had cut these as well, causing George and Andrew to drift south into Maine's Penobscot watershed, where white pine was more abundant, for the time being. Furthermore, England had ceased preferential tariffs on New Brunswick timber while U.S. demand skyrocketed. By midcentury the timber industry shifted westward toward Pennsylvania and upstate New York, and the two states soon replaced New England as the top lumber producer in the United States. Additionally, the Civil War caused a labor shortage, raising wages substantially, and Pennsylvania lumber companies began actively recruiting in Maine and New Brunswick for workers. After a season in Maine, Andrew and George followed the westward tide to the Susquehanna River in Pennsylvania, where they spent the next year stacking logs for the river drive.[61]

Just as Bangor had become New England's preeminent lumbering center, with the Penobscot River providing access to both trees and markets, so had Williamsport on the Susquehanna become the nucleus of mid-Atlantic trade. By the time the Hammond brothers arrived, in 1866, the city was a hub of lumbering activity, with thirty sawmills and loggers fanning out up the rivers and tributaries. As they moved from New Brunswick to Maine to Pennsylvania, the Hammonds traveled from the protocapitalist stage of the lumber industry to one that was on the cusp of becoming fully industrialized. But by 1870 Pennsylvania, too, had nearly depleted its white pine, and the industry again looked to the west. Andrew and George turned their gaze west as well but with gold in their eyes, not logs.[62]

Andrew Hammond learned much of what he would later apply to his own enterprises from working as a teenage lumberjack in the north woods. By the time he left New Brunswick, Andrew was well acquainted with the boom-and-bust cycles of the timber industry and how debt and overextended finances could lead to bankruptcy. Indeed, he built his career upon the realization that only the well capitalized could survive in a resource-based economy. Compared with many other industrial capitalists, Hammond was exceedingly conservative, unwilling to take financial risks, and scrupulously avoiding debt—factors that ensured his long-term success in an inherently volatile industry. No doubt, these traits resulted in part from his father's failure at the hands of an unstable commodity.

When Hammond began his commercial dominance over western Montana, he replicated his grandfather's model of building a business network based upon extensive kinship ties. But recognizing the limitations of agriculture, Hammond avoided investing much time or money in farming or ranching. He also developed a disregard for environmental costs; in his mind such impacts were simply unavoidable consequences on the road to riches. Furthermore, the tradition of lawlessness and independence that pervaded the timber colony of New Brunswick formed much of Hammond's subsequent behavior and outlook. In short, A. B. Hammond was a product of the frontier that shaped his worldview, ethics, and business practices. But it was the frontier of New Brunswick rather than that of the American West.

CHAPTER 3

The Woodhawk War

The upper Missouri River was a strange place to find a lumberjack.[1] For Andrew Hammond, the country seemed to be more sky than earth, and the earth was more like an endless sea of grass than anything resembling a forest. Unlike the thick woods of New Brunswick and Maine, the northern plains appeared hostile and empty. The vast horizon likely filled his heart with both a young man's yearning and dread of the unknown. This was a land where winter temperatures plummeted to forty degrees below zero, where winds exceeded eighty miles per hour and could drive a person insane with their relentless howling. Summer heat often topped one hundred degrees in the shade, except there was no shade. In July, the "whole prairie was dry and yellow, the least motion, even of a wolf crossing it raised the dust," wrote the traveling German Prince Maximilian of Wied in 1833.[2] Nightfall provided little relief, for that was when hordes of mosquitoes descended.

Despite the "sea of grass" cliché, the northern plains were not nearly as monotonous a landscape as they first appeared. They were more like a badly rumpled blanket than a tight sheet stretched across the midsection of the continent. Hills and buttes broke the skyline, while coulees filled with chokecherry, sumac, and wild rose incised into the horizon. Unbelievably, thousands of bison, huge 1,500-pound beasts, could vanish into a dip in the blanket as if swallowed up by the earth. Conversely, hundreds of mounted Indians could suddenly appear, arising from a hidden draw.

This blanket tilted up slightly, almost imperceptibly, to the west, from where the great rivers flowed: the Platte, the Niobrara, the Yellowstone, the Musselshell, and the "Grandmother" River, the Missouri. Near the

Missouri, the land fell away and the sky shrank back to its proper size. In places, the plains broke off abruptly into barren badlands—crumbling hills of gray-green bentonite clay that eroded into domes and minarets. In other places, beige cliffs stair-stepped up from the river, and just enough water trickled through parapets in the sandstone to support a scattering of juniper and ponderosa pine. But for much of its course, the Missouri wound a sinuous path across a floodplain that supported a wide, forested valley of cottonwood, ash, and oak.

These bottomlands provided refuge for both humans and animals from the wind, heat, snow, and cold of the plains above, and this was where a nineteen-year-old logger from New Brunswick was busy chopping wood during the winter of 1868. Tall and lanky, with a physique more befitting a cowboy than a woodcutter, Andrew Hammond cast a wary eye about him for Lakota warriors every time he ventured beyond the confines of his hastily constructed cabin at the confluence of the Missouri and Milk Rivers. While the Aroostook War was mostly bluster and bravado, Hammond had inadvertently found himself in the middle of a real war zone.

The previous winter, Andrew and his older brother, George, had been logging on the Susquehanna River in Pennsylvania. After long days of heavy labor, sawing trees by hand, skidding the logs by horse, and dumping them into the river, George began to speak glowingly of the gold fields of Colorado, urging his brother to join him in heading west. Sitting around the wood stove in the cramped, damp bunkhouse infested with lice and smelling of wet wool, the brothers had likely digested the mining propaganda issued by gateway communities such as St. Joseph, Missouri, hoping to attract immigrants on their way to Colorado.[3]

After tucking away enough money for the trip, the Hammond brothers indeed journeyed to St. Joseph, where they narrowly missed the newly completed Union Pacific train to Julesburg, Colorado. Along the bustling riverfront, however, they received news of recent gold strikes in Montana. Making an impulsive decision, the brothers booked passage on one of the last steamboats of the 1867 season bound for Fort Benton, 2,100 miles upstream. They had no idea what they were getting into.[4]

Although he had come to the upper Missouri seeking his fortune in the Montana gold fields, Andrew Hammond stumbled into a conflict over forest resources that would frame the rest of his life. The species would change from white pine to cottonwood to ponderosa pine to Douglas fir to coastal redwood. His adversaries would change as well, from Indians to white settlers to government bureaucrats to labor leaders to conservationists, but the struggle over control and access to the forests of the West would remain constant for the next six decades.

In the 1860s white Americans were still newcomers to the upper Missouri, a landscape that had been under contention for centuries, and they scarcely understood the dynamics of the shifting coalitions of native tribes. Just as the French, British, and Spanish empires had grappled over natural resources and political power in eighteenth-century Louisiana Territory, the Lakotas, Assiniboines, Crows, and Blackfeet engaged in a similar struggle for land, resources, and power on the upper Missouri in the nineteenth. The lucrative fur trade, the Montana gold rush, white settlement to the east, a shrinking bison range, and Lakotas expansion all combined to turn this incredibly game-rich region into a war zone with multiple groups contending for diminishing natural resources. The delicate détente established during the fur trade era unraveled, slowly at first, then with increasing momentum, culminating with the extirpation of both bison and the nomadic horse culture that had developed around it.[5]

The United States regarded intertribal conflict over the shrinking bison herds as the primary obstacle to ordered settlement and security and attempted to assert federal control over the region. Both the Fort Laramie Treaty of 1851 and the Lame Bull Treaty of 1855 designated "common hunting grounds" carved from buffer zones that Indians had previously negotiated through ritual warfare, intermarriage, and temporary alliances. The treaties attempted to define boundaries between tribes and establish territories while permitting the federal government to build forts and roads through Indian-controlled land. Although the United States laid claim to the region on the map, the land itself was in de facto control of various groups jockeying for positions of power. While the Fort Laramie Treaty stipulated the land between the Musselshell and the Yellowstone as Assiniboine territory, this area was actually a contested hunting ground between the Blackfeet, Gros Ventres, Crows, and Lakotas.[6]

The fur trade brought Indians of the upper Missouri into the orbit of the global market. Whichever tribe dominated the rich hunting ground of the upper Missouri and Yellowstone Rivers also controlled access to European and American trade items, including guns. Since whites needed furs and Indians needed guns, "the fur trade demanded mutual reliance among all participants," according to historian Barton Barbour. Indians became willing partners, trading bison robes for "luxury and utilitarian goods that would otherwise have been impossible to procure."[7] To keep robes flowing to the center of the trade at St. Louis, traders advanced desirable goods to Indians on credit. Rudolph Kurz at Fort Union noted that by encouraging Indians to make robes, "luxuries will become necessaries and they will be compelled to remain on the Missouri in order to procure them."[8] This transformed the upper Missouri into a year-round production center, concentrating both whites and Indians and intensifying the impacts on the limited riparian areas. In addition, bison became a commodity. Without ownership claims, and without government or community restrictions, the animals were doomed.

The Missouri River itself served as the gossamer thread linking the northern plains frontier with the global market via St. Louis. Mackinaws, or keelboats, powered by human muscle had plied the Missouri River since the days of Lewis and Clark. Then, in 1832, a steamboat reached Fort Union at the confluence of the Yellowstone and wedded the industrial revolution to the fur trade. For Indians, steamboats were a mixed blessing; they brought valuable trade items and annuities but also diseases and white settlers.

As the northern plains fur trade developed, Indians experienced devastating waves of smallpox. In 1837, the steamboat *St. Peter's* brought a load of annuities to the tribes of the upper Missouri. An unintentional cargo had stowed away among the coffee, blankets, kettles, knifes, tobacco, and sugar. By the time the boat reached Fort Pierre, smallpox raged among the crew. Rather than return and forsake his lucrative government contract, the captain continued upriver to Fort Union, spreading the disease along the way. The concentrated populations of the Mandans, Hidatsas,

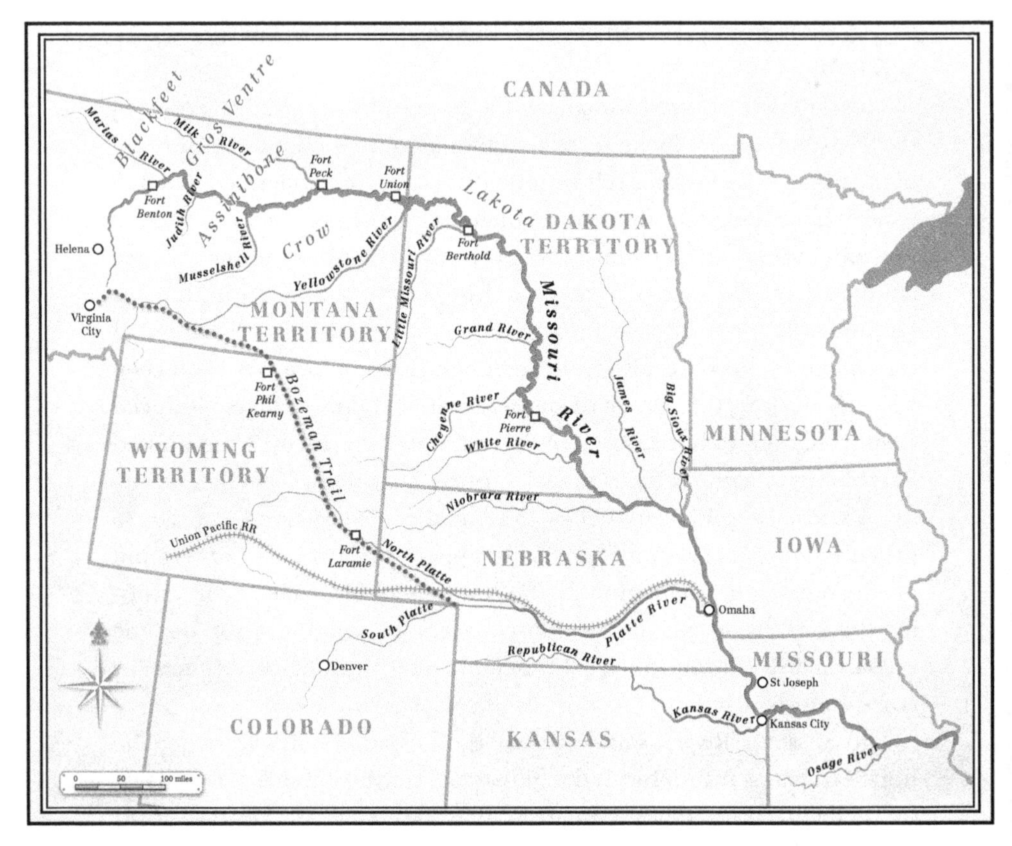

The Missouri River. Map drawn by Gerry Krieg. Copyright © 2014 by the University of Oklahoma Press.

and Arikaras along the Missouri experienced the brunt of the epidemic. Many killed themselves and each other to avoid the horrors of the disease. The Blackfeet also suffered, with a loss of three-fourths of their population, and the Assiniboine never fully recovered.[9]

The Lakotas, however, had received a warning from Indian Agent Joshua Pilcher, who told them to avoid the trading posts that summer. Doctors also succeeded in vaccinating more than one thousand Lakotas the next spring. Furthermore, the Lakotas' nomadic lifestyle helped them avoid the brunt of the epidemic, as they were widely dispersed. With diseases decimating other groups, Lakotas expanded into the vacuum, and their population increased from five thousand in 1804 to twenty-five thousand in the 1850s. They soon became the dominant power on the upper Missouri. But as bison numbers continued to decline, competition between Lakotas and other groups for diminishing resources led to increased territorial conflict. The fur trade and Lakotas expansion coalesced to transform the region into a war zone. In the middle of this swirling vortex sat the Fort Peck trading post, where Andrew Hammond was busy chopping wood.[10]

As the energy source for the steamboats, wood became one of the contested resources. The cottonwood bottomlands, which provided crucial habitat for the bison, deer, elk, turkey, and other wildlife that Plains Indians depended upon for subsistence, lay at the heart of the conflict. These ribbons of forests set into the Great Plains allowed both people and animals to combine the advantages of very different resources and habitats. Furthermore, winter survival for Plains Indians depended upon finding shelter in the bottomlands. Army officer Richard Dodge noted that "a day which would be death on the high Plains may scarcely be uncomfortably cold in a thicket" and Indians, settlers, and animals all "fly to shelter at the first puff."[11]

In addition to shelter, the bottomlands provided security for humans and forage for horses. When camped on the open plains, Indians had to travel farther and farther to find water and grass for the horses, exposing them to attack and weather. The drought cycles and harsh winters of the northern plains could devastate horse herds. Historian James Sherow noted that the characteristic short grasses of the open plains lose half their protein in winter. To compensate, Indians brought their horses into the bottoms where tall bunchgrasses produced up to twenty times more volume per acre. Horses could also eat cottonwood branches and seedlings. Often Indians cut the tops of the saplings which then sprouted more branches in the spring. Lieutenant J. W. Abert reported in 1845, "We were astonished at seeing great numbers of fallen trees, but afterward learned that the Indians are in the habit of foraging their horses on the tender bark and young twigs of the cottonwood."[12]

For Indians on the Great Plains, horses were the fundamental means of procuring food, as well as bison hides, which yielded increasingly desired trade items. Horses had also become a measure of wealth and status. By the 1850s Indians on the high plains averaged five to six horses per person, with some bands exceeding fifteen per capita. After several decades, large horse herds began degrading the bottomlands. Even a small camp with just thirty-eight horses spending the winter could consume all the trees and brush for a quarter mile.[13]

Indians saw the conversion of cottonwood into steamboat fuel as directly threatening their survival. James Morley, traveling up the Missouri in 1862 aboard the steamboat *Spread Eagle,* noted that "much fallen cottonwood lay about, having been cut by the Indians to subsist their horses." Seizing the opportunity, the men on the steamboat spent two hours loading the downed wood.[14] After the steamboats had passed through in spring and summer, Indians returned in winter to find the bottomlands stripped of timber they needed to survive the winter.

Steamboat traffic increased dramatically in 1866 when Red Cloud's War effectively closed the Bozeman Trail, the overland route to the Montana gold mines. The Lakotas, upset at the flood of white gold seekers through their territory over the previous four years, continually harassed both soldiers and wagon trains on the trail. Upon reaching St. Joseph, Missouri, in 1867, Andrew and George wisely decided on the longer but safer journey by river. Safety, however, was relative, as Lakotas began attacking steamboats as well.

The initial surge in steamboat traffic began with the discovery of gold in Montana in 1862, which brought hordes of gold seekers and more than fifteen thousand tons of freight, including mining equipment, shovels, sawmills, and saddles, along with staples like sugar, flour, coffee, and, most important, whiskey. The mining trade soon eclipsed the Indian fur trade, at least in terms of tonnage at Fort Benton. In addition to furs, the boats returned to St. Louis laden with gold—1,225 tons in 1867 alone. At the peak of the gold rush (1865–68) $42 million in minerals left the territory, nearly all via the Missouri River. Each steamboat hauled anywhere from $20,000 to $1 million in gold, not counting the passengers' own booty, which likely exceeded that of the official load.[15]

Normally the steamboat season on the Missouri ran from May to July, but the high-water year of 1866 demonstrated the feasibility of shipping goods and passengers throughout the summer, and the money to be made lured investors into attempting to extend their profits. A two-hundred-ton cargo, for example, could net $40,000 on an upstream trip, plus another $10,000 in passenger fares. Hauling gold, hides, and passengers downriver added even more. With profits that easily covered the cost of the boat, owners often took excessive risks commissioning vessels unsuitable for the treacherous journey.[16]

One such boat was the *Imperial,* a massive 286-ton stern-wheeler built for the St. Louis-New Orleans commerce but pressed into service on the upper Missouri at the tail end of the 1867 season when low water made passage precarious. A cabin passage to Fort Benton on the *Imperial* cost a princely $300, but young adventurers like the Hammond brothers could shell out $150 each to sleep on the deck amid the cargo. George and Andrew thus joined 2,200 argonauts and seventy other steamboats in traveling to Fort Benton in the peak year of steamboat travel on the upper Missouri.[17]

A jovial crowd began the voyage on the *Imperial.* Andrew and George soon discovered, however, why the Missouri River was called the "Harlot": it changed beds often. The deepest channel could switch direction, heading into a bank or a sandbar. Sometimes water rushed over a sandbar, creating a wave train that could swallow small boats. High winds could kick up at any time and shove the boats sideways or even push them back downstream. Shipwrecks and fires spelled an all-too-common end to a journey. By 1895 the Missouri had claimed 295 shipwrecks, the majority caused by running into submerged logs.[18]

Three themes pervade travelers' accounts: the monotony of the expedition, the constant delays from getting stuck on sandbars or loading fuel wood, and the profusion of wildlife. Coming up behind the *Imperial* on the *Zephyr,* Stephen Spitzley repeatedly penned in his diary, "Nothing of interest. We are making very slow progress." He noted with frustration that it took his steamboat seven days to travel from Fort Rice to Fort Berthold when the Indians walked there in just two. In a letter to his family, Peter Koch wrote, "Had I known that it would take this long to go up to Montana, I never would have taken this route."[19] No doubt Andrew and George—neither of whom left a record of the trip—felt the same.

The size of the *Imperial* soon proved problematic, as the boat kept running aground on the many sandbars exposed by the dropping water level of late summer. The sinuous passage and frequent delays as the boat lurched from sandbar to sandbar proved tedious. When the boat ran aground, the crew first tried "sparing," employing two huge poles on either side of the boat angling forward. The crew would loop hefty lines around the poles and around the capstan; pulling the lines taut lifted the boat up and over the sandbar like a giant grasshopper. If that did not succeed, they buried a cottonwood log deep into the bank and tried winching the boat free. Failing that, the captain called for volunteers among the passengers to get in the water on each side and drag a chain back and forth under the boat to get the sand moving. Able-bodied young men like Andrew and George would have leapt to the task. The last-gasp endeavor was to unload all the freight and try again. Sometimes, despite all efforts, the boat remained stuck until rising water levels carried it off the sandbar. This could result in a twenty- to thirty-hour setback. Flushing out the silt from boilers and clogged water lines also caused frequent delays. But on good days, the ship might log fifty miles.[20]

Nearly daily, the need to procure fuel wood interrupted the journey. Steamboats consumed a prodigious quantity of wood, some thirty cords every twenty-four hours. Settlers, as well as Indians, quickly capitalized on this situation and established wood yards along the lower Missouri. The Métis (descendants of French voyageurs and American Indians) and Indian women ran most of the woodlots, charging from $7.50 to $20 per cord, depending on the desperation of the steamboat. As a young lumberjack with a keen eye for detail, Andrew surely noted these exorbitant prices and would have calculated how he might earn a tidy profit by establishing his own woodlot.[21]

Above St. Joseph, the rich bottomlands of cottonwood, hickory, and oak stretched out for a mile or more on each side of the river. But as the steamboats chugged upriver the scenery became more varied, with high bluffs flanking the forested bottomlands. Past Sioux City the wood yards petered out, and the steamboats had to rely on the crew and deck passengers like Andrew and George, who earned extra money by collecting and cutting cord wood. Upriver, however, the Lakotas readily attacked and drove off steamboat crews attempting to load wood. When a boat needed fuel, men leaped ashore and loaded as much as they could as quickly as possible, leaving the large pieces to be cut on board ship. To avoid detection by Indians, crews often loaded wood at night.[22]

Above the Niobrara River, the Missouri River corridor was little more than an ephemeral strand of American presence through vigorously defended Lakota territory. Regarding the steamboats as an intrusion, Lakota Little Elk informed Lieutenant Henry Maynardier at Fort Berthold in 1860, "We don't want to see any white people or any steamboats, because the goods the steamboats bring up make us sick."[23] A week later, 250 Lakotas rushed Fort Union, killing twenty-five cows and setting fire to the lumberyard, outbuildings, and two mackinaw boats.[24] When the preeminent fur trader Charles Chouteau and Indian agent A. S. Reed presented annuities accompanied by a patronizing speech to several hundred Lakotas at Fort Pierre, they were stunned by the response. The Indians refused to accept the goods and stated that they, in fact, did not "appreciate the guardianship of the Great Father." The whites were invading their country, they said, and "the time was close at hand when they would put a stop to . . . traveling up and down the Missouri River." Lakota leader Standing Elk then accused the whites of property infringement by killing bison and cutting down timber. He informed Chouteau that the trading days were over and "you may consider us your enemies."[25] Indeed, Lakota attacks on steamboats increased in frequency, the years 1866–69 marking the height of conflict on the upper Missouri.[26]

Although the Lakotas rejected the white presence, the Assiniboines, Gros Ventres, Crows, and Blackfeet continued to engage in the fur trade. Fearful of the Lakotas, all but the Blackfeet sought protection from the U.S. military. The army, however, was incapable of providing security even for whites. As one observer noted, "The soldiers in all these forts on the river are kept penned up by the Indians like so many cattle."[27] The situation had become so hazardous that steamboat captains began armoring their wheelhouses with sheet metal to fend off arrows and bullets.[28]

As with the forts along the Bozeman Trail, the Lakotas resented the military presence on the Missouri. At Fort Union, Edwin Denig reported that Hunkpapa Chief Little Bear threatened to burn the forts, forgo the robe trade, and "return to their primitive mode of life."[29] While Denig attributed the conflict to "bad council," the Indians themselves pointed toward the steamboats, which brought disease and whiskey, depleted game populations, and cut down trees for fuel. Throughout the winter of 1866–67, eleven camps of Lakotas, including Sitting Bull's, sporadically harassed Fort Union with potshots at anyone venturing too far afield. Twice Lakotas captured the fort's sawmill but were subsequently driven off by the troops.[30]

The marked graves of those killed by Indians on the banks of the Missouri provided sober reminders to Andrew and others on the *Imperial* of the perilous nature of their journey. When the boat stopped to pick up travelers, the new arrivals would entertain and beguile the existing passengers with stories of Indian attacks and frontier tales. As the boat steamed past legions of mounted warriors lining the bluffs, the stories became increasingly compelling. Despite repeated warnings against walking onshore, young men like Andrew and George often grew restless and would disembark for short hunting forays, cutting off the big oxbow bends and meeting the boat on the other side. At the Great Bend, passengers walked the four miles across while the river wound forty miles around. One unlucky hunter on the *Imperial's* return journey failed to return to the boat and was found scalped, mutilated, and filled with arrows.[31]

With some ten thousand argonauts traveling the Missouri in 1867, the Lakotas became adamant about ceasing steamboat traffic. Bull Owl, a Lakota leader, expressly tied the white intrusion to game depletion. General John Sanborn, however, dismissed his concern, telling the Lakotas, "The steamboats that run on the river do not disturb your game."[32] Passengers, however, filled their diaries with accounts of steamboats plowing into herds of bison as they swam across the river, the paddle wheels often striking and wounding the animals. More bison deaths came from gunshots. With the first sighting of the great beasts, symbols of the untamed West, Andrew and George likely scrambled for their guns along with the rest of the men. The crack of gunshots and the acrid smell of gunpowder punctuated the air. Captains even used their steamboats to pin bison against the riverbanks to prevent their escape as men unloaded their guns into the herd. Aboard the *Imperial,* John Napton reported, "The wheel was reversed, in order to hold the boat amongst them, and everybody commenced shooting with pistol, shot gun or rifle and the buffalo swimming frantically in every direction to get away."[33]

As the *Imperial* steamed through Dakota Territory, the voyage began to mirror the vast sameness of the endless plains that extended in every direction. As the novelty wore off, tedium set in. Each day was the same as the previous—the same scenery, the same passengers, and the same poker game. Even running aground became less of an event and more toil, drawing out the journey even more. Brief excitement came from an occasional pronghorn, or when the ship passed a memorial ground where "several dead Indians were seen in the trees nearby and a great many bones in all stages of decay were scattered around."[34]

Above the Yellowstone, the landscape changed. The candy-striped Dakota badlands of bentonite clay yielded to gently rolling grassy country with cottonwood bottomlands. But the hot August sun remained, and Andrew, George, and the other passengers aboard the *Imperial* sweltered in the heat and battled mosquitoes all night. The prairie grasses had long since dried, and dust clouds from distant bison herds rolled across the sere landscape. Wildlife abounded above the Yellowstone. Wolves and grizzly bears followed the bison herds; elk and deer grazed the river bottoms, and bighorn sheep scrambled across the white sandstone cliffs. Within thirty years, however, Audubon's bighorn would be hunted to extinction and the bison, wolf, and grizzly would be extirpated from the upper Missouri.

The steamboats and mackinaws returning from Fort Benton brought reports that the best gold diggings had all been taken, adding to the brothers' discomfort. The Montana gold rush had already peaked, and they passed several keelboats each day as miners hurried back to the states with unrecorded thousands of dollars in gold. These accounts, combined with the journey's boredom, and his own restless and impatient nature, caused Andrew to reconsider his plan. Between them the brothers had only thirteen dollars, and likely they realized that arriving in a mining camp without a grubstake was unwise. Perhaps Andrew had read Washington Irving's *Adventures of Captain Bonneville* and imagined the romantic life of a fur trapper. There was also substantial money to be made as a woodcutter or "woodhawk" running a wood yard on the river. A hard-working fellow could clear $200 off a single steamboat. Whatever the reason, when the boat pulled up to Fort Peck, Andrew decided he had had enough of river travel. He bid farewell to his brother and enlisted as a fur trader. If he had any idea of the explosive situation of the upper Missouri, he likely would have persisted on to Fort Benton, another three hundred miles.[35]

The *Imperial* continued its struggle upstream. Finally, on September 11, the boat lurched to a standstill at Cow Island, 130 miles below Fort Benton. By this late in the season, the river had dropped so low that it was impossible for the big steamboat to continue, so the *Imperial* disgorged its cargo and passengers. Awaiting teamsters then transferred the loads to ox-drawn wagons that hauled the cargo and people the rest of the way across the broken plains to Fort Benton. When George finally arrived at the bustling port, he, too, eschewed the gold fields and obtained a fourteen-month contract to supply wood to the U.S. government.[36]

Unfortunately for Andrew, the fur trade had also peaked and was in decline. Since the late 1820s, the Chouteau family, through the American Fur Company, had dominated the Missouri River fur trade, but the loss of government contracts in 1865 prompted Charles Chouteau to close his business. Several new outfits rushed in to fill the void, including Durfee and Peck, an upstart company that would control the Missouri River fur trade for the remainder of its days. Unlike the American Fur Company, which relied primarily upon Indians for acquiring furs and hides, the new company employed about a hundred white men who, in addition to staffing trading posts, actively hunted and trapped for hides themselves. The Indians saw this as an infringement upon their economic livelihood and became increasingly confrontational.[37]

Fort Peck, at the junction of the Missouri and Milk Rivers, was a highly desirable spot for both steamboats and Indians. Perched on a ledge above the river, the post was ideally placed for rear-wheel steamboats, which could dock right at the fort to load wood and furs. Fort Peck was a crucial spot for loading fuel wood for the long badland stretch upstream, where the hills were "entirely bare, looking like fresh earth as thrown up on public works." The confluence also sat on a bison migration route, which appealed to the Gros Ventres and Assiniboines. While Fort Peck was a favorable location for steamboats, the Durfee and Peck trading company could not have chosen a worse site for safety. With no federal military presence for the four hundred miles between Fort Buford and

Camp Cooke on the Judith River, whites in this region placed themselves at suicidal risk.[38]

Autumn was a slow time for the fur trade. With the passing of the last steamboat in September, Andrew and his companions scattered across the landscape to eke out a living by cutting fuel wood, trading for hides, and poisoning wolves for their pelts in anticipation of the arrival of steamboats the following spring, when they could cash in. Looking to join the woodcutters, two men from Fort Buford were fired upon by Indians and fled back to the river. Unnerved, they flagged down the *Miner* and caught a ride back downriver. Receiving these reports of the escalating "Woodhawk War" that was sweeping through the upper Missouri, George became concerned about his brother's safety—and with good reason.[39]

Andrew and his companions had just pitched their camp when a lone Indian arrived. Thinking he posed little threat, the woodcutters let him into their tent. Before they realized what was happening, he pulled out a tomahawk concealed under his blanket and killed one of the men before making a quick escape. Understandably, this put the woodcutters on guard the rest of the season. The men lived in constant fear of attack and never went anywhere alone. Even traveling between their winter cabins, they carried guns. The objective of the Indians, however, seemed to be a campaign of harassment in an effort to close the Missouri to steamboat traffic. Describing a typical encounter, Peter Koch wrote, "As soon as the Indians were discovered they went away after exchanging a number of shots, whether anyone was killed on either side or not."[40]

Winters were hard on the men. Isolated in small clusters once the ice jammed the river and prevented traffic, they had little to eat but catfish and game. Koch summed up his life in his diary: "twenty-five years old and poor as a rat." Winter, however, provided an opportunity to harvest wolf pelts that they could sell for three dollars apiece. Working in pairs or small groups, Andrew and his partners became "wolfers" and set out for buffalo country but avoided known Indian hunting grounds. Upon discovering a herd of bison, a wolfer shot one, taking care not to startle the herd. He then dipped a porcupine quill in strychnine and blew it into the veins of the dying animal. The still-beating heart pumped the poison throughout the body. A mile or so away, the wolfers killed another bison and repeated the process until they had thirty or forty poisoned bison carcasses in a great circle. Returning a few days later, the men stacked the dead, frozen wolves that had come to feed on the carcasses into a big pile to keep magpies from ruining the pelts while they waited for milder weather to skin them. Where Indian presence proved too dangerous, wolfers returned only in the spring to collect the dead wolves, sometimes getting more than a hundred off a single bison.[41]

Even the Gros Ventres and Assiniboines, who were on friendly terms with the traders, had little regard for the wolfers and their wanton destruction of bison. Indians witnessed their dogs poisoned by the bait. Even greater was the loss of their horses when they ingested the grass coated with strychnine that the wolves had vomited up before dying. In retaliation, the Indians "cut up the wolf skins whenever they had a chance and annoyed the wolfers in every possible way." The Crows, for example, would frequently steal the wolfers' horses and their belongings, forcing the men to walk back to the post, which could be dangerous on the open plains exposed to winter blizzards. Lakotas, however, would just as soon slay a wolfer as a woodcutter, as they were often one and the same. A few years after his woodcutter winter, Peter Koch wrote, "Since leaving the place I have often wondered that we were not all killed."[42]

Nearly all Indians had special contempt for the woodhawks as well, whom they regarded as depleting a valuable resource. The Lakotas specifically blamed woodcutters for "spoiling our game."[43] In his council with the Hunkpapa band, Father Pierre DeSmet received an earful of complaints. Sitting Bull expressed his opposition to wood cutting along the Missouri River, while his companion, Black Moon, justified his hostility toward whites, stating, "They ruin our country; they cut our timber with impunity."[44]

White society also disparaged woodhawks as degenerate, filthy, ignorant, shiftless, conniving, cheating extortionists. Many were misanthropes and malcontents who fit into neither white nor Indian society. The most infamous woodhawk fitting this depiction was one of Andrew's upstream neighbors, "Liver-eatin'" Johnson, who maintained his woodlot for more than a quarter century at the mouth of the Musselshell River. As one of the last old-time mountain men, Johnson reportedly killed eighteen Crow warriors one by one. Not only did he scalp them, but he eviscerated their bodies and ate their livers raw in front of the survivors. Some say he just pretended to gobble the organs, rubbing them against his beard. Either way, he gained a reputation among Indians, who left him alone. His collection of grimacing, white skulls on stakes pitched along the riverbank created quite a stir among the more genteel steamboat passengers when they stopped for fuel at Johnson's woodlot. Most woodhawks, however, were restless young men like Andrew Hammond and Peter Koch. Seeking cash and adventure, they spent a season, rarely more, bucking firewood to feed the insatiable appetite of the steamboats.[45]

Spring brought anticipation and apprehension to the woodcutters. With the break-up of river ice in April, Andrew could expect the arrival of the first steamboat the following month and sell the stock of wolf hides and fuel wood he had amassed over the winter. On the other hand, the warmer weather brought an increase in Lakota hostilities. J. A. Wells, a woodhawk on the Musselshell, counted fifty-eight whites killed between the Yellowstone and Judith Rivers in 1868. The Woodhawk War came to a head the next spring when fifty woodhawks and wolfers gathered at the mouth of the Musselshell to await the first steamboat of the season. When the Lakotas attacked, the woodhawks took advantage of their unusual numbers and new breech-loading rifles. A bloodbath ensued, resulting in the death of thirty-three Indians.[46]

The open warfare sweeping across the western plains led the United States to create a peace commission to negotiate safe transportation corridors to the Montana mining camps. The commissioners agreed to close the Bozeman Trail in exchange for peace on the Missouri. Woodcutting, however, remained a sticking point for Lakota leaders who regarded the cottonwood bottomlands as tribal property. Unlike Black Moon and Sitting Bull, these bands regarded the timber as an economic asset that they wished to control. One Lakota leader, The Grass, told the commissioners, "I want us to be paid for that wood which is being taken out of our country." Both he and Two Bears, another leader, insisted that their people should choose who could cut wood, essentially granting concessions to men "who have lived in our country ten years, that have Indian families whom they are supporting."[47] Acknowledging the Lakotas property claims, General John Sanborn agreed to let them take charge of supplying wood for steamboats or at least receive monetary benefits.

Congress, however, failed to ratify the Sanborn agreements, and the Lakotas came to view this as yet another empty promise.

Despite the Lakotas' military strength and negotiating skills, events far beyond their control changed the dynamic on the upper Missouri. The waning of the Montana gold rush and the completion of the first transcontinental railroad in 1869 created a precipitous drop in steamboat traffic, rendering profits from woodcutting negligible. The next year, steamboat traffic plummeted to a mere eight vessels, and over the next few years Montana's annual mineral production dropped to $4 million. Out of the several hundred cords that Peter Koch and his partner bucked during the winter of 1869, they sold fewer than twenty-five. Economics, more than Indians, drove the woodhawks off the Missouri.[48]

Although mining declined in the 1870s, rampant industrialization in the East increased the demand for bison, not as Indian-processed robes but as dried hides to be manufactured into tough leather belts for factories. In 1868 Durfee and Peck took in 25,000 to 30,000 buffalo robes. Ten years later, the company shipped 75,000 bison hides down the Missouri. By 1884 the bison were gone from the Great Plains.[49]

Many woodhawks, including one of Andrew's companions, lost their lives during the Woodhawk War, but Hammond, Peter Koch, and Liver-eatin' Johnson all escaped from the Missouri quagmire. Koch headed to Bozeman, where he eventually became an officer of the First National Bank. Johnson took his reputation to Red Lodge, Montana, where he became sheriff. He later went on to fame when his name was changed to "Jeremiah" and Robert Redford provided a new visage.

Uncommonly ambitious, Andrew Hammond had no wish to be another Liver-eatin' Johnson, but he was not about to abandon all his hard work during the winter of 1867–68. With the last big steamboat year, Hammond reaped a tidy profit from his woodlot and wolf hides, but by late summer he was ready to leave and boarded a boat for Fort Benton. The Montana gold fields were still producing, and he might just strike it rich. Besides, anything was better than another winter as a woodhawk on the upper Missouri.

In later years, Hammond attached a high degree of importance to this period of his life. This was the first time he had done anything significant on his own, without George or his other siblings. The experience infused Andrew with a sense of self-reliance and confidence that would remain with him the rest of his life. The desperately hard work for little reward left him with little patience for young men who complained of long hours and low pay. Andrew also drew a metaphorical lesson from witnessing his companion's murder. He often told his managers, "Never let an Indian in your tent," meaning never take someone into your counsel who might not be entirely on your side. Indeed, for the rest of his life, Hammond drew almost exclusively upon his relatives to compose his inner circle of managers. Most of all, Hammond saw the experience as an initiation, a badge of Old West legitimacy that set him apart from those who came afterward, and he would use it to establish a degree of authority. In lieu of a formal education, this became his diploma.[50]

CHAPTER 4

To Hell's Gate and Back

In the spring of 1868, Andrew Hammond stepped from a nine-month wilderness exile and life-threatening ordeal into a maelstrom of commercial activity. Disembarking the steamboat at the Fort Benton levee, Hammond picked his way among barrels, boxes, towering bundles of buffalo robes, and closely guarded gold ore awaiting shipment downstream. Between the far end of town and the levee, sixteen-mule teams pulling covered wagons stood ready to haul freight bound for the gold fields and trading posts. Even more important than the barrels of whiskey being loaded on the wagons was the machinery for the ore-crushing quartz mills. The mills signified the end of the placer mining era, in which independent miners with little more than picks, shovels, and gold pans could work the surface and streams. To get at the ore underground and mill it into a form that could be transported back to the states required massive capital investment, which was readily supplied by European and Wall Street financiers and corporations.[1]

Young men seeking adventure in the rough and tumble town found it, for better or worse, in the saloons and back alleys. Miners, coming and going, shared whiskey with fur traders and woodhawks. Gamblers in tailored suits fleeced cowboys and teamsters fresh off the trail. Indians of many different tribes traded furs for knives, guns, kettles, and beads. Prospectors were often stuck in Fort Benton waiting for overland transport to the Helena area mines, while others waited for downstream transport back to the states. With little to do, men frequented the saloons, gambling houses, and brothels, all of which were open twenty-four hours a day. In Benton everything was legal and all sorts of vices were available to separate a miner from his gold. Outright robbery and murder for a poke of gold were not uncommon. Termed the "bloodiest block in the west," Fort Benton averaged three murders per month in 1867–77.[2]

Fresh off the boat after nine months of chopping wood, Andrew Hammond, not surprisingly, headed for the nearest saloon. George, who had long given up his brother for dead, nearly choked on his whiskey when Andrew walked through the door. Happily reunited, the brothers recapped their adventures and planned their next move. Visions of gold and striking it rich still danced in their heads, but news from Helena indicated that the placer mines had played out. Furthermore, George still had three months left on his government timber contract, so Andrew took a job at a local lumberyard.[3]

As soon as George's contract expired, the Hammond brothers headed for the gold mines to assess their prospects, however discouraging. Passing through Helena, which was little more than a collection of log cabins at the mouth of Last Chance Gulch, they discovered that they had indeed arrived too late and the best claims were already taken. Disheartened, the brothers decided to continue west, following Mullan Road to the trading outpost of Hell Gate near the confluence of the Clark Fork and the Bitterroot River.[4]

Conveniently located between the gold fields and the Oregon country, Hell Gate served wagon trains, settlers, and prospectors. Indians passed through on the road to buffalo, while settlers in the Bitterroot Valley sought supplies and an outlet for their produce. Traders brought supplies from Walla Walla to the west and from Fort Benton to the northeast to redistribute to mining camps and settlements that seemed to be springing up everywhere. To the north, the Kootenai gold rush of 1864 created "a demand for all kinds of supplies, and everything sold at war prices." The Higgins and Worden trading post raked in $64,000 in two years, prompting Frank Worden and C. P. "Cap" Higgins to build a sawmill four miles to the east, where Rattlesnake Creek could provide power. In 1865 they installed a flourmill and moved their store from Hell Gate to the mill. David Pattee set up a hotel and carpenter shop nearby, and the town of Missoula was born.[5]

Without much to support it other than the gold rush, however, Montana's entire economy began to fizzle in 1868. From the Bitterroot Valley, John Owen observed, "Things look gloomy throughout the length and breadth of the Territory. Unless some New diggings are found I fear we are all Bankrupt." The territory's white population had topped out two years earlier at 28,000, dropping to 20,595 by 1870. American Indians still outnumbered the settlers by nearly 10,000.[6]

The lack of prospects in Montana prompted the brothers to continue west. George, who was something of a horse trader, picked up several dozen horses in Hell Gate with the intention of selling them on the West Coast. Traveling through the Cascades, the brothers were, no doubt, astounded at the size and quantity of trees they passed. Endless forests of towering Douglas fir that dwarfed the white pine of New Brunswick unfolded before them. Arriving in Puget Sound, they saw timber camps nestled into nearly every cove. Loggers had cut most of the trees near the bay but continued to haul enormous logs out of the forest and into the water, where they assembled them into rafts that could be towed to one of the many mills that ringed the sound. A pall of wood smoke rising over the forest indicated they were approaching Port Gamble, the signature sawmill of Andrew Pope and William Talbot's Puget Mill Company, the region's largest lumber company.[7]

At Port Gamble, thousands of logs floated in the cove before being sucked into the giant sawmill. On the far side of the mill, rough-cut lumber lay stacked on the dock waiting to be loaded onto one of many ships anchored nearby. On the bluffs above sat a replica of a New England mill town: cottages with white picket fences, a bunkhouse, a hotel, a church, the largest store on the sound, and the two-story house of Cyrus Walker, the mill manager.[8]

Like most lumber companies, the Puget Mill Company preferred to hire Protestant men from Maine and New Brunswick as they had the reputation of being hardworking and sober. Andrew and George quickly landed jobs. They may have lined up employment with Pope and Talbot before their arrival, since they were distantly related to one of the firm's founders. Even if the Hammond brothers had not contacted William Talbot directly, their name and origins were certainly assets. In the timber industry, lumberjacks lay at the bottom of the hierarchy, suffering from hard work, low pay, and miserable conditions. With the Hammond brothers' connections, experience, and education, Walker likely hired the young men to work not in the woods, as they had in Maine and Pennsylvania, but instead in the mill, office, or store.[9]

Despite the relatively good pay at the mill—$30 for an eleven-and-a-half-hour day—after six months, George headed south to San Francisco and then to the Sierras to continue lumbering. Andrew, however, stayed on at Puget Mill, which provided a valuable education on how to run a successful lumber company. From vertical integration to timber poaching to company towns, Hammond followed the model of the Puget Mill Company.

Like most nineteenth-century West Coast enterprises, the Puget Mill Company owed its genesis to the California gold rush. Although the mining camps could obtain lumber from the Sierra Nevada, San Francisco's high demand and port location made transporting lumber via ship more practical than hauling logs from the mountains. While the Pacific Northwest contained abundant and immense forests, the rocky coast offered few anchorages. Puget Sound, however, provided both trees and safe harbor, and in the fall of 1853, entrepreneurs William Talbot and Andrew Pope established a sawmill at Port Gamble. By 1858 Port Gamble boasted the largest mill and largest store on Puget Sound, employing 175 men, while across the bay Seattle housed fewer than one hundred.[10]

On the West Coast, lumbering and shipping went hand in hand, and Pope and Talbot realized they could greatly reduce shipping costs by operating their own fleet. By 1861 the Puget Mill Company controlled eleven ships plying the waters between Puget Sound and San Francisco. While California remained the primary market, it could not absorb all that the Puget Sound mills produced, and before long the Puget Mill Company was exporting lumber to Hawaii, Australia, and South America. Employing "a ruthless approach to competitors," the firm dominated the export trade.[11] Headquartered in San Francisco, Pope and Talbot ran door and window factories as well as lumberyards, dabbled in real estate, and were considered the "lumber kings of the Pacific Coast."[12] In the early twentieth century, however, the Hammond Lumber Company would usurp their crown.

As was standard practice, Pope and Talbot, along with other lumber companies, simply cut whatever they wanted regardless of land ownership. Noting the "continued trespass," U.S. Attorney John McGilvra declared in 1861 that the mills in Puget Sound "are almost wholly supplied with logs from the government lands." McGilvra secured indictments against several lumbermen, but evidently lumbermen could avoid prosecution with a cash payment to McGilvra. By the late 1870s, lumber companies, primarily the Puget Mill Company, had cut illegally an estimated $40 million in timber from public lands around Puget Sound. Ten years later, however, attitudes toward timber poaching shifted dramatically, and it was Andrew Hammond who became the target of the federal government.[13]

Besides outright theft, lumbermen also purchased timberlands from the federal government for absurdly low prices. Congress had intended the bargains to promote settlement, not land consolidation, but lumbermen formed a substantial power block within Washington's territorial government and could thus promote the selling of timberlands that the federal government had granted to the territory to finance the state university. Using military scrip, the company bought an additional 17,398 acres from the federal government for $1.25 per acre. Through such methods, by 1874 Pope and Talbot had amassed some 186,000 acres to become the largest private holder of timberlands in Washington, with the exception of the Northern Pacific Railroad.[14]

While Hammond would follow Pope and Talbot's lead in vertical integration, overseas marketing, and timber poaching, perhaps the deepest impression Hammond received was his experience living in Port Gamble, the quintessential company town. At Port Gamble, the Puget Mill Company was the only employer, owned all the houses, owned the only store, and controlled transportation in the sound by requiring all ships to use their tugboats. Historian Robert Ficken noted that resident managers ruled over life in mill towns "like little kingdoms," with near-absolute authority. The isolation of Puget Sound insulated the company towns from government interference, industry competition, and labor unions. Ficken wrote, "The lumber companies managed affairs as they saw fit, expelling workers who caused trouble, and through the company store, monopolized trade with employees and local farmers."[15] Decades later, Hammond applied this same model in building the mill towns of Bonner, Montana, and Samoa, California.

In the meantime, however, Andrew continued to fantasize about striking it rich. At twenty-two, he was still restless, and after eighteen months at Puget Mill, news of a new gold strike in Montana rekindled his interest. Quitting his job at Pope and Talbot, Hammond returned to Montana, heading for the Cedar Creek mining district on the lower Clark Fork River, sixty miles west of Missoula. When Andrew arrived in the spring of 1870, he found three thousand men crowded into the narrow gulch and the placer mining nearly exhausted. Once again, he had arrived too late. From the mining to lumbering frontiers, Hammond was just behind the first wave of prosperous pioneers.[16]

Stymied by his mining prospects, Andrew continued upriver back to Hell Gate and took a job with George White, who was in dire need of a competent clerk for his general store. White's health was failing, however, and he died the following August of "abdominal dropsy." Andrew stayed on to help Mrs. White and her three young children, until she decided to close shop. By now, residents had dismantled most of Hell Gate and used the wood from its buildings for new construction in Missoula or for firewood. Without a store and post office, Hell Gate, once a vibrant frontier town with shootings, vigilante hangings, and fistfights between judges and defendants, quickly faded away. With money in his pocket and a solid reputation as a competent clerk, Hammond drifted a few miles east to Missoula, where the town's population had tripled to three hundred from the previous year as a result of the Cedar Creek strike. Countywide, the first census in 1870 yielded 2,554 white inhabitants, of whom 2,084 were male. While economic opportunity was higher on Hammond's agenda than romance, he would find both in the booming new town.[17]

By 1871 Missoula consisted of some fifty buildings housing two stores, a flourmill, two hotels, two blacksmith shops, a sawmill, a public school, a Catholic Church, a new courthouse, a Chinese laundry, a brewery, two lawyers, two physicians, livery stables, several saloons, a Masonic lodge, and even a bookstore. The businesses were open on Sundays as that was when people from the surrounding countryside came to town. Emma Slack, Missoula's first schoolteacher, noted disapprovingly, "There was no Sabbath observed then, stores wide open, often town full of Indians horse racing." Although it was a supply town, mining strikes could quickly empty the town. But when prospectors found the Kootenai gold rush a bust or arrived at the Cedar Creek mining district to find others had already snatched all the claims, they filtered back to take up more steady prospects such as farming, ranching, and merchandising.[18]

The valley's social and economic activity swirled around the new adobe mercantile of Bonner and Welch on the corner of Front and Higgins. At this favorite gathering place, men discussed politics and whispered rumors of new gold strikes while women debated the quality of cloth with fashionable clerks and stocked up on staples such as flour, sugar, and canned goods. Five years earlier, E. L. Bonner, a former clerk at New York's prestigious Lord and Taylor, had arrived in Missoula with a pack train of 125 mules and horses loaded with merchandise. Bonner and his partners, D. J. Welch and Richard Eddy, soon hit upon a successful merchandising strategy: bring the goods to the customer. Realizing the scarcity of supplies in mining districts around Missoula, they began setting up temporary stores in tents. These "branch stores" required little capital investment and could literally fold up shop and move on once the diggings were exhausted. The temporary stores required a central warehouse and distribution facility, and Missoula offered an ideal location.[19]

With its mercantile web, Bonner and Welch served as the glue between the hinterland and the emerging national market. Farmers in the Bitterroot and Missoula Valleys grew wheat and produce, which they sold to Bonner and Welch for distribution to mining camps. In turn, settlers purchased their staples and winter groceries from the mercantile. Everything from apples and bacon to shovels and boots arrived via mule train from either Walla Walla or Fort Benton. In 1869 the completion of the Union Pacific brought the railroad to within six hundred miles of Missoula, and goods could now be sent via rail instead of steamboat. Once goods reached the railhead at Corinne, Utah, however, mule or oxen teams still had to haul supplies to Montana, an arduous ten- to fourteen-day overland journey that they could make only during the summer and fall, since mud and snow blocked the routes for months.[20]

Despite the railroad connection in Utah, Missoula remained a remote outpost. Only two stage lines served the valley: one to Fort Benton, taking four days, and the other to Walla Walla, which took five. The long months of cold and isolation led many Montanans to winter in the states, despite the distant and harrowing journey to Corinne. Residents hoped the seclusion would soon end. As early as 1867, the *Helena Weekly Herald* reported that the route of the Northern Pacific would run through Missoula. Three years later, Bonner returned from the east with exciting news. He reported that the railroad had already laid three hundred miles of track from Minnesota, would complete another eight hundred the following year, and should arrive in Montana by 1872. Bonner's prognosis proved overly optimistic. The railroad would not reach Missoula until 1883, leaving the region relatively isolated from the rest of the country for another ten years.[21]

Bonner's attention was also drawn to a young man with a solid reputation as an honest and competent clerk, and Andrew Hammond landed a job with Bonner and Welch. Following the death of George White in 1871, Hammond had shown an aptitude for financial matters and become known throughout the community for his firm and impartial transactions. For Bonner, Hammond's fluency in French was an additional asset as it allowed him to conduct business with the region's sizable Métis population. In October 1872, Bonner and Welch dissolved their partnership and Hammond moved into the void, becoming head clerk. Bonner, Eddy, and Hammond soon established a working relationship that capitalized on their respective talents.[22]

An avid outdoorsman, Eddy preferred to spend his time hunting and fishing rather than working in the office, and he was virtually absent during the fall hunting season. Bonner, on the other hand, "liked to strut around the Waldorf-Astoria" and enjoyed cavorting with financiers and politicians on the East Coast.[23] He thus became the buyer for the company, often spending as much as eight months at a time in New York. Bonner also served as a political liaison when the firm became embroiled in disputes with the federal government over access to timber resources. Early on, the Missoula press recognized Bonner's influence but perhaps overstated his importance, printing that "Mr. Bonner is now in Washington helping Congress to regulate the affairs of the nation."[24]

In Bonner, Hammond discovered a mentor and father figure who suited his serious and ambitious nature more than his footloose and carefree brother, George, did. Hammond looked to Bonner as the consummate businessman and began to model himself after him, even cutting his hair and beard in an identical style; their photographs bear an uncanny resemblance to each other. As was the fashion among gentlemen, such as E. L. Bonner, Hammond replaced his first name with his initials: A. B.

Whenever Bonner left town on business or Eddy went hunting, neither man had any doubt that when they returned every penny would be accounted for and not a candy cane missing. When Bonner and Eddy realized the competence and skill of the young man they had hired as their clerk, they quickly made Hammond the store manager and considered it a stroke of good fortune.[25]

Despite his ability as "a knight of the yard stick and scissors," after a few years Hammond grew restless again.[26] With the arrival in Montana of his uncle Valentine "Walt" Coombes and his brother Fred, he took leave of storekeeping and joined his relatives in a ranching venture near Flathead Lake. The open air seemed to agree with him, and in his spare time A. B. began writing poetry. Thinking he might have finally found his true calling, when he returned to Missoula Hammond tried out his verses on newspaper editor Chauncey Barbour. Barbour informed his readers, "Andrew's smiling phiz has a better effect on a person than a dose of laughing gas. But after hearing one verse of Hammond's poetry, Barbour suddenly remembered he "had urgent business" elsewhere.[27]

Poetry, apparently, was not in Hammond's future. Nor was ranching, as he eventually sold off his entire herd. His next experience, as a traveling snake-oil salesman, likely convinced him that the mercantile business held the most promise. In the spring of 1876, Hammond joined H. C. Myers in traveling up and down the Bitterroot Valley by wagon peddling cure-alls. This time, however, A. B. returned much less buoyant. Barbour noted Hammond's "green and melancholy look."[28]

Perhaps Hammond had simply ingested a bit too much of his own remedy. Nevertheless, this marked the first indications of the poor health that would plague him for the rest of his life. The following winter, Hammond fell "dangerously ill" with an unspecified ailment. The snake oil having failed, he turned to another health fad—electric shock—and hooked himself up to a "galvanic battery." Unfortunately, he could not control the electric force and thus received a severe shock that laid him out for a few days.[29] Whatever the cause, Hammond's poor health cured his wanderlust, and he settled into his role as a storekeeper.

Although less romantic than finding a gold mine or building a cattle empire, mercantilism proved a far more successful path toward wealth in the American West. Countless thousands of gold seekers died destitute, while the mundane mercantile begat great fortunes. Copper magnate William A. Clark and railroad builder Collis P. Huntington started as storekeepers.

By July 1876, Hammond's business acumen had so impressed Eddy and Bonner that they made him a full partner and reincorporated as Eddy, Hammond and Company, while Bonner opened another store in Deer Lodge, eighty miles to the east. With Eddy off hunting and fishing, Hammond soon became the firm's point man and general manager. The company expanded, opening branch stores in the nearby Bitterroot and Flathead Valleys. The partners also replaced their old adobe trading post at the corner of Higgins and Front with a new brick building that would eventually become the venerable Missoula Mercantile Company. Although Hammond had no way of knowing it at the time, becoming a store clerk for Bonner and Eddy was the single most important event in his life. His career, his fortune, and his political connections were all forged inside the mercantile.[30]

In late July 1877, most of Missoula's residents gathered at the nearly completed Eddy, Hammond and Company building. With nervous anticipation, many in the crowd glanced southwest over the Clark Fork River, their eyes following the Bitterroot Wagon Road, which sliced diagonally across the valley toward the mouth of Lolo Canyon. From the rooftop of the building, armed men stood guard while women and children sought refuge in the basement. The tension in Missoula hinged on events fifteen miles away. A. B. Hammond, Chauncey Barbour, and about two hundred other volunteers of the ad hoc "Montana Militia" had entrenched themselves in Lolo Canyon in anticipation of a clash with approaching Nez Perce Indians.[31]

On the heels of Custer's defeat at the Little Bighorn the year before, trepidation about a general Indian uprising pervaded the territories. The army, spurred on by settlers, had attacked several Nez Perce camps in Idaho during this wave of fear and hysteria. With the Sioux wars to the east and now the Nez Perce problems to the west, Missoulians felt particularly vulnerable. Newspaper accounts contributed to the anxiety, presciently assuming the Nez Perce would flee Idaho, head over the Lolo trail, and pass through Missoula or up the Bitterroot following their traditional routes. Insisting that there was a "splendid chance to strike the Hostiles by way of the Lo Lo Pass with 4 or 500 well armed men," "Cap" Higgins rallied a volunteer force that included Hammond, who joined Captain E. A. Kenney's company. Higgins, on the other hand, declined to take part in the action, claiming that his life insurance policy prevented him from doing so.[32]

Coming off Lolo Pass, the Nez Perce bumped into the Missoula volunteers ensconced behind their hastily built blockade. Chief Looking Glass proposed that if they let them pass unmolested, the Indians in turn would "shed no blood in the Bitter Root Valley."[33] No doubt this sounded like a reasonable proposition to the volunteers—farmers and merchants concerned about their homes and families. Captain Charles Rawn insisted upon unconditional surrender. But when he returned from the council with the Nez Perce, Rawn found the Bitterroot volunteers preparing to return home. Upon hearing Looking Glass's proposal, they had no wish to provoke a violent confrontation with the Nez Perce. Hammond and the other Missoulians, however, remained.[34]

The next morning, the men behind the barricades noticed a long line of Indian women and children leading their horses up a nearby ravine and over the mountain. Warriors flanked either side and fanned out across the grassy hillside above, affording them easy shots into the barricade, yet the Indians held their fire. Reported one eyewitness, "Upon looking up we discovered the Indians passing along the side of the cliff where we thought a goat could not pass, much less an entire tribe of Indians with all their impediments. The entire band dropped into the valley beyond us and then proceeded up the Bitterroot. . . . They were good-natured, cracked jokes and seemed very amused at the way they had fooled Rawn" by slipping past the barricade. Outflanked and outsmarted, Rawn decided discretion was the better part of valor and let the Nez Perce pass without resistance, a decision for which Higgins, Barbour, and others chastised him, dubbing the nonevent "Fort Fizzle."[35]

With no sign of pursuit, the Nez Perce believed they had left the war behind. They moved leisurely up the Bitterroot, trading horses and purchasing supplies from the settlers along the way. Stores did a thriving business with the Nez Perce, who paid exorbitant prices for coffee, sugar, and tobacco with gold dust and greenbacks. Although Charlo refused to join or aid the Nez Perce, on more than one occasion the Salish stepped in to prevent outbreaks of violence between the Nez Perce and the settlers. "A goodly number of our friendly Flathead [Salish] Indians armed with rifles such as they had, gathered in to protect us," recalled one storeowner.[36]

Ten days after Fort Fizzle, while camped in the upper Big Hole Valley, the Nez Perce awoke to volleys of bullets being fired into their tepees. Despite the surprise attack, the Nez Perce turned the tables and repelled Colonel John Gibbon's force. The Nez Perce fled toward the Yellowstone country, taking the U.S. Army on one of the most epic campaigns of the Indian wars. Finally, on September 30, a full month after Fort Fizzle, General Nelson Miles, who had come up the Missouri by steamboat, pinned down the Nez Perce on the flanks of the Bear Paw Mountains. After thirteen battles and 1,700 miles, and with Looking Glass and the other chiefs dead, Chief Joseph surrendered, just forty miles short of reaching Canada and safety.[37]

The Missoula volunteers had long returned to their daily routine, but they continued to follow the saga of the Nez Perce with excitement mixed with sympathy and respect for Joseph and his band. The flight of the Nez Perce marked the end of the Indian wars and the passing of the Old West. Although many in Missoula ridiculed the episode at Fort Fizzle and chastised Bitterroot settlers for trading with the Nez Perce, the incident illustrated the spirit of capitalism that was sweeping the frontier. The rhetoric of the Fort Fizzle debacle highlighted the divide between the young, professional merchants and industrialists intent upon amassing capital and the old guard of Indian fighters, independent traders, fur trappers, and citizen militia. As a veteran of the Old West, Higgins deplored the actions of Rawn and the volunteers. As a businessman, however, Hammond saw little to gain by combat and much to profit by trade. Montana thus began to shift from a frontier/agricultural economy to an industrial one.

Similarly, 1877 underscored the national transition to industrialization. The Panic of 1873 had set off unprecedented business failures and threw more than three million people out of work. By 1877 the national unemployment rate stood at 27 percent. Starving, impoverished, and desperate, millions crowded into the urban centers of New York, Baltimore, Chicago, and Pittsburgh. Under financial strain, the Pennsylvania Railroad cut its already low wages, igniting a series of spontaneous and violent railroad strikes. When the strikes finally subsided, over one hundred people lay dead and a thousand were imprisoned. This first nationwide labor strike set the stage for a multigenerational conflict between two emerging forces in America—labor and capital—with the federal government often intervening on the behalf of capitalists.[38]

Montana's relative isolation insulated it from the worst aspects of the longest depression the nation had yet experienced. Although both Gold Creek and Cedar Creek mines began to peter out in 1873, silver mines in Butte and Phillipsburg kept the Montana economy afloat. The white population of Missoula County dipped only slightly, suggesting a stable but stagnant economy. Perhaps most telling, from 1872 to 1880 only one additional business—a saloon—opened in the city. Eddy, Hammond and Company, however, continued to expand. Following Fort Fizzle, Hammond supervised the final construction of the new building and also took charge of the Missoula express office. Under Hammond's leadership, the firm soon became a regional economic and political powerhouse.[39]

Shortly after Hammond began his clerking job, Florence Abbott, a precocious twelve-year-old, arrived in Missoula to further her schooling. With her parents still in Oregon, she moved in with her sister and brother-in-law, Edwina and Richard Eddy. Over the next five years, while Hammond was trying to find himself—writing poetry at Flathead Lake or hawking snake oil between bouts of clerking—Florence became one of Missoula's top students and grew into a beautiful young woman. She soon attracted Hammond's attention.[40]

Florence's father, Lorenzo Abbott, however, disapproved of the ardor developing between the seventeen-year-old student and the twenty-nine-year-old clerk. In 1878, when the Eddys left for San Francisco, where Edwina would spend the winter, Florence accompanied them. Having graduated from high school, Florence wished to continue her education, and no doubt her sister also enjoyed the companionship. Nevertheless, when Florence returned to Missoula in May, she and A. B. renewed their courtship. After discovering them in a compromising position one evening inside the store, Florence's father marched into Hammond's office and informed him, "You *will* marry her."[41]

Hammond's marriage to his partner's sister-in-law the following February solidified both his business and personal life. The newlyweds moved into Eddy's new house, a modest two-story residence that the two families shared for the next twenty years. The Hammonds celebrated Christmas 1879 with the birth of their first child, Edwina, named in honor of Florence's sister. Over the next eight years, they would have five more children.[42]

The Eddy-Hammond house became the scene of many of the town's social gatherings. To celebrate the couple's first anniversary, "a party of friends gathered at the house of A. B. Hammond . . . and chased the hours with flying feet." A. B. and Florence soon became Missoula's preeminent couple. Youthful Florence, with her thin lips and rosy cheeks, was the paragon of high society. Fashionably dressed and sporting the latest coiffure, she hosted lunches with Montana's elite. Hammond, meanwhile, continued his involvement in community affairs and became active in local politics.[43]

As his fortunes increased, however, Hammond gradually withdrew. Gone was the genial young sales clerk and eligible bachelor who readily volunteered for civic duties ranging from serving on the Christmas tree committee to fighting the Nez Perce. A growing intolerance won him few friends, while his tenacious oversight irked many of his business associates. Leaving the socializing to others, Hammond turned his attention to business affairs with a resolute and boundless energy. He was gradually becoming the uncompromising, unpopular, and vilified businessman that would define his reputation for the rest of his eighty-five years.

CHAPTER 5

Working on the Railroad

Upon settling at Hell Gate in 1856, Frank Woody purchased a barrel of whiskey that he promised to open "for the drinking public to imbibe" when the Northern Pacific Railroad reached the Missoula Valley. He was that confident the railroad survey of two years before would soon yield a transcontinental line. Little did he know that he would have to wait twenty-seven years. But when it finally arrived, the railroad transformed Missoula from a small town into a regional commercial center and changed Hammond from a store manager to a millionaire.[1]

Eight years after Woody's optimistic proclamation, Abraham Lincoln signed into law the largest land grant in U.S. history, bequeathing forty-two million acres—more than 2 percent of the landmass of the lower forty-eight states—to the Northern Pacific Railway Company. The railroad company argued that since it was building a line through such a sparsely inhabited region, the federal government should subsidize the endeavor as freight receipts could not possibly cover the construction costs. But instead of a cash subsidy, Congress offered land as the inducement, assuming that the railroad would sell the land to settlers to recoup its outlay. With an eye to the future, however, the railroad selected valuable timberlands from which it, and its subsidiaries, would reap immense profits. In addition to the two-hundred-foot right-of-way for construction of the railroad itself, the railroad received alternate square-mile sections along a forty- to eighty-mile-wide swath from Minnesota to the Pacific coast. In Montana the Northern Pacific received 14.7 million acres, sixteen percent of the state's total area. The checkerboard ownership that resulted from such railroad grants would create a future land management nightmare.[2]

Shortly after approving the Northern Pacific grant, Congress began having second thoughts in the wake of public opposition to such generosity. Nonetheless, in 1868, Congress granted the Northern Pacific an extension until 1877 to complete the line and qualify for the grant. A series of organizational and financial setbacks delayed construction, but finally, in August 1870, the Northern Pacific began laying track outside Duluth, Minnesota, leading E. L. Bonner to predict that the railroad would arrive in Montana within two years. By June 1872, however, the railroad had barely crossed into Dakota Territory.[3]

Bonner was not the only overly optimistic booster. Chief among the railroad promoters was Samuel Hauser, a civil engineer who arrived in Montana during the Civil War and had invested heavily in a silver smelter that failed in large part because of high transportation costs. With outside financing, Hauser started the First National Bank of Helena and continued his interest in both railroads and mining. An unabashed speculator, Hauser quickly jumped into the most spurious of schemes with boyish enthusiasm and often found himself grossly overextended. A greater contrast to Hammond's conservative business approach could scarcely be imagined, and yet the two men soon became business partners. Indeed, Hammond owed much of his success to Hauser's effusive personality. It was the railroad that brought them together.[4]

The Northern Pacific, now part of Jay Cooke's financial empire, also engaged in a massive publicity campaign. To entice settlers, the railroad produced a pamphlet declaring Montana's climate as having "the mildness of Southern Ohio." Such optimism encouraged Cooke to overextend the railroad's credit. This, combined with speculation, sluggish bond sales, the Franco-Prussian War, the Crédit Mobiler scandal, and inflated reports of Indian hostilities, led to Cooke's financial collapse in September 1873. Cooke's failure took out the stock market and sent the nation into a six-year depression, the greatest economic collapse America had yet experienced. The Panic of 1873 literally stopped the Northern Pacific in its tracks, and the railroad fell into receivership.[5]

The 1870s marked a period of wildly fluctuating economic conditions. While the nation spiraled downward financially, recurring gold and silver discoveries buoyed the western economy. In Montana, mining camps provided a market for local agricultural and lumber products, yet the lack of a railroad prevented export of such bulky commodities. Furthermore, full exploitation of the ore bodies was economically viable only with railroads, which brought the heavy machinery required to crush the ore and ship the concentrate to eastern markets. Transportation, or the lack thereof, defined the economy of Montana Territory.

Dependence upon placer mining, a single, soon-exhausted resource, triggered the rise of numerous towns and cities—Virginia City, Bannack, Louisville. These all passed into oblivion nearly as fast as they sprang up. In comparison, the smaller trading centers like Missoula, surrounded by prime agricultural lands, maintained their tenuous grip on existence. But as the gold mines began petering out in the 1870s, it seemed Missoula would forever be a small town catering to local farmers and ranchers. Location, nonetheless, proved to be the crucial underlying factor ensuring its future. The same geographical factors that made the Missoula Valley a nexus of American Indian activity also contributed to its viability as a commercial center despite having only a fraction of the population of the mining towns of Helena and Butte. Although it housed only 440 people in 1880, Missoula served as a ganglion of civilization with nerves radiating into five valleys. To the north lay the ranchlands of the Flathead Valley, to the northeast the Blackfoot River corridor provided timber, to the east were the gold fields of the upper Clark Fork basin, to the south lay the farms and orchards of the Bitterroot, and to the west spread the wide Missoula Valley and the lower Clark Fork. Indians, trappers, and settlers traveled up and down the three rivers converging in Missoula. Farmers and ranchers brought their wheat, produce, and cattle into town, where merchants distributed all of it to the mining camps. More important, the valley was tributary to two-thirds of Montana's timber resources.[6]

At the center of this commercial activity sat Eddy, Hammond and Co., which despite the national depression continued to expand slowly and steadily throughout the 1870s, adding a new building in 1878, then a stable across the street to accommodate mule teams. Beginning a pattern he continued for the following sixty years, A. B. Hammond used the economic downturn to buy up other businesses, this time the Fahey store in Stevensville, thirty miles to the south. By the end of the decade, Eddy, Hammond and Company was pulling in $15,000 a month and had become Missoula's primary business.[7]

While all of this economic activity occurred without a railroad, it was predicated on the promise of its eminent arrival. Construction resumed in 1879 with the reorganization of the Northern Pacific. Progress west of the Missouri, however, proved sluggish at best, and Montanans cast about for other options. Hauser, with his political, social, and economic connections, led the push to grant a territorial subsidy to the Utah and Northern, a branch of the Union Pacific that could be extended to the Montana mines.

With most of Montana's white residents clustered along the Continental Divide in Butte and Helena, it mattered little whether rail access came from the south or the east. Moreover, the Utah and Northern was making steady progress toward Montana after Jay Gould took over the line in 1878. For the Utah and Northern, Hauser wanted Helena, where he lived, to be the terminus. However, Jay Gould informed Hauser that extending the line past Butte to Helena would cost an additional $500,000. The farther the railroad progressed, the more citizen support for a subsidy diminished. Without the subsidy, the Utah and Northern halted at Butte, arriving on a bitterly cold night in December 1881.[8]

In 1870 Butte's population was 350; ten years later it exploded to 5,000, to become the most prosperous mining camp in the world. During that decade William A. Clark built a mill and smelter, and Marcus Daly acquired the Alice and Anaconda mines. In addition, Congress passed the Bland-Allison Act, which compelled the federal government to purchase silver, stimulating silver production. A forward-looking railroad executive such as Jay Gould would naturally focus on the up-and-coming Butte rather than the older mining town of Helena.[9]

Having failed to entice the Utah and Northern to his town, Hauser turned back toward the Northern Pacific, still inching its way across eastern Montana but without a clear route through the mountains. The surveyors had narrowed their choices down to three possible passes over the Continental Divide: Deer Lodge (south of Butte), Mullan (west of Helena), and Pipestone (east of Butte). Each had its drawbacks. Mullan was the shortest but most expensive route, requiring a long tunnel under the summit of the Continental Divide; Deer Lodge did not require a tunnel but was forty miles longer, while Pipestone had the highest summit. Regardless of where the railroad crossed the Divide, it still had to pass through Hellgate Canyon and Missoula one hundred miles farther west.[10]

In 1881 financial wizard Henry Villard gained control of the Northern Pacific (NP), and his representative, Thomas Oakes, became vice president in charge of operations. Oakes happened to be an old friend of Hauser, who quickly moved to secure a commitment from the NP to use his Helena bank for its Montana accounts. Meanwhile, Hauser also lobbied General Adna Anderson, the NP's chief engineer, to guarantee the road would run through Helena. Anderson's predecessor had recommended Deer Lodge Pass, which from an engineering standpoint was the most cost effective. However, this route bypassed the mining districts of both Butte and Helena. E. L. Bonner, now established in the town of Deer Lodge, also met with Anderson and no doubt promoted the Pipestone or the Deer Lodge route, either of which would run through *his* town. As late as July 1881, Oakes favored the Pipestone route. Nonetheless, Hauser, with his grandiose promises and wildly optimistic propositions, prevailed, and Anderson agreed on Mullan Pass. With the route now decided but not yet publicly announced, Anderson, Hauser, and Oakes all purchased land in Helena, indicating that Hauser probably suggested how they might personally profit by running the line through his town. Such profiteering by company executives plagued nineteenth-century railroads and contributed to their financial instability.[11]

A few months later the Utah and Northern reached a "pooling" agreement with the Northern Pacific whereby the NP would bypass Butte and Gould's lines would stay out of Helena. Although this agreement came after Anderson's decision, the principals may have discussed it previously, and that would have played a role in determining where the NP crossed the Rockies.[12]

After his meetings with Hauser and Bonner, Anderson spent a day in Missoula. While the route along the Clark Fork River was obvious, the location of the depot, yards, and shops remained unresolved. Would they be in Missoula or farther down the valley? The town's very existence hung on the outcome. Should the railroad decide to locate its facilities elsewhere, Missoula would wither and die, much the way that Hell Gate disappeared when Higgins and Worden moved their mill. To entice the railroad to stop in Missoula, "Cap" Higgins, Frank Worden, and lumberman Washington McCormick donated hundreds of acres along the north end of town for a depot and rail yard. Despite this largesse, Anderson bypassed these town elders and spent the day with A. B. Hammond, who accompanied the engineer to Flathead Lake. When he returned, Hammond had in hand an exclusive contract to supply all of the railroad's construction lumber from Mullan Pass to Thompson Falls, a distance of 175 miles. The massive timber requirements included three thousand cross ties for every mile, plus lumber for tunnels, trestles and bridges, for a total of twenty-one million board feet. In addition, Eddy, Hammond and Co. would supply the railroad with all its provisions: clothing, blankets—everything the railroad needed save steel. This contract enabled the firm to go from an annual business of $180,000 in 1880 to $450,000 two years later and continue growing by about half a million a year for the next ten years.[13]

While the size of the deal was certainly remarkable, what was especially surprising was that it gave a mercantile firm with neither construction experience nor sawmills an exclusive contract to build the roadbed and provide all of the construction lumber. Anderson's deal with Hammond also contrasted with the railroad's previous practice of accepting bids for its lumber contract. Furthermore, not only had Higgins, Worden, and McCormick provided the Northern Pacific with prime real estate; Higgins and Worden owned a sawmill, and McCormick had the largest lumber company in the valley. Naturally, these men felt cheated, and a rift developed within the Missoula business community that grew into a hostile feud between Higgins and Hammond over the ensuing years.[14]

We may never know for certain what persuaded Anderson to grant Hammond such a lucrative contract. While Hauser and Bonner had likely recommended Hammond to Anderson, surely the young man's poise, demeanor, and confidence helped persuade the Northern Pacific official. No doubt, Hammond also relied on his previous lumbering experience, exaggerating his background and ability. However it came about, this massive infusion of outside capital catapulted Eddy, Hammond and Co. from a general store into the county's largest employer and economic entity, quickly outpacing their main competition, the more established Worden and Higgins. For his part, Hammond rocketed from being a general manager of a local mercantile to one of Montana's wealthiest and most powerful men.

The partners wasted little time. Eddy began looking for a site to build a giant sawmill, Bonner traveled east to purchase the latest mill machinery, and Hammond scouted the construction route. Within a month, Eddy had men cutting trees along the Blackfoot River for railroad ties and bridge timbers, while Hammond sent crews west to work on the more difficult sections. Bonner succeeded in getting one sawmill shipped west before winter and lined up two more to arrive in the spring. Although winter prevented the shipment of machinery, it hardly slowed production as it made for easier logging: logs could be skidded on the snow and hauled on horse-drawn sleighs. By May 1882, the company had five sawmills in place and three more on order. Some of these mills produced a staggering (for Montana) twenty thousand board feet per day and before long had eliminated much of the white pine from the region.[15]

From his office inside the Eddy, Hammond and Co. building, Hammond supervised the complex operation with the precision and authority of a field marshal. He hired loggers, mill workers, laborers, and mule skinners—anyone who wanted work and was more or less sober. He subcontracted for grading and clearing the railroad right-of-way, purchased timber and lumber from independent loggers and mills, and sent Bonner, who remained in New York, specifications for sawmills and other machinery. Hammond issued tents, groceries, tools, and other supplies for the workers up and down the line. He also continued to run the store, although he now had the assistance of C. H. (Herb) McLeod.

When Hammond became partner in Eddy, Hammond and Co. in 1876, he began casting about for trustworthy and sober assistants. In the frontier communities of Montana, however, these were difficult to find. Most westerners dreamed of becoming cattle barons, striking it rich in the mines, or at least starting their own business or farm. Hammond had little use for employees who walked off the job every time a new gold or silver strike came along or who failed to show up to work because they were too hung over. In frustration, he asked relatives in New Brunswick to send hardworking young men to Montana, where they could make twice their previous income. The ability of many in New Brunswick to speak both French and English was a bonus.

In a classic example of chain migration, many answered Hammond's call. Thomas Hathaway came first, arriving in 1878 to work as a bookkeeper in the store, and his family followed the next year. Herb McLeod showed up in 1880 with ten dollars in his pocket to work as a clerk and eventually became the general manager of all of Hammond's Montana businesses. With McLeod running the store and Hathaway keeping the accounts, Hammond sent word to New Brunswick requesting more capable, honest men who shared his work ethic and around whom he could build an empire. In the 1880s John M. Keith and his brother Henry C. Keith came to Missoula to work for Hammond as clerks. Before long, John Keith became cashier of the First National Bank, and Henry managed the branch store at Kalispell. Also from New Brunswick came a host of others who worked as clerks, loggers, or mill hands. By 1892, out of the company's sixteen store and department managers, only three were not from New Brunswick.[16]

The financial opportunities in Montana looked so promising with the arrival of the railroad that Hammond encouraged his entire family to relocate to Missoula. Hammond hired his brothers Fred, George, and Henry to supervise the mills and logging operations, while his uncle Walt Coombes oversaw freight operations. In 1886 the four Hammond brothers were joined by their sisters, Mary and Sarah, along with their husbands, and their mother, Glorianna. Sarah had married her cousin, Charles Beckwith, from one of the few other English families in Madawaska. Seventeen years later, Mary married George Fenwick, a cousin of Herb McLeod, who eventually married Clara Beckwith, the daughter of Charles and Sarah. Herb's cousin George McLeod, who also came west to work for Hammond, married Clara's younger sister, Emma, while John Keith married another Beckwith sister, Harriet. Thus, by the late 1880s, A. B. Hammond had fixed himself as the nucleus in a swirling medley of family and financial relationships. Whether related to A. B. by blood or marriage, nearly all of these families became financially dependent upon Hammond, and the men formed a web of intensely loyal and dependable managers of the various Hammond operations.[17]

The railroad contract quickly made Eddy, Hammond and Co. the largest mercantile in the region, and it announced the opening of a new store in Butte and the purchase of another in Frenchtown, bringing the total to eight stores. In addition, supplying the railroad with timbers temporarily made the firm the largest employer in the territory. Hammond was hiring anyone who could swing an axe, yet he was still short of men and lumber. To make up the shortfall, he subcontracted and repeatedly advertised for "10 million feet of sawed lumber and five hundred thousand hewed ties."[18] As he had with clerks, Hammond imported "shipments" of Acadian loggers from Madawaska, descendants of the same families that his great-grandfather, John Coombes, had pushed out of Kingsclear. With such a labor force and infrastructure, the firm also supplied cordwood, ties, and piling to the Northern Pacific's Pend d'Oreille Division in Idaho to help with its massive timber requirements for bridges and trestles.[19]

It was an ideal situation for the three partners. The trees were free for the taking (or so they believed), while outside capital supplied the funds, funneled through the First National Bank of Missoula, of which Bonner, Hauser, and Hammond were primary stockholders. Full employment provided the mercantile with customers flush with cash. Before long the partners' stores began receiving weekly shipments of staples like coffee, sugar, and flour, along with sewing machines, plows, saws, silk, buttons, carpets, and dresses to sell to their employees. Not only was the firm supplying the Northern Pacific crews with housing and goods, but it also established branch stores along the route to sell such nonessentials as whiskey, tobacco, and sweets. The company successfully merged the "wangan" concept from Hammond's native New Brunswick with Bonner and Eddy's camp stores to supply Montana's railroad workers.[20]

The construction of the railroad was a monumental feat. Traversing dense forests and steep mountainsides, the 130-mile section along the Clark Fork was the most difficult and most expensive of the entire Northern Pacific line. At the western end of the section, Chinese laborers hung from rope ladders to place dynamite to blast a roadbed from sheer cliffs. East of Missoula, Mormon crews, receiving twice the pay of the "Celestials," graded the roadbed with horse teams. Meanwhile, Hammond sent his brawny New Brunswick loggers deep into the woods to cut railroad ties. But rather than paying them wages, Hammond contracted to pay them by the tie. Swinging a double-bladed axe, an independent "tie-hack" could produce about forty ties per day. Upstream from Missoula, the sinuous Clark Fork required ten separate bridges, but the greatest challenge lay just west of town after the route left the river. Spanning 860 feet and 226 feet high, the Marent Gulch trestle was the highest bridge in the United States at the time and was among the tallest wooden structures ever built. Farther downstream, workers had to slice the route out of the narrow Cabinet Gorge, where, after they had finally laid the tracks, a landslide wiped out several hundred feet, leaving a 1,300-foot chasm.[21]

Despite the setbacks, Eddy, Hammond and Co. completed their section well ahead of schedule. The first train entered Missoula from the west on June 22, 1883, six weeks before the formal celebration at Gold Creek that marked the completion of the nation's second transcontinental railroad. When the train arrived in Missoula, one eyewitness recalled, "Citizens came to the scene in droves, some of them to behold for the first time in their lives a locomotive." To mark the occasion, Frank Woody rolled out his twenty-seven-year-old barrel of whiskey, and "all that day there was rejoicing in Missoula."[22]

Not everyone was so celebratory, however. The railroad sliced across the south end of the Flathead Reservation, and the Indians knew it would bring even more settlers. By linking their isolated, rural community to the world at large, some Missoula residents worried that the railroad would increase dance halls and prostitution, lead to an influx of gamblers, vagabonds, and criminals, and contribute to overall moral degradation. The railroad also shattered the town's relative tranquility, spewed coal dust and soot, and imposed a standardized time. Even those who were enthusiastic about the benefits expressed concern over the growing corporate power that the railroad embodied. By the late nineteenth century, railroads were the largest commercial enterprise in the

United States and freely flexed their economic and political muscle. In short, the railroad brought industrialization, with all its problems as well as advantages. While the railroad brought more goods at lower prices and helped integrate Montana into the national economy, it accelerated the exploitation of natural resources and allowed eastern and European capital to determine development patterns, at least initially.[23]

Although Hammond, Hauser, and Bonner profited immensely from their railroad contracts, for Henry Villard, president of the Northern Pacific, the completion of the line marked the beginnings of his financial troubles. Construction expenses had grossly exceeded the estimates, and by October 1883 the railroad faced a deficit of $9.5 million. Villard issued an additional $20 million in mortgage bonds, bringing Northern Pacific's indebtedness to more than $61 million. Stock value plummeted, and Villard resigned in 1884 under financial and physical stress. Five years later, Villard regained control of the Northern Pacific and served as chairman until the Panic of 1893. In the final analysis, it appears that that Villard was simply a conduit of capital; it was the regional entrepreneurs such as Hammond, Hauser, and Bonner who actually built the railroads and in doing so transformed the American West.[24]

The completion of the Northern Pacific Railroad in 1883 stimulated a decade-long burst of economic activity in Montana. Predicated on mining and the railroad's ability to transport ore, banking, lumbering, merchandising, and agriculture all boomed. Montana's population skyrocketed from 39,000 to 132,000. Similarly, the population of Missoula County jumped from 2,500 to 14,000, and the number of farms in the county quadrupled. Nearly every month it seemed a new mine, mill, or smelter appeared, needing workers, fuel, and timber. Ranchers added more cattle to the open range, now cleared of both bison and Indians. Farmers reaped high wheat yields and harvests from the Bitterroot orchards. With the railroad, Montana producers could export cattle, wheat, fruit, wool, and lumber. And, of course, silver, lead, and especially copper ores could now be shipped economically, stimulating even more development. Manufactured goods such as wagons, plows, harnesses, clothes, and furniture all flowed into Eddy, Hammond and

Co. and became available to consumers throughout the region. Montana was riding the tide of a national boom, and speculation ran rampant.[25]

Montana's premier speculator was Samuel Hauser, and he relied heavily upon his First National Bank of Helena for his investments. In an age when character counted for more than ledgers, banks in New York allowed Hauser to make overdrafts since he personally guaranteed repayment. Hauser's bank often advanced money to his friends, frequently allowed overdrafts, accepted inadequate loan securities, and hesitated to collect debts. When his bank did foreclose, it often found itself saddled with worthless mining property. When eastern banks called for repayment, Hauser was often hard pressed to cover the demands. Nevertheless, Hauser's bank became the second largest in the Pacific Northwest.[26]

Although Hauser and Hammond shared investments, they harbored radically differing ideas about finances. While both had experienced the Panic of 1873 and the fluctuating economic conditions of the 1870s, Hammond understood the lessons of New Brunswick and the consequences of its colonial dependence upon a single extractive industry. The volatile nature of the lumber industry made him acutely aware of the dangers of speculation. Nevertheless, in addition to railroads and lumbering, Hammond became involved in real estate development, construction, mining, livestock, flour mills, water and power companies, politics, and a newspaper. In fact, during the 1880s, Hammond had his hand in so many different projects that the *Missoulian* put a reporter on a separate beat just to cover his activities.[27]

The economic boom brought on by the railroad stimulated the need for more capital, and local businessmen began casting about for investors. Thus, Cap Higgins, founder of the First National Bank of Missoula, approached Hauser, who by the early 1880s was the majority stockholder. Hauser, in turn, invited Hammond and Bonner, flush from their railroad contract, to pump their excess cash into the bank. With the infusion of capital, in September 1882 the bank directors elected Hammond as vice president to represent his and Bonner's interests, while Higgins retained his title as bank president.[28] A struggle over control of the bank soon erupted. The conflict between Higgins and Hammond eventually split the town into two factions, with competing banks, stores, newspapers, political parties, and real estate companies.

Hammond was beginning to develop a modern business practice, one that made sense financially but conflicted with tradition and the social ethic of small-town America in the nineteenth century. Indeed, he struggled with the subtleties and social obligations of conducting business in such an environment. While many residents appreciated Hammond's financial acuity and exactitude, others deplored his hard-nosed approach, which offered little flexibility for the exigencies of frontier life. The core of the Higgins-Hammond feud was the debate over which took precedence—the community or fiscal responsibility.

Hammond had no intention of squandering his investment in the First National Bank and immediately took an active role in its management, putting him on a collision course with Higgins and trapping Hauser between two prima donnas. In accepting the investment offer, Bonner and Hammond were well aware of the lax business practices of the First National Bank. Higgins, for example, allowed overdrafts of $75,000 or more, many of which came from his own bank account. Even Hauser considered his friend Higgins inept as a bank manager but declined to address the issue. Hammond suspected that Higgins, like Hauser, approved questionable loans to friends. When Hammond sought to examine the bank records, Higgins denied access, raising Hammond's suspicions as well as his ire. Concerned that bad loans would reduce the value of the bank stock, Hammond demanded changes. Holding Hauser, as majority stockholder, ultimately responsible for supporting incompetent management, Hammond told him, "You place parties in control and then expect me to quarrel with them about their blunders. I was willing to do all I could to save the bank but now the time has come for a new deal." Hammond further suggested that Hauser's friendship with Higgins clouded his obligations.[29]

Hauser must have wondered about this impertinent young man whom he had helped obtain the Northern Pacific contract and invited into his banking enterprise. Hammond was now telling him, the president of the largest financial institution in the territory and soon-to-be territorial governor, how to run his business. Despite his propensity for the good-old-boy network, Hauser recognized Hammond's business sense and competence. Attempting conciliation, Hauser suggested that Hammond and Bonner buy out Higgins and appoint a new president. But

Hammond insisted that Hauser buy them out instead. Agitated at Hammond's intransigence, Hauser replied that as vice president, Hammond was fully within his duties to deny loans and bear some responsibility himself.[30]

Seeking the rest of the story, Hauser asked Higgins about the dispute with Hammond. Hoping to downplay his own complicity, Higgins replied, "I do not know of any hostility or disagreement or any cause for even a misunderstanding."[31] Clearly, Hauser was not getting a straight answer, so he turned to his cousin, Ferdinand Kennett, who had been cashier since 1877 and whose loyalty to Hauser was unquestioned. In a long letter Kennett apprised Hauser of the situation. Taking issue with Hammond's meticulous oversight, Kennett wrote, "Hammond will not be satisfied with the present arrangement. . . . As far as I am concerned if Hammond is to be the controlling director I wish to step out and will do so." Although Kennett stated, "Much of Hammond's fault finding is utterly without foundation," he admitted in his postscript, "There are some bad debts which we cannot ascertain exactly as yet the status of."[32]

Kennett summed up the situation as he saw it: "Hammond is a dangerous man and would I think hurt the business of the Bank by being too severe with parties who are now hard up if he were to have full swing."[33] While Kennett and Higgins clearly had a personality clash with Hammond, many of their differences lay in values and what they saw as the fundamental purpose of the bank. Kennett saw the bank as providing a service for the community and worried that they would alienate friends and neighbors should they press too hard for debt collection. For Hammond, the bank's fiscal health and protecting his own investment superseded all other concerns.

After six weeks of wrangling, Hammond came to an agreement with Hauser, Higgins, and Kennett. Higgins could remain as president but only a committee could approve loans. The agreement gave Hammond access to the accounts, and he discovered that the bank had carried $50,000 in bad debts as assets. Chagrined, he told Hauser, "Men have borrowed money without collateral security who have never been able to pay 50 cents on the dollar since they have been in this country."[34] Bonner had a more colorful assessment of the situation. He compared the bank loan policy to the equivalent of letting people use sunken railroad ties in the Blackfoot River as collateral, adding, "I would as soon think of loaning money on a school of mackerel in Boston Harbor."[35]

The détente lasted less than two years. The bank directors lined up along two poles, with Higgins and Kennett on one end and Hammond and Bonner on the other. Bonner, still splitting his time between the East Coast and Deer Lodge, had full confidence in his junior partner. Kennett was convinced that Hammond would "never be satisfied until he is out or the head of the Bank." Kennett also complained to Hauser that disputes between Higgins and Hammond not only made for unpleasant working conditions but were also "seriously affecting the business of the Bank." By August 1887 the situation had deteriorated to the point where Kennett declared that if he could sell his house, he "would be glad to take even some subordinate position at a much smaller salary than to stay here."[36] Hauser, however, ducked confrontation and ignored Kennett's complaints, hoping they would go away.

Hauser became even more enmeshed with Bonner and Hammond, enlisting their help to build five of the seven Montana branch lines for the Northern Pacific. As early as 1881 the NP's chief engineer, Adna Anderson, began laying plans for a series of feeder lines into mining districts and the Bitterroot Valley. The Northern Pacific's charter, however, prevented it from building such branch roads. To circumvent this provision, railroad officials enlisted local businessmen to build the lines, which the NP then purchased for substantially more than construction costs. This proved to be a financial windfall to the local entrepreneurs and their backers, who included Villard, Oakes, and Anderson. With company officials profiting by endorsing purchases of subsidiary branch lines that they themselves had financed, it was hardly any wonder that the NP was in such desperate financial condition. This chicanery, however, required the assistance of compliant and malleable local businessmen.[37]

To build the Montana branch lines, Anderson enlisted Hauser, who proved an ideal front man, throwing his enthusiasm and reputation into the projects. Meanwhile, less than a month after the gold spike ceremony, Hammond, convinced that the Bitterroot would be a viable feeder line, began to organize his own railroad from Missoula to Salmon, Idaho, with a branch over the Sapphire Mountains to Butte. The Northern Pacific, however, opposed "the building of any lines connecting with our main lines unless we have control thereof and absolute ownership." Nevertheless, Hammond's crew began surveying a route, prodding the NP into backing more branch lines in Montana.[38]

More threatening to the Northern Pacific was James J. Hill's Great Northern, which would connect St. Paul, Minnesota, to Helena and Butte. The Union Pacific also announced it was considering building a line into the Bitterroot from Philipsburg to secure the vast timberlands and agricultural production of the valley. Such pressure from the competition compelled the NP to sponsor branch lines, eventually contributing to the railroad's financial meltdown.[39]

Unrealistic mining speculation factored heavily in charting NP's Montana branch lines. A Philipsburg branch would tap the Granite Mountain Mine, one of the country's largest silver mines, while the Bitterroot line would serve the Curlew silver mine, in which Hammond and Hauser were among the primary investors. Even the Yellowstone Park line looked toward the Cooke City mines as a source of potential traffic. In nearly every case, Hauser promoted branch lines to mines where he had financial interest. He convinced the NP that these would be valuable additions to their system and would soon pay for themselves. Indeed, the NP based its purchase price on Hauser's overly optimistic traffic projections, which, except for the Bitterroot line, fell dramatically short. Hauser, however, got railroads to his mines while making money on the construction.[40]

Hammond's dogged pursuit of a Bitterroot line paid off, and in January he partnered with Hauser, Bonner, Anderson, and Frank Worden to form the Drummond and Philipsburg Railroad and the Missoula and Bitter Root Valley Railroad. As with building the Northern Pacific, the mercantile, and other enterprises, Hauser, Bonner, and Eddy supplied the capital and served as titular heads, while Hammond as vice president oversaw the day-to-day operations and functioned as the CEO. Although Hauser was the president of both the Drummond and Phillipsburg and the Missoula and Bitter Root Valley Railroads and Anderson was the chief engineer, Hammond insisted on personal oversight of construction expenses. With his insistence upon written contracts and timely payments, Hammond exasperated Hauser and Anderson, who were used to conducting business with a wink and handshake. In February 1887 Hammond told Hauser that the severe winter would cause heavy cattle losses, making for a tight money market, and thus they should ask the NP for half the payment in bonds up front. He wrote to Hauser that it would be unwise to assume "the responsibility of building these Roads with local money without some assistance in the way of bonds, no matter how great the profit we might expect to realize in the operations." The Northern Pacific conceded the point and thereafter purchased half the bonds in advance to speed construction.[41]

In April Hammond negotiated a separate contract with the NP to supply ties and timbers as well as grade the roads and, in the case of the Bitterroot branch, lay the tracks. To protect himself, Hammond insisted, "No charge for stumpage shall be made by the Northern Pacific Railroad Company for timber that we may take from their land for use in the construction of roads." Hammond essentially wanted a written guarantee that the NP grant lands would supply the timber free of charge. Hammond also haggled with Anderson over prices and insisted upon immediate payment for a recently constructed bridge. Hammond was quickly gaining a reputation as a shrewd and uncompromising businessman. While his dogmatic approach may have been less than socially palatable, he was establishing practices that would pay financial dividends in ensuing years.[42]

Building two railroad branches simultaneously proved challenging. Although Hammond began work on the Bitterroot line, in July he had to divert work crews to the Phillipsburg branch, as it promised more immediate results from the mining industry. So keen were the partners to get this line finished that they subcontracted grading to anyone with a team of horses. Finally, in September, the Philipsburg line was completed and the construction crews resumed work on the Bitterroot.[43] In less than six months Hammond had completed the railroad as far as Victor, thirty-five miles up the Bitterroot, and invited the press, politicians, businessmen, and other guests on the first train. The *Missoulian* reported, "Mr. Hammond with pardonable pride showed his visitors the works at this place [Florence, Montana, named after Mrs. Hammond] and the wonderful rapidity with which logs can be made into lumber."[44] Although the severe winter had delayed construction, by September 1888 Hammond had completed the railroad from Missoula to Grantsdale, fifty miles upriver, and turned it over to the Northern Pacific.

In building the Bitterroot railroad, Hammond and his partners received $50 per acre for clearing, $0.40 each for ties, and from $0.30 to $1.90 per yard for excavating. Hammond then subcontracted the work, paying $35 per acre for clearing, $0.25 each for ties, and from $0.17 to $1.33 for excavating. Ultimately, the partners (including Anderson, who still worked for the NP) charged the Northern Pacific $134,235 more than the actual construction cost of $693,079, giving them a net profit of just over 16 percent on an eighteen-month investment. Hammond also profited from his separate timber contracts and town site development in Victor.[45]

The other Northern Pacific branch lines followed a similar procedure. Through his First National Bank of Helena, Hauser, who was now territorial governor, along with Bonner, Hammond, and other investors, fronted the initial expenses by mortgaging the road for $20,000 per mile and issuing fifty-year bonds; the Northern Pacific bought half of these, as Hammond had insisted, and guaranteed the interest payments on the remainder. Upon completion, the NP would take possession, rebond the road for $25,000 per mile, and pay off the local investors. The NP intended to use the traffic receipts, which it split with the local investors, to make semiannual installments to redeem the bonds. Subsequent events, however, would dictate otherwise.[46]

Not surprisingly, Hauser's buoyant traffic projections from his mining ventures fell woefully short. Furthermore, the branch lines were worth far less than the $20,000 per mile it supposedly cost to build them. An auditor's report of 1896 noted that for the Phillipsburg branch, "construction was light and probably did not cost to exceed $9,000 per mile." The report also noted that the Bitterroot line was of "comparatively easy construction, costing from $12,000 to $15,000 per mile. The line is only in fair condition." Another estimate pegged the entire Montana branch system as worth $7 million "which is considerably less than the construction cost . . . or the bonded debt." With the Panic of 1893 and the collapse of Montana's mining industry, the Northern Pacific found itself with a system of branch lines that were for the most part "unproductive and valueless."[47]

Perhaps the most egregious example of the profiteering of individual investors at the expense of the parent company was the case of Rocky

Fork Railway and Coal Trust. In 1888 Hammond, Bonner, Hauser, and Marcus Daly partnered with Henry Villard and Thomas Oakes of the Northern Pacific to form the trust. They built a branch railroad and founded the town of Red Lodge, Montana, to supply coal for the Northern Pacific. By 1891 it was the largest fuel producer in the state. The trust then flipped the Rocky Fork Railway, which had cost $800,000 to build, to the Northern Pacific for $1.4 million. When the railroad went into receivership, in 1893, Brayton Ives, one of the major stockholders, charged mismanagement and brought its directors to trial. Hauser could make no precise accounting of why the railroad paid nearly twice as much for a branch line as its construction cost. Nevertheless, he and Oakes were eventually cleared.[48]

Hammond's Bitterroot railroad proved to be the exception. The 1896 auditor's report noted, "This branch is one of the best of the Northern Pacific and Montana properties and is a good feeder to the Northern Pacific Main Line." While the Curlew mine failed to live up to expectations, the Bitterroot Valley continued to produce agricultural goods and ship produce, livestock, grain, and lumber throughout the depression and even provided "a very handsome margin after payment of interest."[49]

As the Bitterroot railroad neared completion, the bank dispute between Higgins and Hammond flared into a management crisis. By October 1887, even Kennett admitted that Higgins was "no use to the bank." Kennett, however, could not abide Hammond's micromanagement. Kennett admitted that Hammond "is undoubtedly a good business man and deserves credit for his success and what he has done for Missoula but at the same time I don't and can't respect him." Over the winter, Hammond finally maneuvered Hauser to his side and forced Higgins's resignation. While Kennett agreed to serve as president temporarily, he informed Hauser that if Hammond gained control of the bank, he should "send someone here in my place." In the shuffle, the directors elected Hammond's nephew John Keith as cashier to look after the daily operations while casting about for someone to buy Higgins's stock and serve as a figurehead president.[50]

Despite being deposed, Higgins, as Kennett complained, "still occupies the back room and hangs around all the time." Kennett also groused that "the situation is far from pleasant," and other stockholders were weary of the quarrel and wanting to sell their stock. Finally, by September, Kennett had had enough of Hammond's fastidious oversight and manipulation. Declaring, "I want to get out of his clutches at once," Kennett offered to sell his stock to Hauser below market value.[51]

Hammond, meanwhile, headed east with his wife for a two-month business vacation. Upon his return, Hauser finally agreed to buy both Kennett's and Higgins bank stock, splitting it with Bonner and Marcus Daly, who was becoming interested in Missoula and would be a perfect proxy president. With the Bitterroot branch certified and turned over to the Northern Pacific, Hammond took his share of the final payment from Hauser in stock in the First National Bank of Missoula, making him the institution's largest stockholder and cementing his position as the financial kingpin of the five valleys.[52]

Workers attempting to free a logjam on a river in New Brunswick. While river drives were the primary method of getting the timber to the mill, they caused substantial ecological damage. Hammond worked similar log drives on the Susquehanna River in Pennsylvania. Note the log brand "H" visible on the end of some logs. (Provincial Archives of New Brunswick, Ole Larsen fonds: P6–197)

In 1867 Andrew and George Hammond ascended the Missouri River in a steamboat much like this one. While George continued on to Fort Benton, Andrew spent the winter of 1867 in Fort Peck cutting firewood to fuel the steamboats. Note the stacks of cordwood on deck. (83–0141, Archives and Special Collections, Mansfield Library, The University of Montana)

Stacked high with supplies for the mining camps, the Fort Benton levee in the 1860s was bustling with activity whenever a steamboat landed. Having survived his ordeal as a woodhawk, Andrew Hammond arrived in Fort Benton in 1868 and reunited with his brother George. (W. Hook photo, Montana Historical Society Research Center)

The residents, mostly woodhawks and wolfers, in front of Fort Musselshell, circa 1870. The Woodhawk War, which came to a head here in the spring of 1868, resulted in the death of thirty-three Lakota Indians. This photo shows stacks of fuel wood awaiting the steamboats, which by 1869 had diminished to a trickle after the completion of the first transcontinental railroad. (Photo courtesy of James Burst)

The Pope and Talbot mill at Port Gamble, circa 1861. In 1853, entrepreneurs William Talbot and Andrew Pope established this sawmill. By 1858 Port Gamble boasted the largest mill on Puget Sound. Seeking work, Hammond arrived here in 1868 and learned the many advantages of a company town that he would later employ to great effect. (OrHi 39485, Oregon Historical Society)

Andrew B. Hammond at about the time he became a sales clerk in Missoula, 1871. (84–0075, Archives and Special Collections, Mansfield Library, The University of Montana)

The Missoula Mercantile Company, the successor to Eddy, Hammond and Company, circa 1885. At the corner of Front and Higgins, the building served as the social and economic hub of the Missoula Valley. (70–0113, Archives and Special Collections, Mansfield Library, The University of Montana)

A. B. Hammond, circa 1890. At forty-two years old, he was at the height of his Montana career and could have retired as a multimillionaire, but he dreamed of bigger things. (88–0008, Archives and Special Collections, Mansfield Library, The University of Montana)

In 1890 Hammond added a second floor to the Missoula Mercantile Co. (the Merc). Today, the exterior remains largely unchanged. (70–0099, Archives and Special Collections, Mansfield Library, The University of Montana)

The grocery department was one of several housed inside the Merc, circa 1900. Other departments included men's and women's clothing, furniture, produce, hardware and tack, and liquor. (76–0440, Archives and Special Collections, Mansfield Library, The University of Montana)

Higgins Avenue, Missoula, circa 1905. Facing South, this photo shows the nucleus of Hammond's Montana empire. The Merc is in the left foreground; behind it, the First National Bank building sports a turret. Across the street is the castle-like Hammond Building, and in the right foreground is the Florence Hotel. The Merc is the only original building still standing. (76–0164, Archives and Special Collections, Mansfield Library, The University of Montana)

Chief Charlo in 1910. As his proud poise and elaborate dress suggest, Charlo refused to accept defeat, despite the forced relocation. (78–0225, Archives and Special Collections, Mansfield Library, The University of Montana)

The Bitterroot Salish at Stevensville shortly before their exodus from the Bitterroot Valley in 1891. (76–0376, Archives and Special Collections, Mansfield Library, The University of Montana)

Editorial cartoon from the *Missoulian,* 1891. Hammond's head and "M. M. Co." (Missoula Mercantile Co.) are at the center of the "Missoula Octopus," while its tentacles include towns, political parties, and businesses. Harrison Spaulding, editor of the *Missoulian,* especially resented Hammond's control over the county's Republican Party. Although Hammond dominated Missoula's political and economic landscape, claims that he controlled the "Higgins Estate" and "Church" were absurd. (72–0701, Archives and Special Collections, Mansfield Library, The University of Montana)

A. B. Hammond and family, circa 1892. (88–0023 Archives and Special Collections, Mansfield Library, The University of Montana)

The Marent Gulch trestle northwest of Missoula. When finished in 1883, it was the world's largest wooden structure, standing 226 feet high, spanning 860 feet, and containing 800,000 feet of lumber. (MHS H-1068, Montana Historical Society Research Center)

Celebration of the completion of the Northern Pacific Railroad at Gold Creek, Montana, in 1883. Hammond received the contract to supply the railroad with all of its construction lumber from Mullan Pass to Thompson Falls, a distance of 175 miles. Although the man on the train is unidentified in the photograph, he bears a strong resemblance to A. B. Hammond. (University of Washington, Special Collections, UW 594.)

Christopher P. Higgins, founder of Missoula and Hammond's rival for control of the First National Bank. (82–0161, Archives and Special Collections, Mansfield Library, The University of Montana)

Completed in 1888, Hammond's Missoula and Bitter Root Valley Railroad provided easy access to the valley's timber, nearly all of which went to the Big Blackfoot Mill at Bonner and, in turn, to supply the Butte copper mines with timbers and cordwood. (75–6014, Archives and Special Collections, Mansfield Library, The University of Montana)

Copper magnate Marcus Daly, alternately Hammond's best customer and business rival, political coconspirator and antagonist, and cofounder of the Hammond Lumber Company. (79–0001, Archives and Special Collections, Mansfield Library, The University of Montana)

Logjam on the Blackfoot River above the Bonner Mill during the flood of 1908. Such logjams inflicted substantial ecological damage to rivers and riparian corridors. (Mtg0001474, Archives and Special Collections, Mansfield Library, The University of Montana)

The Bonner Mill on the Blackfoot River. Hammond built the mill in 1886 and sold it to the Anaconda Copper Mining Company in 1898. The pylons or cribs in the middle of the river anchored the log booms, which caught the logs coming downriver so they could be directed into the log pond for milling. The dam generated electricity for the mill and town. The cylinder on the far right was used to burn the mill scraps. (D-II_a-04, Archives and Special Collections, Mansfield Library, The University of Montana)

The young George B. McLeod, hired by Hammond in 1900. In 1946 he became president of the HLC and thereafter guided it through some of its most prosperous years. McLeod oversaw the transfer of the HLC to Georgia Pacific, giving him the distinction of being the company's first employee as well as its last. (88–0020, Archives and Special Collections, Mansfield Library, The University of Montana)

Charles Herbert (C. H.) McLeod, Hammond's right-hand man in Missoula. Born in 1859 in New Brunswick, C. H. arrived in Missoula in 1880 to work for Hammond's Missoula Mercantile Company. In 1885 McLeod became general manager of the MMC, and he ran it until he retired in 1940. (UM archives 88–0014, Archives and Special Collections, Mansfield Library, The University of Montana)

CHAPTER 6

The Cramer Gulch War

With railroads, mining, and lumbering forming an interlocking triumvirate, Montana's economy during the late nineteenth century depended upon easy access to timber. This essential natural resource was highly contested, however, with individuals, corporations, and government all vying for control. While the federal government ostensibly held title to the forests, the ideology of privatization fostered the Great Barbeque, whereby the government dispensed land to states, corporations, and individuals. While many favored retaining forests in government ownership, most echoed Congressman Jacob Collamer's contention that the job of the government was to "get rid of the public lands. . . . I do not care where they go." By 1883 the federal government had purged itself of 620 million acres, of which 11 percent, or 155 million acres, went to the railroads. Privatization, however, did little to curb the long-established tradition of timber theft as both individuals and corporations regarded the forests as available for exploitation regardless of ownership.[1]

Industrialization exacerbated the problem as the railroad linked the natural resources of the West to the industrial centers of the East, quickly integrating a preindustrial society into the global market. In supplying the ties and timbers for the construction of the Northern Pacific Railroad across Montana, A. B. Hammond established portable sawmills and logging camps along the route, moving both as the railroad progressed and as the supply of timber became exhausted. By 1885 Eddy, Hammond and Co. had cut much of the merchantable timber along the Clark Fork, primarily for railroad construction. Not only had the railroads themselves created a demand for timber, but they also made the development of large-scale mining profitable. Mining, in turn, accelerated the demand for more timber. While the first industrial revolution revolved around steam power, the second depended upon electricity—electricity that traveled through copper wires.[2]

The copper mines in Butte had an insatiable appetite for wood. As miners burrowed deeper into the earth, they needed more timbers to support the tunnels. The demand for timber during the 1880s was so great that Missoula County alone supported at least a dozen independent mills and more than a thousand workers. Then in 1884 Marcus Daly built the world's largest copper smelter at Anaconda, which required prodigious quantities of cordwood for fuel. The following January Daly contracted with Hammond's firm to supply ten million board feet (bf) of lumber, with the option to increase if necessary. The next month Hammond agreed to supply Northern Pacific with thirty thousand cords of wood. Then in April the Gregory Mining Company ordered five hundred thousand board feet of lumber.[3] Never one to turn away business, Hammond continued racking up orders that far exceeded his company's capacity. He would simply hire more men and build more sawmills.

Before long, however, Hammond's firm denuded the "splendid forests of Hell Gate Canyon" and began casting about for more trees. In August 1885, the company built a new sawmill along the banks of the Clark Fork at Bonita, thirty miles upstream from Missoula "in the midst of some of the finest timber seen in Montana."[4] Hammond dispatched his uncle Robert Coombes to run the new mill and establish a logging camp up nearby Cramer Gulch. When Hammond's logging crew arrived the next fall, they discovered fifty Métis loggers employed by Bill Thompson, a mill owner from the Butte area, busy cutting timber. Felling the big trees as quickly as possible, Thompson's crew stamped his brand on the logs, a common practice when in contested terrain.

Hearing of Thompson's incursion, Coombes, armed with a revolver, hurried from the mill with a group of men intent on driving out the invaders. Not intimidated, the Thompson crowd refused to leave, and the two crews brawled over possession of the logs. In typical lumberjack fashion, Thompson's foreman offered to fight Coombes or any man in his crowd over the right to the timber. Coombes backed down but sent for a "fighting lumberjack" to take charge of the Hammond crew. Bill

Harris, who had previously skirmished with Thompson's men over timber in nearby Rock Creek, arrived, along with his reputation. Harris, flanked by two men armed with peaveys (long poles topped with a steel hook for rolling logs), announced he would fight any two of Thompson's crew—no, make it three—if there were any who dared face him. No one took the challenge.

With no clear demarcation of ownership, total mayhem broke out. Both crews rushed to cut and brand the logs before the other. Theoretically, the logs became the property of whoever branded them, but the race degenerated into who could physically take control of the logs, even as they were loaded. "They would act like crazy men," reported one eyewitness. Amid the trees crashing and the smell of damp bark, loggers shoved and swore at each other, cut log chains with axes, dumped the other crew's loads, and engaged in annoyances and petty tricks, but they avoided actual bloodshed. Their corporate timber bosses were not worth dying for. The Hammond team had the advantage, as it did not need to saw the logs on site but could simply hitch and drag them to the yard at the mouth of the gulch. Behind the scene, both Hammond and Thompson knew neither had a legal right to the timber and reached a compromise, agreeing to let each crew work unmolested. Even under this détente, it was still a contest as to who could log faster. Ultimately, "there were few gulches in Montana which were stripped of their timber faster than was Cramer Gulch that winter."[5]

Far from a unique occurrence, Cramer Gulch illustrates the struggle over America's natural resources during the nineteenth century. Central to this conflict was who held the rights to access and cut timber on the seventy-three million acres of public domain. Although legal uncertainty played a role in the conflict, the greater factors were ideological and cultural. As historian Thomas Cox noted, English immigrants "brought with them ideas about the poor mans' customary rights to common lands" and thus saw forests as open for exploitation regardless of ownership. By 1800 timber trespass had become so established in America that it ceased to be a crime in the public mind.[6]

The ecology of the northern hardwood forest also encouraged widespread timber poaching. In New Brunswick and Maine, white pine grew in scattered groves among the beech, maples, and spruce. With little economic incentive to own timberland, lumberjacks simply cut the localized pine before moving on. To Hammond, the situation in Montana appeared to be no different and the forests readily available. Unlike the dispersed white pine of New Brunswick, however, the ponderosa pine of Montana occurred in nearly pure stands that stretched for miles across the valleys and lower mountainsides. Such concentration facilitated the development of a mechanized lumber industry and encouraged exclusive land tenure. Hammond and other lumbermen thus quickly moved to establish control over vast sections of terrain.

Unbeknownst to Hammond, the extensive park-like groves of large ponderosa pines were a result of deliberate Indian management. Written and oral histories of the Salish Indians indicate that they intentionally burned both the valleys and the forests to improve hunting, increase berry production, and facilitate gathering. The Salish purposely maintained open stands of timber, knowing that frequent, low-density fires prevented insect and disease outbreaks as well as fuel buildup. Fires also recycled soil nutrients and stimulated production of game browse.[7]

While many scholars now recognize fire management as a type of land tenure, during the 1855 treaty negotiations Washington's territorial governor, Isaac Stevens, also implicitly acknowledged Salish ownership of the forests and user rights. Although Stevens sought to extinguish Indian title and institute a reservation system, the U.S. government recognized exclusive Indian territories but saw natural resources, such as fisheries and hunting grounds, as held in common among different tribes and even suggested these resources be shared with settlers. The Indian conception of territory, in contrast, regarded the land itself as a commons with rights of access shared equally. While this held true for forests with scattered resources, like tepee poles and firewood, particular bands or family groups held highly productive resources, such as fishing sites or berry patches, for their exclusive use.[8]

European settlers also understood the notion of a forest commons. An ancient concept, the term "commons" originates from medieval England, whereby village communities managed and held in common uncultivated lands, especially forests and pastures. With the advent of the feudal system, the lords assumed ownership of the land, yet villagers retained the right to use forest and pasture as a commons and continued to manage them on a cooperative basis. Over the ensuing centuries the lords gradually whittled away at the commons through fenced enclosures, transforming communal Saxon villages into a society constructed around private property. The tradition of the English commons and the German *almende,* a similar concept, combined with resentment toward enclosures, caused New World immigrants to view the fish, game, and forests of America as a commons instead of private property, in much the same way as did American Indians. Often settlers were the very individuals who had been excluded from access to these resources in the Old World and thus resisted both privatization and government control.[9]

In New England, where society developed around concentrated villages, commons retained this traditional meaning. Where land settlement was more scattered, however, such as in New Brunswick, the backwoods of Maine, and the American West, the term "commons" assumed a different meaning. It referred to a no-man's-land with resources free for the taking, what the eighteenth-century economist Thomas Malthus termed "open-access."[10]

In nineteenth-century America, the federal government appropriated the concept of community commons, employing terms such as "public domain" to refer to government property. Such "public lands" vacillated between open access and government control but could scarcely be considered as actual commons. Furthermore, the federal government operated under conflicting laws and administrations, and even within administrations there were opposing views on what to do with the public estate. The biggest questions were how best to dispose of these lands and to whom, and how much, if any, should be retained by the government and for what purpose? The nation deliberated whether public lands should be managed as open-access resources, controlled by government regulations, or privatized. During the peak years of western settlement, the core question for policy makers concerned the privatization of timberlands. Would privatization foster their conservation and wise use? Should they be distributed to citizens in the manner of homestead acts? Or should forests be maintained under government control?

Much hinged on this debate, for the outcome would determine settlement and land use patterns, the distribution of wealth and power in the West, the integrity of forest ecosystems, and the relationship Americans had with their forests. Despite the high stakes, the situation remained in flux for decades. While the nation deliberated the fate of its timberlands, lumbermen like A. B. Hammond quickly took advantage of this state of uncertainty.

The transition from an agricultural to industrial economy intensified the debate over forests. For centuries farmers had regarded forests as an open-access resource. In supporting the 1862 Homestead Act, for example, Speaker of the House Galusha Grow insisted that settlers were entitled "to a reasonable amount of wood-land," from the commons for building, fencing, and firewood.[11] It was not clear, however, if the "right" to firewood and building materials extended to public lands or only to lands acquired through preemption and homestead laws. Although Grow and others in Congress suggested that timber was to be used only for subsistence, they failed to make a legal distinction between settlers and corporations. Meanwhile, the rise of laissez-faire capitalism favored the privatization of resources.

Grow's assertion contradicted the legal status of federal timberlands. As early as 1817 Congress authorized the secretary of the navy to reserve timberlands for shipbuilding purposes and imposed penalties for commercial exploitation. The ineffectiveness of protection led to legislation against illegal cutting of timber from public land in 1831. Sixteen years later, the Supreme Court broadly interpreted this act to include not just timber reserved for naval requirements but all timber on public land. Thus, by 1847, the government had established that reserved timberlands were subject to federal control and ownership.[12]

Protests ensued, however, when the federal government began asserting control over timber. In 1852, when agents from the General Land Office began investigating timber thefts, seizing timber cut from public lands, and selling it at public auction, Representative Ben Eastman of Wisconsin complained of "a whole posse of deputies and timber agents appointed by the President without the least authority of law." Eastman told Congress, with a bit of hyperbole, that lumbermen "have been harassed almost beyond endurance with pretended seizures and suits, prosecutions and indictments until they have been driven almost to the desperation of an open revolt against their persecutors."[13] Eight years later, R. W. Raymond, commissioner of Mining Statistics, noted the continuing conflict over forests but thought that "the entire standing army of the United States could not enforce the regulations against cutting timber upon government lands."[14] Nonetheless, in 1872 Willis Drummond, Land Office commissioner, appealed for congressional aid in combating timber depredations.[15]

In the western territories, with limited government oversight and extensive forests, settlers simply cut timber from public lands, unencumbered by regulations. The first settlers in the Missoula Valley, for example, constructed sawmills to supply lumber for their immediate needs. These, however, were water-powered sawmills, limiting production. At a capacity of two thousand board feet, the Higgins and Worden mill was no larger than the mill built by Archelaus Hammond in New Brunswick a half century earlier. Lacking railroads, Montana lumber could be used only for local consumption, not exported, curtailing the impact upon the forests. But in 1883, the railroad linked the natural resources of Montana to markets in the Midwest, turning forests from community resources into a commodity, creating a series of "Cramer Gulches" along the way. As timber shifted from being a common necessity—akin to air, water, and soil—to a valuable product, so did the discussion over the fate of the forests.[16]

While open access to timber was the de facto situation on the frontier, America also had a long tradition of conserving forests. In the seventeenth century, the New England colonies had enacted multiple ordinances governing the cutting of timber from the commons. William Penn recognized the forest as a valuable resource when he stipulated that in his colony, for every five acres cleared, one acre should be left as forest. Regulating the timber commons was not confined to the East, either. In 1851 the newly established Mormon community of Utah imposed a $100 fine on "anyone who should waste, burn, or otherwise destroy timber in the mountains."[17] The publication of *Man and Nature* by George Perkins Marsh in 1864 fostered an increasing public awareness of the ecological value of forests. Americans began to recognize the role forests played in erosion control and watershed protection. Stripping the forests bare accrued benefits to the few while the community bore the costs.

The profligate waste that accompanied logging galled many observers. In 1875, Land Commissioner Samuel Burdett noted the connection between logging practices and fire, reporting that only the best part of a tree was used while "the rest is left to decay or to add to the general destruction caused by the periodic mountain fires." Lumbermen, however, complained that fires, not logging, were the real threat to forests—yet their practices created the conditions for the catastrophic forest fires that would sweep through the region over the next century.[18]

Alarmed at the destruction, Commissioner Burdett attributed wasteful logging practices to the open-access regime of the American West whereby lumbermen "strip the lands of their timber and thereupon abandon them." Since this was government land, the lumbermen had no incentive to conserve or use the timber economically. Burdett pointed out the conundrum to Congress: since the lands in the West were not surveyed, settlers could not obtain title and therefore those who needed timber for survival "are compelled to become themselves depredators on the public lands."[19]

Foreshadowing the establishment of the Forest Service by thirty years, Burdett believed that the best solution was for the federal government to retain title, selling only the right to timber and "leave sufficient quantity standing to secure the shade and moisture necessary." But with a nod to his boss, Secretary of Interior Columbus Delano, who favored privatization, Burdett also noted that placing lands in private ownership might also promote conservation, since it would be in the landowners' own best interest to protect their tracts.[20]

The debate within the Grant administration over whether to privatize timberlands spilled over into Congress, with proposals playing out in accordance with the sectional differences among the West, East, and South. In agreement with Delano, Representative Charles Clayton of California responded to Burdett's report by introducing a bill allowing for public sale of timberlands. Whereas Clayton argued that privatization would better protect timberlands, Senator George Boutwell of Massachusetts offered an amendment that allowed for appraisal and sale of timber while retaining government ownership as a sort of commons. Senator Albert Howe of Mississippi responded, "When he [Boutwell] calls upon us to embark very heavily in the protection of generations yet unborn, I am very much inclined to reply that they have never done anything for me." Clayton's bill passed without Boutwell's amendment and opened up timberlands to privatization and subsequent exploitation. Although the Clayton Act applied primarily to the South, western lumbermen looked forward to its extension into their region.[21]

America appeared to be on the path toward privatization of all timberlands. A. B. Hammond's reentry into the timber industry in the 1880s, however, coincided with a protracted, and often resolute, shift toward increased government management. The contradictory and ambiguous decisions emanating from Washington, D.C., during this treacle-like adjustment period provided Hammond with opportunity, but they also ensnared him.

The departure of President Grant in 1877 marked a notable change within the Department of Interior. Seeking to establish legitimacy after the highly contested election of 1876, Rutherford B. Hayes selected the indomitable reformer, Carl Schurz, as secretary. While his predecessor had advocated privatization, Schurz favored government retention of timberlands and enforcement of existing laws. Born and educated in Germany, Schurz subscribed to the European concepts of centralized, state-controlled forest management and regarded timber trespass as a threat to federal authority. In his annual report to Congress, Schurz made clear his intentions to crack down on illegal timber harvest from public lands, calling for criminal prosecution in "arresting the evil." Citing predictions of a coming timber shortage, Schurz stated, "The rapidity with which this country is being stripped of its forest must alarm every thinking man." To show he was serious, Schurz dispatched timber agents throughout the country, targeting large-scale commercial operators, many of whom operated in the Great Lake states and western territories.[22]

In Montana, Marshall Wheeler, acting under instructions from Schurz, seized 250,000 bf of lumber from two of the territory's most established lumberyards in December 1877. "This action of the Washington authorities has caused a good deal of annoyance to dealers as well as builders," reported the *Weekly Missoulian.* The paper also printed a telegram from the attorney general stating that the government was willing to settle such seizures in exchange for a stumpage fee of two dollars per thousand board feet (mbf), about 10 percent of retail value.[23]

The timber agent for the case assured Schurz that the new policies had brought about a change in attitude. He wrote, "The lumber dealers now inform me that they considered the tax fair and equitable, and were willing to pay it, believing that the government should protect the public lands from depredations, and receive something for the timber cut therefrom." Schurz, however, disliked this practice of charging a stumpage fee in lieu of prosecution. He believed it merely encouraged the practice of timber depredation and asked the attorney general to discontinue the practice. Clearly, the Hayes administration was divided over the issue.[24]

Schurz faced an untenable situation. By rapidly disposing of lands to the private sector, the General Land Office (under the Interior Department) was instrumental in expanding industrial capitalism and rapid settlement of the West. Industrialization, however, depended upon wood products, and most of the forests remained in federal ownership. Logging on government lands was illegal, but it was difficult to procure timber of any substantial quantity otherwise. Schurz, thus, implored Congress to correct the matter with legislation. Congress responded by passing a pair of timber laws, which, ironically, exacerbated the conflict and created even more confusion.[25]

The acts of June 3, 1878, took two forms. The Free Timber Act allowed residents of the Rocky Mountain states to cut timber on *mineral lands* for building, agricultural, mining, or other domestic use, while the act sanctioned the sale of federal timberlands in California, Nevada, Oregon, and Washington. In 1891 Congress extended the latter provision to the Rocky Mountain states as well. These two laws would have profound and lasting impacts upon the western landscape. As economist Shirley Coon has observed, the Timber and Stone Act, in particular, "appears to have been nothing more than an unscrupulous scheme for the private acquisition of the publicly owned timber lands."[26]

The scheme was readily transparent. Upon their passage, Land Commissioner J. A. Williamson regarded both laws as "equivalent to a donation of all the timber lands to the inhabitants of those states and territories. The machinery of the Land Office is wholly inadequate to prevent the depredations which will be committed." Schurz agreed; sounding a dire note, he stated, "It will stimulate a wasteful consumption beyond actual needs and lead to wanton destruction . . . enforcement of the regulations will prove entirely inadequate, and . . . in a few years the mountainsides in those states and territories will be stripped bare."[27]

To curb potential abuse, Schurz interpreted the Free Timber Act narrowly and issued a set of regulations that explicitly restricted its application to "lands which are in fact mineral." Schurz noted that the act applied only to "bona fide residents" and forbade export to other states and territories. He concluded that this precluded railroad corporations from cutting timber. Additionally, Schurz expressed his intent to "preserve young timber and undergrowth . . . to the end that the mountainsides may not be left denuded and barren." Enforcement of these regulations was practically nonexistent, however, as lack of congressional funding forced the Department of Interior to discontinue special agents in seven states and territories, including Montana.[28]

Furthermore, the incoming Arthur Administration quickly backed away from Schurz's policies. While President Chester A. Arthur expressed serious concerns over destruction of the nation's forests, he nonetheless appointed Senator Henry Teller, a Colorado mine owner and railroad lawyer, as secretary of interior. Amenable to western development, Teller firmly believed in privatization of timberlands, did not consider timber harvests on public lands illegal, and often dismissed cases of timber poaching. Suggesting that all of western Montana could be considered a mineral district and thus open to timber cutting, Teller modified the Free Timber Act to allow lumbermen and railroads to cut timber on federal lands for commercial purposes and export. Teller's loose interpretation of the timber laws had profound consequences for Montana and for Hammond in particular.[29]

Upon acquiring the contract to supply the Northern Pacific Railroad with ties and lumber, in 1881 Hammond and his associates began cutting timber with impunity. Intending to form a large lumber company to better exploit the railroad land grant, the following year Hammond, Bonner, and Eddy joined Marcus Daly and Washington Dunn, the superintendent of construction for Northern Pacific, to incorporate the Montana Improvement Company (MIC). Bonner served as president and Hammond became the treasurer and general manager of his first corporation. Reportedly backed by two million dollars and armed with a twenty-year contract with Northern Pacific to supply all the railroad's lumber and fuel needs from Miles City, Montana, to The Dalles, Oregon, the MIC was well poised to become western Montana's dominant business—except that its activities were largely illegal.[30]

As an incentive to build the transcontinental railroad, the federal government had granted the Northern Pacific the odd-numbered sections while retaining the even-numbered sections along the railroad's route. Both the government and the public expected the railroad to subdivide and sell its grant land to defray construction costs. The Northern Pacific, however, discovered that it would be more profitable to retain its grant as timberlands. In addition, the charter legislation permitted the railroad to take timber from "public lands adjacent to the line" for construction purposes. Teller extended this privilege by allowing railroads and their agents, such as the MIC, to cut timber "wherever the companies desired, the word 'adjacent' being interpreted to mean practically anywhere in the United States," reported Land Commissioner William Sparks a few years later, noting that Teller's actions "promoted rather than checked timber depredations."[31] Not only had the railroad acquired more than one hundred million acres of timberland, but under Teller the culturally embedded practice of timber poaching on public lands acquired legal sanction—but only for corporations.

Teller's reluctance to enforce regulations must have frustrated his land commissioner, N. C. McFarland, who recognized that the system permitted "capitalists to indirectly obtain great bodies of public land" and control the timber supply. To minimize Teller's generosity, McFarland pronounced that the railroads, after their original construction, were prohibited from cutting public timber. McFarland, thus, interpreted the Free Timber Act differently than Teller did, declaring that mines could cut timber from adjacent public lands. But he muddled the issue by adding "but not for their private gain or commercial purposes," failing to explain what noncommercial mining might be. He maintained that it was "unlawful for mill men or others to cut timber on the public land for sale." The inconsistencies between Teller and McFarland proved awkward for Hammond and his associates.[32]

Even more problematic for the MIC was an 1875 Supreme Court ruling that held railroad grant lands still belonged to the government until surveyed and patented. Accordingly, McFarland informed his special timber agent in Montana, William Prosser, that since the Northern Pacific grant lands were not surveyed, it was not possible to determine the odd and even sections and therefore it was the "duty of the Government to protect the timber thereon." Yet under the lax Teller administration, nothing was done.[33]

In addition, the Northern Pacific granted the MIC rights to timber they did not own by allowing the MIC to log both railroad grant lands and adjacent public lands. As both Schurz and McFarland had specified, timber on these "adjacent" lands could be used only for railroad purposes. Regulations prohibited logging on all public lands except "mineral lands," and on these the export and commercial sale of lumber was forbidden. Furthermore, since the grant lands remained unsurveyed, the United States retained title, and in logging these sections the MIC could not avoid cutting federal timber.[34]

Government regulations, however, were little more than minor annoyances to Hammond. By 1883 the MIC had already cut 6,000,000 bf and was sending another 250,000 bf (equal to thirty railroad car loads) of lumber each day to markets outside Montana in violation of federal policy. In addition, the mines in Butte chewed up 50,000 bf daily. With seven sawmills in continuous operation, the MIC employed more than two thousand men and was advertising for hundreds more. Throughout 1884 Hammond continued to expand his operations, establishing logging operations and commercial lumberyards throughout Montana, Idaho, and Oregon.[35]

While the federal government struggled to assert its authority in the hinterland of Montana, the Northern Pacific Railroad and the MIC moved to restrict access of individuals to the forests. The railroad established dominion over the grant lands by requiring stumpage fees and limiting the timber harvest, relying upon the MIC to patrol the forest to keep other wood cutters out.[36] The special timber agent, Prosser, reported that not only did the MIC dominate all the railroad grant lands along a thousand-mile stretch; it also "claims control of all the timber on *government lands.*" Citizen complaints soon reached the secretary of the interior. Upset at being excluded from what he considered the commons, Montana resident S. H. Williams reported on the activities of the

MIC, stating, "They have from two to three thousand men here, steadily chopping the government timber, and sawing it up into lumber and shingles for their own benefit, and pocketing the proceeds themselves; and if anybody else wants any to fence with, or use on their place, or for firewood, they make a terrible fuss about it, and threaten to put them in states-prison."[37]

In Missoula, Prosser confronted Hammond with these charges. Hammond told the agent that Bonner had met with Secretary Teller and "had received permission and authority from him to cut all the timber they might require from government land—at least where the land was not surveyed." Although Hammond could have been bluffing, Teller—given his record—may have indeed granted this permission. However, his successor, Lucius Lamar, found no record of such authorization. In condemning both MIC and Teller in a single stroke, Lamar stated, "He [the secretary] is not himself at liberty to violate the law, nor can he authorize any one else to violate the law."[38]

The activities of the MIC were so blatant that they attracted the attention of the U.S. attorney general, Benjamin Brewster. Brewster noted that the company had apparently obtained permission from Teller to cut timber and build sawmills on the Flathead Indian Reservation. Teller, however, had stipulated that the MIC was to leave the reservation as soon as it had completed the construction of the railroad. Brewster noted, "The road has long been completed, but the firm insists on keeping their mills on the reservation. They are running night and day. . . . They are cutting out all the available timber." Meanwhile, Teller had forbidden Indians themselves from logging on their reservations.[39]

While the MIC patrolled the forests and kept residents from cutting timber, the railroad took action against higher-level timber poachers on what it regarded as its land. When Northern Pacific Vice President Thomas Oakes discovered that Montana Territorial Governor Samuel Hauser had logged some of the railroad sections and refused to pay stumpage, he dashed off this angry note: "Read this over carefully and let me know if you intend to take this position in reference to our timber interests. If we have no rights in this property you will respect, I shall at once withdraw our deposits from your bank, put the Wickes Br. [Hauser's railroad] on a strictly local basis and in every other respect make things so hot for you, you will think the Devil is after you." Hauser immediately backed down, replying, "I am willing to be bulldozed by men in position and power and stand it like a little man," clearly demonstrating the supremacy of the railroad.[40]

Although Hauser had enticed and accommodated the railroad, he was now subject to its power—his mines needed the railroad more than it needed him. Hammond, however, had recognized that the railroads needed timber for construction, for fuel, and as a steady export. While the Cramer Gulch war resulted from Montana lumbermen seeing the forest as an open-access resource, Hammond and the Northern Pacific regarded the public lands as their exclusive territory, exacting stumpage fees and threatening "trespassers" with imprisonment. This attempt by corporations to mimic the power of the state by asserting access and control over a given geographical area recalled the feudal enclosures of the commons and underscored the emerging corporate power. Given the laissez-faire attitude of multiple administrations, the Northern Pacific and the MIC anticipated a move toward privatization of the public domain. Yet an unexpected shift began as the public rebelled against the power of corporations, especially railroads. Faced with a choice between government and corporate control, citizens often preferred the former. This sentiment coincided with the rise of federal power in the United States at the end of the nineteenth century, putting the federal government and A. B. Hammond on a collision course.

CHAPTER 7

Sparks Lights a Fire

The Montana gold rush of 1863–70 coincided with an exodus from the South devastated by the Civil War. The resulting influx of southern settlers into Montana tilted the newly formed territory toward a Democratic majority. As a Canadian immigrant, A. B. Hammond entered Montana with no particular political affiliation, but under the influence of his Dixie business partners, Bonner, Eddy, and Hauser, he embraced the Democratic Party. So it was with great occasion that the Hammond-Eddy household celebrated the election of Grover Cleveland in 1884, the first Democratic president since James Buchanan, twenty-four years earlier. Cleveland appointed Hauser territorial governor, and it appeared that Hammond's interests would be well represented at both the territorial and national levels. Before long, however, Cleveland's land policies would cause Hammond to switch parties and reconfigure Montana's political landscape.[1]

The presidential campaign of 1884 was particularly nasty and contentious. The Republican Party split over the nomination of Senator James Blaine, who was an ardent supporter of big business, railroads, and the widely reviled tycoon Jay Gould. In disgust, the Mugwumps, a group of reform-minded Republicans led by the esteemed former interior secretary Carl Schurz, abandoned the Republicans and supported Grover Cleveland, who campaigned against the power of railroads and government corruption. Although many in the business world feared Cleveland's rhetoric of reform, he received infusions of cash from railroad magnate James J. Hill and others. Upon his narrow victory, Jay Gould wired the president-elect, conceding, "The vast business interests of the country will be entirely safe in your hands."[2]

Regarding land use issues, however, Cleveland began sweeping house almost immediately, much to Hammond's dismay. Hammond had predicated his lumber business on free access to government timber, and the Cleveland Administration threatened to assert federal authority over and control of the nation's forests. Forests became the political battleground between private industry and the federal government.

Hammond's troubles began with a pair of Cleveland's appointments. Upon taking office in March 1885, Cleveland selected former Confederate Colonel Lucius Quintus Cincinnatus Lamar as secretary of interior and tapped former U.S. Representative William Andrew Jackson Sparks as his land commissioner. Although Lamar and Sparks were both committed to land reform, the two men had marked differences in their personalities. A chivalrous, old-school Mississippi gentleman, Lamar was highly conciliatory toward rebuilding the Union. Even northerners considered the stately and deeply pensive Lamar "noble-hearted, honorable, and generous." In contrast, Sparks was an explosive crusader. He possessed the very qualities Cleveland looked for in a land commissioner: stubborn, combative, and meticulously honest. During his years in Congress, Sparks had earned a reputation as a tireless reformer and staunch advocate for curbing the abuses and power of the railroads.[3]

Cleveland's top priorities included reforming the Land Office and tackling the fraud, speculation, and environmental destruction resulting from the railroad land grants. Riddled with corruption and graft, the Land Office was also understaffed and increasingly saddled with responsibilities. Congressmen whose election campaigns were financed by railroads, land speculators, and lumbermen deliberately kept the Land Office underfunded.[4]

Sparks swept into the Land Office like a rat terrier, intent on ferreting out evil and corruption wherever he found them, and the timber poachers received his special notice. Lumber companies aroused the ire of reformers like Sparks and those in the nascent conservation movement, not only for fraudulently acquiring vast tracts of the public domain but also for the waste and destruction that accompanied the cutting of timber.

Eight days after taking office, Sparks suspended most land claims, especially those under the Timber and Stone Act, citing reports of "widespread, persistent, public land robbery." As the flood of land patents dried up, Congress demanded an explanation. Lamar patiently informed Congress that not only did Sparks have legal authority to issue such a moratorium but he fully supported the action. In his defense, Sparks stated that the "public domain was being made the prey of unscrupulous speculation and the worst forms of land monopoly through systematic frauds." An examination of land entries over the next year vindicated Sparks, as his office discovered that 2,223 out of 2,591 cases were indeed fraudulent.[5]

Sparks recapped his first three months in office in a scathing condemnation of the condition of the General Land Office under the previous administration. He informed Congress that the public domain had been "wasted under defective and improvident laws and through a laxity of public administration astonishing in a business sense if not culpable in recklessness of official responsibility." Aghast at the general attitude of his department toward land disposal, Sparks stated, "The vast machinery of the land dept. appears to have been devoted to the chief result of conveying the title of the US to private lands upon fraudulent entries."[6]

The majority of these frauds involved claims on the West Coast under the Timber and Stone Act. In the rest of the country, lumbermen simply cut whatever they wanted from federal lands. This widespread practice of timber poaching was as abhorrent to Sparks as land fraud was. In his first year, Sparks dispatched twenty-one special agents from Mississippi to Washington to investigate 1,219 cases of timber trespass involving more than $9 million in stolen timber. Filing 137 criminal and civil suits, Sparks recovered only $101,085, leading some to question the worth of his efforts. The lone agent for Montana investigated twenty-six cases involving nineteen million board feet and 1.5 million railroad ties taken from public land. The largest violator, by far, was Hammond's Montana Improvement Company.[7]

Sparks's campaign against timber poaching ignited controversy across the country, probably just as he had hoped. Believing he could dictate immediate policy changes, Sparks issued a flurry of regulations, clarifications, retractions, and modifications of the timber laws. Lamar's more subtle approach was to persuade Congress to rewrite the laws. Lamar desired a legal method of selling timber from public lands while retaining government title. Regardless of motivation, both men wished to establish a concrete policy and distributed a circular across the country clarifying the provisions of the Free Timber Act.[8]

In contrast to his predecessor, Henry Teller, Lamar issued a strict interpretation of the act and limited timber cutting on public lands to those "distinctly mineral in character" and restricted harvest to individual residents. In a provision that appeared to be targeted directly at the MIC, the circular stated that neither a railroad company nor its agents were allowed to cut any timber "from the public lands, *for sale or disposal*, either to *other companies* or to the *public* or for *exportation*." This regulation essentially prohibited commercial logging on all public lands. Montana's mining and lumber industries, however, had grown up during the laxity of the Teller administration.[9]

Predictably, the territory's business leaders howled their opposition to Lamar's policy and flooded the press with prophecies of total economic collapse as a result of the regulations. E. L. Bonner warned, "Our mines in 1884 produced twenty million dollars. If we are not allowed to cut timber for these mines production will cease at once. Mining is the great pursuit of Montana."[10] The Butte *Inter-mountain* joined the chorus, calling the regulations "unjust and ruinous to the material interests of Montana." Attempting to equate Sparks with the growing conservation movement, the newspaper noted, "Sparks must be of the opinion that timber is one of the most sacred products of nature, not to be defiled by the rude hand of man but intended by God to grow and die and rot, safe from the profanation of the axeman's stroke, and that it were sacrilegious to use it for fuel, building or mining purposes."[11]

Sparks, however, was no John Muir. He genuinely wished to provide individual settlers with access to federal timber. Like the Progressive Era conservationists who would follow, Sparks saw the forest issue as embedded within a larger social agenda: curtailing the rise of corporate power. Linking capitalism to deforestation, Sparks reported, "The struggle to accumulate great private fortunes from the forests of the country has reduced forest areas to a minimum."[12]

Both Sparks and Lamar singled out the Montana Improvement Company as the most flagrant abuser of the timber laws. Forwarding the reports from the Arthur administration that charged the MIC with cutting timber on government lands while excluding and threatening locals, Lamar reiterated to Sparks the instructions of the previous land commissioner: "the purpose of the Government is to prevent the unlawful taking of timber from *all* Government lands *until the title to such lands has actually passed from the United States.*" Unlike his predecessor, however, Lamar chose to act and directed Sparks to take "prompt and vigorous measures to ascertain the amount of timber already cut" by the MIC. Lamar also requested the attorney general to order the MIC to cease its operations on all unsurveyed lands—essentially all of western Montana and northern Idaho.[13]

The graft, corruption, and political influence of the railroads were particular targets for Cleveland's reform team, and unfortunately for Hammond, Lamar tied the MIC to the Northern Pacific Railroad, insisting that the NP controlled 51 percent of the MIC's stock, with Eddy, Hammond and Co. holding the rest. According to Lamar, not only was the MIC preventing other lumber dealers from accessing raw materials; it also received half-price shipping rates. Sparks thus viewed the MIC as a railroad-controlled monopoly that intentionally violated laws. Unlike other lumbermen, who readily confessed their crime and paid a fine, often in lieu of a stumpage fee, Hammond refused to admit any wrongdoing. Sparks found this particularly annoying and repeatedly referred to the MIC as "bold, defiant and persistent depredators on the public domain."[14]

Based on the MIC's association with the Northern Pacific, in July 1885, Sparks filed a formal complaint against the two companies for cutting federal timber. The government charged the MIC with claiming control of the forests on railroad *and* government sections between Miles City, Montana, and Wallula, Washington, a distance of 925 miles. Legally, this timber was only for construction of the railroad, yet the MIC continued to log and operate mills long after its completion, exporting lumber as far as Utah and Minnesota. Both Hammond and Bonner insisted that Secretary Teller had sanctioned their activities.[15]

Especially objectionable to the government was the report of a twenty-year contract between the MIC and the Northern Pacific giving the MIC the right to remove all merchantable timber from the railroad grant lands between the eastern end of Hellgate Canyon and Sandpoint, Idaho. The railroad agreed to withdraw these lands from sale, build sidings for the MIC at each of its mills, and guarantee low freight rates in return for railroad ties and bridge timbers at a minimal price. Both companies resolved to "keep trespassers on government timber lands from cutting more than the actual wants of settlers for mining and other domestic uses," although, in reality, they excluded everyone but themselves. Not only did Hammond deny that the railroad held any stock in his company, but he also refuted the contract's existence. Furthermore, Hammond insisted that MIC cut only from the odd-numbered (railroad grant) sections and did not knowingly cut from even-numbered (government) sections. How his logging crews were able to differentiate between the two, considering the entire region had yet to be surveyed, he neglected to mention.[16]

The connection between the Northern Pacific and the MIC was the linchpin in the government's case, and Bonner and Hammond worked quickly to decouple it, using their influence to redirect the public discourse. Just three days after its initial account, the St. Paul *Pioneer Press* issued a retraction of its previous condemnation of the MIC, calling Sparks "impetuous" in prematurely filing charges. Both Hammond and Bonner pinned the blame on Special Timber Agent M. J. Haley, who became "an exceedingly reckless and defamatory character" in the newspaper report. The paper also noted, "We have reliable information that not a dollar of this [MIC] stock was ever transferred to the Northern Pacific Company." To this day, the exact relationship between the two companies remains hazy. Nonetheless, the MIC enjoyed a cozy affiliation with the railroad that pushed the bounds of legality.[17]

The partners also succeeded in manipulating the Montana press. The *Missoulian* compiled dispatches from newspapers around the country under the headline "Unrelenting Prosecution of the Improvement

Company." The editor felt no compunction about expressing his sympathies, writing with pointed exaggeration, "Judging from reports the entire power of the United States government and of Victoria, Queen of Great Britain and Ireland and Empress of India is being brought to bear on the Montana Improvement Company."[18] While the Butte *Intermountain* referred to the charges against the MIC as "absolutely ridiculous," the paper admitted that if the government intended to quash a monopoly, "then the aggravation will be but temporary, and may result in some benefit." Even the boosters, it appeared, opposed a single corporation controlling access to natural resources.[19]

As the largest employers in Missoula County, Hammond, Bonner, and Eddy wielded considerable influence. They maintained a near-monopoly on the wholesale and retail trade in western Montana, selling groceries, clothing, and hardware on credit. Hammond's bank held most of the homeowner mortgages in Missoula. As the primary real estate developer, he determined the downtown business rents and location of building lots. Hammond even controlled the city's water supply. Nearly everyone in the county was indebted in some fashion to Eddy, Hammond and Co. Furthermore, the partners were the second-largest taxpayers in Missoula County, after the Northern Pacific. Two years later, the tax burden of Hammond's enterprises would double even that of the railroad. Besides their economic supremacy, the three men were widely respected, generally well liked (thus far), and active in community affairs. They could, thus, frame the government accusations as an assault against the entire community.[20]

Nonetheless, by October 1885, Special Agent Haley had accumulated enough evidence to file thirty-one separate counts against the Northern Pacific and the MIC for stealing $613,402 of public timber from 1881 to 1885. This included an enormous forty-five million board feet of lumber (enough to fill 5,500 rail cars), 84,700 railroad ties, fifteen million shingles, thirty-two thousand cords, and twenty thousand posts. While this constituted only a fraction of what the MIC had cut in four years, it reflected the quantities that the government believed it could prove were exported from the territory and not used for railroad construction. The charges of timber theft against other lumbermen in Montana paled in comparison, ranging from $9,092 to $66,805.[21]

While Bonner and Eddy appeared nonplussed, Hammond, meticulous in his accounts, was indignant at the charges of impropriety. At the helm of the largest business in western Montana, Hammond considered himself an upstanding leader of the community. Concluding that Sparks had targeted him as a national scapegoat, Hammond stewed. Experienced in the subtle dynamics between politics and business, Bonner, on the other hand, counted on his political connections and cronyism to resolve the matter. Confident that once he clarified the demands of lumber and mining interests to Lamar, Bonner wrote that "the rulings will be reversed." The easygoing Richard Eddy simply moved toward early retirement—hunting, fishing, and wintering among his newly acquired orange groves in Southern California.[22]

Although he was not directly implicated, the new territorial governor, Samuel Hauser, also had good reason to be worried. He shared many financial and political interests with Bonner and Hammond. Furthermore, Hauser was a major purchaser of lumber for his mines and cord wood for his smelters. He had already run afoul of the Northern Pacific for poaching its timber, and now his major supplier, the MIC, was about to be cut off.

Hauser began pulling strings. He implored Montana Delegate Joseph K. Toole to straighten things out in Washington. Anticipating the outcry, Toole confidently informed the governor, "I had called on Lamar before your telegram came and had all suits discontinued against MIC." Toole had an inflated sense of his influence, and his notice to Hauser was a bit premature. Hammond, who was engaged in a lively and rather caustic dialogue with Hauser over banking affairs, dashed off an uncharacteristically conciliatory note thanking Hauser for his efforts, noting that he would "not soon forget it." After his own meeting with Lamar, Bonner, too, mistakenly believed that he had convinced him to rein in Sparks and rescind the charges.[23]

Hauser also enlisted the help of Marcus Daly, one of the original incorporators of the MIC. Daly, in turn, gathered a petition to Lamar signed by all the leading businessmen of Butte asking for resolution of "the timber question." Daly recommended that Hauser "appoint a Commission to go to Washington and have a personal interview with Sect. Lamar" and suggested that George Hearst, as one of the Anaconda owners, "would probably be willing to make one of such a Commission." Hauser also called on an old friend, Senator George Vest of Missouri, to intervene, but to no avail.[24]

Although far from Washington, Hammond had a more realistic assessment of the situation than his partners did. He insightfully realized that the railroads, especially the Northern Pacific, had fallen out of favor and lost access to the Cleveland administration. Privately, Hammond told Hauser, "Sparks is the bad man." He complained that Sparks seemed "to be a hawk on the subject of reform" and used the press to make "all kinds of wild statements that have not the shadow of truth simply to make political capital."[25]

With so much commotion over the issue, Lamar clarified the regulation by stipulating that export and commercial harvest were still illegal but mining companies could cut all the timber they needed. To exploit this provision, the MIC became an agent of the mining companies, at least on paper.[26]

Seeing through this ploy, Sparks issued an even more restrictive circular. The new regulations limited timber cutting on public lands to areas in which it was consumed, forbade manufacturing of lumber, and held that a settler could not pay someone for timber cutting but must do the logging himself. Hammond noted that the Sparks circular "virtually stops the lumber business." Normally self-possessed, Hammond confessed his worry to Hauser: "I am at a loss what to do. . . . It now looks as if this fellow Sparks was going to shut down the whole country."[27] Sparks had no intention of shutting down the whole country, just the MIC. But this time he had gone too far, and Lamar distributed a statement that these new provisions had been issued "inadvertently."[28]

Lamar, however, also sought to rein in the MIC and insisted the company cease logging until its case was resolved. In flagrant disregard, the MIC continued to expand its operations. There was simply too much money to be made to worry about fine points of legality, and the company already controlled "nearly the entire lumber trade of Western Montana," with a sawmill production capacity of twenty-five million board feet a year, ten million of which were destined for the Anaconda Copper Mining Company. In clear violation of the government decrees, the MIC anticipated exporting fifty thousand tons of lumber in 1885. In addition to the continued construction of a mammoth mill at the mouth of the Blackfoot River, Hammond opened a mill at Bonita to access the timber of Cramer Gulch and the Clark Fork. The company also ran its mill in Wallace, Idaho, night and day.[29]

The conflict became a contest of wills between Sparks and Hammond. If Sparks was going to shut down his lumber business, Hammond threatened to shut down Montana, for without wood neither the mines nor smelters of Butte would function, throwing thousands out of work and severely restricting the nation's copper supply. In December 1885, Hammond announced he was shutting down the Wallace mill and would soon close the others. By Christmas, only one mill remained in operation. Hammond cited the combination of low lumber prices, the new requirement of having to pay government stumpage rates, and—hoping to mobilize public opinion in favor of the MIC—the impending lawsuits as reasons for the closures. That winter, however, with improved market conditions, the MIC quietly reopened its timber camps and advertised for more loggers.[30]

Upon lining up sufficient evidence, the Cleveland administration brought suit against the MIC in March 1886 for $1,164,000, the estimated value of timber the company had cut from government land. The U.S. district attorney for Montana Territory also filed an injunction prohibiting the MIC from logging on public land while the suit was pending and convened a grand jury. Normally, Sparks deplored the practice of early settlement of timber poaching cases, believing that it encouraged depredations, as lumbermen simply regarded the low fines as stumpage fees. But in this case, he was willing to settle in order to finally put the matter to rest. Indignant at the government's treatment, Hammond and Bonner, however, had no intention of settling out of court. Failing to reach an agreement, the government carried the suits over until the next court term.[31]

Despite the suits and injunction, the MIC reopened its mills and continued logging but now concentrated on the surveyed railroad sections along the Blackfoot and Clark Fork Rivers. In July 1886, the company finished building the much-anticipated Blackfoot Mill, the region's largest, and A. B.'s younger brother, Henry, took charge of the operations. In its first month, the Blackfoot Mill produced 55 thousand bf per day, and within two years it churned out over 100 thousand bf daily. With the mills cranking up, the MIC began hiring all the workers it could find. Exhausting the local labor supply, it advertised in St. Paul for more men. Opening five logging camps on the Blackfoot River, Henry anticipated cutting 20 million bf during the winter, 13 million more than the previous year. Although the mines still consumed a vast amount, much of this timber went to the railroad branch lines that Hammond, Bonner, and Hauser were building.[32]

Between the two court terms, Hauser presented a litany of modifications to Sparks, maintaining that all of western Montana should be considered mineral lands and thus open to logging under the Free Timber Act. Hauser claimed that "every stream in this region contains placer gold." He further stated that enforcement of the regulations would shut down nine-tenths of the small mills. Despite obvious evidence to the contrary, Hauser tried to convince Sparks that the timber cut from government land was not exported but used exclusively by the mines.[33]

Sparks was not fooled. He rejected Hauser's recommendations, pointing out to Lamar that "the interests of the settler and the lumbermen are not in this matter identical, as the lumbermen ship lumber as far as St. Paul." Sparks noted that all of the signers of Hauser's petition were "*lumber dealers,*" as if this were condemnation enough. Immersed in populism, Sparks viewed himself as looking out for the individual settler against the rapaciousness of big business.[34]

More susceptible to the pressures of business interests, Congress demanded an accounting of Sparks' actions. Now on the defensive, Sparks provided Congress with scores of letters from western settlers who supported the "efforts of the Interior Department in its determination to hold the public domain in the interest of honest settlers." Others complained of the undue power, influence, and heavy-handedness of both the Northern Pacific and the MIC. Two prominent Missoula businessmen thanked Sparks for enforcing the law against the MIC.[35] Particularly annoying to Montanans were the high lumber prices the MIC charged because of its monopoly-like control of the forests. High prices were an unintended consequence of Spark's regulations as they prevented small and honest lumbermen from cutting on the public domain. The MIC, on the other hand, continued to enjoy exclusive access to the Northern Pacific land grant, the only timberlands now legally available for exploitation.

The suits resumed in the fall, but by now the government had difficulty locating witnesses. Special Timber Agent M. J. Haley reported that the MIC destroyed evidence and persuaded witnesses to leave the area or remain silent. Haley asserted, "I found that no reliance could be placed on statements made by the officers of the Montana Improvement Company."[36] Frustrated, Sparks wrote to Lamar that the directors of the MIC "are bold, unscrupulous and persistent violators of law and will leave no means untried to defeat the government in the suits now pending against them."[37]

Faced with both criminal and civil suits, Hammond and his partners engaged in a shell game to protect their considerable assets. In contrast to their former proprietary partnerships, by forming corporations the men became stockholders and thus could not be held individually liable. In August 1885, Eddy, Hammond and Co. dissolved its partnership and incorporated as the Missoula Mercantile Company. While the partners stated that their intention was to allow longtime employees the opportunity to invest in the company, after incorporation employees held only 4 out of 2,500 shares. To further protect themselves, Hammond, Bonner and Eddy incorporated the Missoula Real Estate Association to buy, sell, and lease real estate in the growing city.[38]

In December 1886, the grand jury issued indictments against Bonner, Hammond, and Eddy for timber trespass. They pled not guilty, forcing the case to trial. Meanwhile, in an associated case against the Northern Pacific the following month, the Montana Supreme Court denied the government's claim of $1.1 million in damages, ruling that the government and the railroad were "not tenants in common, and it is impossible to tell on whose land the timber was cut." The court further ruled that the government failed to prove that the lumber was not, in fact, used for railroad construction. Effectively lifting Lamar's injunction against the MIC, the court stated that the government could not impose an injunction against timber cutting until the lands were "surveyed and the odd and even sections specifically designated." In issuing this decision, the

Montana court defied the previous U.S. Supreme Court ruling that held that railroad grant lands belonged to the federal government until title was actually transferred. Clearly something was amiss. Incensed, Sparks sought an appeal to the U.S. Supreme Court.[39]

Meanwhile, the battle moved from the courts to the forests. Under instructions from Sparks, Haley seized railroad ties that the MIC had illegally cut from public lands. Sparks thought Haley should have arrested the offenders because they simply "resumed their timber depredations." Asserting federal authority in the woods was a different matter, as Hammond's attorney advised the MIC employees "to go armed and to resist any attempt to arrest them by the United Sates authorities." In a near-reenactment of the timber conflicts in New Brunswick and New England, the MIC loggers "terrorized" and overwhelmed the deputy U.S. marshal and retook the seized ties.[40]

Furious, Sparks blamed the Montana court decision, stating, "These marauders upon the public timber appear to have been thus emboldened to further disregard of the law." Agent Haley agreed, reporting that the MIC "has never been so willful or defiant in its timber depredations as it has been in the past month. The officials of the company openly boast that the Senate, Congress, and the courts are with them." It appeared that not only had the MIC taken de facto control of the forests, but it had also gained the political high ground.[41]

U.S. District Attorney Henry Hobson also charged the MIC with manipulating the court system of Montana. He reported, "It seems almost impossible in that country, for the government to obtain justice in its own courts." Having fully prepared his case against the MIC, Hobson discovered that Montana Supreme Court Judge Steven De Wolfe, who owed his appointment to Hammond and Bonner, had postponed the hearing until just before the end of the court term. Then De Wolfe suddenly decided that the method of jury selection was illegal, "leaving the government in court with its case, its counsel and its witnesses, but without a jury." Hobson summarized his frustration: "Public sentiment in Montana seems to be against the government; the courts seem to be organized and maintained for the purpose of upholding that sentiment." He believed, presciently, that the government would never again have as strong a case.[42]

After spending more than $2 million in pursuit of the MIC, by 1887, lack of funds forced Sparks to suspend the investigation. Beseeching Congress for more money, Sparks insisted that delay was weakening the government's position, as the MIC was not only continuing its "unlawful depredations on public timber" but actively destroying evidence.[43]

While Sparks fretted, the MIC further thwarted his efforts by disbanding in August 1887. Bonner, Eddy, and Daly sold their interest in the lumber business, while Hammond sought to insulate himself from government prosecution through kinship ties and legal maneuvering. Attempting to conceal his involvement in timber poaching, Hammond created a dummy board of directors, a practice that he used for the next two decades. With Henry Hammond ostensibly in charge, the Blackfoot Mill Company assumed the assets of the MIC. A few months later this morphed into the Blackfoot Milling and Manufacturing Company. With Henry running the mill, A. B. installed his older brother, George, along with his brother-in-law Charles Beckwith and his trusted lieutenant C. H. McLeod, as directors. A. B.'s other brother, Fred, managed the Bonita mill until A. B. transferred it to another brother-in-law, George Fenwick.[44]

Despite Hammond's contention that Sparks operated out of zealotry and retribution, Sparks, like Lamar, saw the timber suits as a means to prod Congress into passing meaningful reforms. Even while he attempted to enforce the Free Timber Act, Sparks advocated its repeal as it invited "great waste and greedy speculation" and made no provisions for the preservation of timber for future generations. Lamar added, "Individual avarice and corporate greed . . . unless checked by wholesome modifications of the law, will soon cause all the mineral lands to be stripped of their forests."[45] With insight into the problem of open-access resources and echoing a common refrain of the era, Lamar noted, "That which is every one's property is no one's care."[46] With this single line, Lamar pierced the heart of the issue. As long as the forests remained an open-access resource, lumbermen would have no incentive to conserve. Lamar was essentially building a case to establish the forest reserves (predecessors to the national forests), although this would not take place for another six years, under a different administration.

Lamar recognized defined ownership as the key to forest conservation. He proposed a survey, appraisal, and demarcation of timberlands, with every fourth section "permanently reserved from sale." On the one hand, Lamar seemed to favor privatization, suggesting that the government sell the appraised sections at public auction. On the other, influenced by Sparks's populism, Lamar advocated a system that would supply timber without "concentration of these lands in the hands of capitalists." He suggested that the Free Timber Act should provide for "the *sale* of timber on mineral lands" while retaining government ownership of the lands themselves.[47]

Lamar's reports echoed previous secretaries in advocating forest conservation to ensure a continuing supply of timber and to prevent erosion. Exasperated at congressional inaction on the issue, Lamar reproached the lawmakers. With the forests "disappearing at a rate that excites grave apprehension," Sparks, too, insisted that "the time for tinkering has passed" and called on Congress to void all public land entry laws except the Homestead Act. Sparks recognized the perils of privatization, especially when it came to the railroads and timber companies. As one solution, Lamar and Sparks implored Congress to establish permanent forest reservations in mountainous regions and at the headwaters of major watersheds.[48]

Although Sparks devoted considerable energy to prosecuting timber poachers, his primary mission was to rescind railroad land grants and bring those lands back into the public domain. Sparks viewed the Indemnity Act of 1874 as especially problematic. Under this law, railroads could claim grant lands up to fifty miles from the track if closer sections were already settled. Realizing that distant timberlands were often worth more than the treeless valleys near the track, railroads employed men to file claims on such "barren and worthless land." This would allow the companies to exchange their grant lands for more valuable parcels. The railroads would also sell land upon which they had yet to receive title and then claim indemnity. With recommendations from Sparks and Lamar, President Cleveland revoked fifty-four million acres of such railroad indemnity lands. In addition, the forfeiture of railroad grant lands and Sparks's cancellation of fraudulent claims brought the total acreage he restored to the public domain to more than eighty-three million acres. Many of these lands eventually were incorporated into the national forest system.[49]

Nevertheless, the land commissioner's impetuous and dogmatic approach ultimately proved his undoing. Upon Sparks's recommendation, Congress passed an act directing the secretary of interior to adjust railroad lands that had been conveyed in error. Under this new law, Sparks submitted an adjustment of the land grant to the Chicago, St. Paul, Minneapolis and Omaha Railroad, seeking to reclaim 245,000 acres. The more-cautious Lamar disagreed and rejected Sparks's calculations. Enraged, Sparks submitted his resignation as a protest. While supporting Sparks in principle, Cleveland recognized the commissioner was a liability in his relationship to the business community and accepted Sparks's resignation. Sparks returned home to Illinois, while Lamar advanced to the Supreme Court. But by the end of 1888, the president, too, would be out of a job.[50]

With Cleveland, Lamar, and especially Sparks all out of office, and with Republican President Benjamin Harrison in Washington, the government lost its enthusiasm for pursuing timber trespass. The suits against Hammond languished in bureaucratic limbo. Although he never denied that government timber had been cut, Hammond refused to accept personal responsibility. At first, he claimed Teller granted permission. Then he claimed that his crews cut government timber only inadvertently since the sections lacked a survey. Finally, Hammond pinned the depredations on his longtime employees and relatives George Fenwick and Henry Hammond, who were the direct supervisors of the operations. Regardless of how it happened, Hammond defended his timber depredations as having "resulted in the material development, upbuilding and growth of the Territory." Not everyone agreed, and many complained that the MIC drove them out of business.[51]

To what extent could the exploitation of natural resources be justified in aiding the material development of the West? Hammond was hardly the only one who regarded laissez-faire capitalism as essential to the nation's material progress. In this view, open access to the nation's forests fueled development, whereas government control hampered progress. But others questioned this and sought some means of popular control over resources and their use. Lacking firm social structures, such as community-regulated commons to manage natural resources, citizens turned to the federal government. Beset with competing interests, however, the government served less as a monolithic actor and more as an arena where this issue would play out.

Rapid western expansion and settlement demanded lumber for buildings, businesses, railroads, and—especially in Montana—mining. This demand, combined with a series of presidential administrations with an ideological predisposition toward privatization, led to relaxed enforcement and poor public policy throughout the 1860s and '70s. At the same time, Congress repeatedly increased the appropriations for timber agents attempting to curb poaching of federal timber. The government, thus, attempted to assert control over forests while promoting western expansion. In doing so it confronted a conflicting objective: how to conserve resources for future generations while providing timber for a rapidly industrializing nation. Addressing this issue would form much of the public discourse about the West for the next twenty years. As it turned out, the activities of A. B. Hammond provided a major catalyst to bringing about the end of the Great Barbeque.

As the railroads propelled the nation headlong into the industrial era, the gulf between those seeking social and government control over resources and those advocating privatization would grow ever wider. Rather than resulting in victory or defeat, this struggle fostered two of the most powerful and influential of modern institutions—the corporation and the federal government—and helped shape the complicated relationship between them.

Although Sparks was out of the picture, the timber suits lingered for another three decades. While the resolution of the timber suits proved anticlimactic, it signified the slow turn from laissez-faire capitalism to government regulation and increased federal power. In the meantime, the struggle for access and control over the nation's forests was heating up.

CHAPTER 8

A Political Animal

During America's Gilded Age, a man's political affiliation formed a large part of his identity. It determined his friends, associates, and business connections. Men identified more with their standing as a good Democrat or a stalwart Republican than with their particular profession. In an age before professional sports, men discussed and debated politics more than any other topic. Although many, including women, were disenfranchised, politics was highly participatory—nearly everyone who could do so voted and almost always voted a straight party line. The Republicans attracted northern Protestant voters who sought to control alcohol consumption and advocated government policies favorable to business and industry. Democrats, often Catholic or from the South, or both, favored a minimal federal presence in both economic and social issues, but supported policies that aided smallholders and agricultural interests. At the national level, such party loyalty led to a nearly equal balance of power between Republicans and Democrats—a situation that in modern parlance would be called "gridlock."[1]

While political affiliation remained an important part of one's social and economic network, in the American West it held a looser grasp than in it did the rest of the country. Nonetheless, like many businessmen, A. B. Hammond became involved in local politics. In 1884 Hammond was elected Missoula city alderman, and later that summer voters chose him as a delegate to Missoula County's Democratic convention.[2]

Unlike his predecessor and partner, Richard Eddy, Hammond took his duties as alderman seriously. Although he was a Democrat, Hammond's activities lay more in line with nineteenth-century Republican policies.

Taking an active role in the weekly meetings, Hammond supported ordinances that regulated saloon hours and prohibited "the keeping of Houses of Ill Fame." Roadwork and bridges drew his special attention. Above all, he sought to keep the streets in good condition. Hammond displayed a keen interest in transforming Missoula from a frontier village into a center of commerce, or at least the sort of commerce that he saw as legitimate.[3]

It was on the national level, however, that Hammond's choice of party was most conflicted. President Cleveland's land policies had, by 1888, alienated Hammond and other Montana Democrats by restricting their access to and control over natural resources. The Cleveland administration's timber suits against the MIC particularly aggrieved both Hammond and Marcus Daly. Not surprisingly, as both a member of Montana's Democratic Committee and a delegate to the Democratic National Convention in St. Louis, Hammond argued strenuously against renominating the president. While Cleveland retained little support in the West, eastern delegates handed him the nomination. Hammond returned from St. Louis discouraged by the prospect of another four years of Cleveland. To protect their economic interests, Hammond and his partners abandoned the Democrats, secretly switched parties, engaged in widespread electoral corruption, and conspired to send their handpicked candidate for territorial delegate to Washington, D.C.[4]

In 1888 Montana stood on the verge of statehood, which would allow it to send two senators and a representative to Washington. In the meantime, a single nonvoting delegate represented the territory. The election of a Republican delegate from a Democratic territory would send a clear message to Washington that Montanans were so incensed over Cleveland's land policies and timber prosecutions that they were willing to switch parties and potentially tip the national balance of power to the Republicans upon achieving statehood. Many of Montana's leading Democrats had suffered embarrassment or prosecution under the Cleveland administration. Above all, they saw Cleveland as hindering the economic development of the territory. The Democrats' nomination of William A. Clark as delegate brought the issue to a head.[5]

Originally from Pennsylvania, Clark arrived in Bannack, Montana, in 1863. He spent a successful year placer mining, sold his claim, and went into business supplying the nearby mining camps. He reaped a small fortune, which he then invested in banking and mining ventures. Historian K. Ross Toole summed up his business: "A dollar never went out but what it returned with another sticking to it." Standing a wiry five feet seven, Clark possessed intense, penetrating eyes. Austere and vain, "there was about him a white starched fastidiousness." Clark "combined a remarkable intelligence and attention to detail" with fanatical ambition and a relentless work schedule, habitually putting in twelve- to sixteen-hour days. Like Hammond, Clark had a reputation for being coldly calculating and uncompromising, both in business and politics. Although he commanded respect, Clark was not particularly well liked. Nonetheless, he began to harbor senatorial ambitions, so much so that it became an obsession into which he would throw his immense wealth.[6]

In contrast, Clark's primary mining rival, Marcus Daly, exuded congeniality, charisma, and generosity. But as historian Michael Malone noted, "Daly's ready wit and relaxed charm masked a keen and ruthless intelligence and an explosive temper." Speaking in an Irish brogue, Daly appeared simple and unpretentious to the thousands of Irish workers who flooded into Butte to work in the mines. Butte held a third of Montana's population, and a majority of those were Democratic-leaning Irish Catholics, most of whom worked for Daly and the Anaconda Copper Mining Company. As Democratic Party chieftain, Marcus Daly controlled a cohesive and substantial voting bloc, one that Clark needed to become senator.[7]

Daly and Clark owned Butte's largest mines and vied for control of the Democratic Party. Like Hammond and Cap Higgins in Missoula, Daly and Clark kept bumping into each other in the quest to dominate Butte economically and Montana politically. Such king-of-the-hill struggles for supremacy permeated the Gilded Age, often to the detriment of local communities and the adversaries themselves. Indeed, the schism between Clark and Daly ripped Montana's Democratic Party in half and provided an opening for Hammond and the Republicans.

The animosity between Clark and Daly began at the Montana Democratic Convention in September 1888. The incumbent delegate, Martin Maginnis, refused a third term, throwing the field wide open. Clark quickly moved to secure the nomination. With statehood on the horizon, Clark viewed his prospective election as a delegate as a stepping stone to the U.S. Senate, his ultimate goal.[8]

Clark's nomination put Daly, Hauser, Bonner, and Hammond in a conundrum. Unlike Maginnis, Clark had wealth that gave him a degree of political autonomy and made him unreliable; the others would have little control over him. The four men recognized that no matter who won the presidential race, Clark would be of little use in protecting their economic interests. If Cleveland won, no delegate from Montana would have much influence with an administration prone to ignore the pleas of western politicians. On the other hand, if Republican nominee Benjamin Harrison won, a Republican delegate would certainly have more leverage than Clark, an independent-minded Democrat. Should the four business partners remain loyal to the Democrats and support an unpalatable candidate, or should they seek an alternative?

Following Clark's nomination, T. C. Power, one of Montana's leading Republicans, approached Daly with a solution: ditch Clark and support a fellow Irishman who was also a malleable Republican: Thomas Carter, a young and largely unknown lawyer. Two days later, in a surprise to everyone except those on the inside, Carter emerged as the Republican nominee. Daly, Hauser, Bonner, and Hammond concluded that if they backed him, a victorious Carter would be indebted to the turncoat Democrats and would gladly do their bidding. Perhaps Carter could even exert some influence in getting the timber suits dismissed should Harrison win the presidency.[9]

The conspirators kept their plans of ditching Clark for Carter quiet. Given Montana's political leaning, Clark banked on an easy victory and acted as if he had already won. He gave a few speeches but otherwise conducted a lackluster campaign. Although Daly, Bonner, and Hauser made no public endorsement of either candidate and appeared neutral on the election, Clark assumed that as fellow Democrats they would deliver their respective counties, which accounted for 50 percent of the territory's population. Absent from Montana from mid-August until late October, Hammond appeared entirely uninterested.[10]

Daly, however, harbored personal bitterness toward Clark. Historian David Emmons traced the Clark-Daly feud to the ancient conflict between the Catholics and Protestants on the Emerald Isle. Emmons argued that for Daly, "who wore his Irish Catholicism like a badge," and the rest of the Butte Irish, Carter was a more attractive candidate than Clark, who was a Scotch-Irish Presbyterian and a Grand Mason.[11] Bonner and Hammond regarded the election not as a deep-seated ethnic conflict, but as simply a business decision in order to have "a friend at court."[12]

In Missoula County, C. H. McLeod became the point man for the plot. Managing the anti-Clark campaign, he sent Missoula Mercantile Company (MMC) employees to canvass the Bitterroot on behalf of Carter while he visited the logging camps, handing out cigars, whiskey, and money as he informed employees that they must vote for Carter or lose their jobs. Montana had yet to adopt the secret ballot, so voters, especially employees, could be easily intimidated. Meanwhile, Hammond was in St. Paul meeting with Northern Pacific Vice President Thomas Oakes to enlist the railroad's support for Carter. Named as a codefendant in the timber suits, the railroad also sought an advocate.[13]

On election night, Clark hosted a grand banquet to celebrate his victory, only to discover that he had lost resoundingly. Reports of voter manipulation quickly flooded the press. The *Butte Miner* (owned by Clark) maintained that Anaconda "employees were given orders as to how they should vote. . . . This was notoriously the case in certain mines . . . where . . . shift bosses led the miners to the polls and saw that the poor men voted just as they were commanded to."[14] However, as David Emmons pointed out, ethnicity trumped party loyalty and the Butte Irish needed little persuasion to vote for Carter over Clark, who had committed several anti-Catholic gaffes during his campaign.[15]

In Deer Lodge and Missoula Counties, where Bonner, Hammond, and the Northern Pacific were the largest employers, voters overwhelming went for Carter, turning in thousands of Democratic ballots with "Carter's name pasted over Clark's."[16] Martin Maginnis, who as delegate had lobbied hard against the timber suits on behalf of Hammond, was aghast at the duplicity involved in the election. He blamed Clark's defeat on "a powerful personal combination against him of men whose support he expected [and who] . . . used their influence with their employees."[17]

Although it was obvious that the Missoula Mercantile and the Northern Pacific had coerced their workers into voting for Carter, Clark held

Daly personally responsible for the loss, and the simmering acrimony between the two men burst into an open conflagration. For the next dozen years, the Clark-Daly feud corroded Montana politics to an unenviable low and cast a long shadow even into the twentieth century. Hammond, however, was more concerned with catching up to the copper kings financially than getting involved in a grudge fight.

For Bonner and Hammond, Carter's election was money well spent. Within a month of taking office, Carter succeeded in convincing the incoming Harrison administration to suspend the timber suits. A staunch conservative, Carter rose rapidly in the national arena, becoming a close friend of Benjamin Harrison. When Carter lost his House seat in 1890, Harrison appointed him land commissioner, putting him in a position to further aid Hammond and his timber operations. In 1892 Harrison made Carter his campaign manager and chair of the Republican National Committee. Carter then served two nonconsecutive terms in the U.S. Senate. As Bonner noted years later, "We did not make any mistake. Carter was our friend in Washington."[18]

Throughout his career Carter failed to shake the reputation that Bonner and Hammond had put him in power. In 1900 the Helena newspaper editorialized that ever since 1888, Carter had been a tool of "those who have been despoiling the public domain of the timber and he has ever been their devoted servant. . . . On the timber question he was elected delegate, on the timber issue he was made commissioner of the federal land office, on the timber issue he was elected US senator."[19] Carter was hardly a docile servant, however. Historian Michael Malone characterized him as "a shrewd, tough, and conniving politician who survived many a battle by adept maneuvering and by consistently allying himself with wealth and power."[20] In reference to his political twisting and crooked deals, Montanans nicknamed him "Corkscrew Tom."

Although Daly, Hauser, Bonner, and Hammond united to support Carter in 1888, the coalition unraveled the following year. Upon achieving statehood in 1889, Montanans needed to ratify a constitution and choose their first state officials and legislature, which would then elect two senators. The outcome of the statewide election could shift the national balance of power, as Republicans held a precarious majority in the U.S. Senate. Hauser and Daly regarded their defection from the

Democrats as an anomaly and quickly returned to the fold. They expected Bonner and Hammond to join them and solidify control over the party in the upcoming election. Daly, however, became increasingly suspicious of Hammond's loyalty to the Democratic Party.[21]

Meanwhile, the editor of the *Missoulian* openly blasted Hammond. Although he identified himself as Republican, Harrison Spaulding was something of a renegade and more aligned with the rising tide of anticorporate populism. Accusing Hammond of both economic and political monopolization, Spaulding told readers, "We are fighting the outrageous practices, especially political, of the Hammond corporation." He charged that the Merc (MMC) had taken over the Republican Party by importing "train loads of men . . . most of whom are not citizens" to control the county convention. Spaulding believed that the MMC was "trying to ferment strife and division in the Republican Party as they want to elect the Democratic ticket this fall." While Hammond was indeed attempting to control the Republican convention, it was not to favor the opposition. Daly's suspicions were far more accurate than Spaulding's.[22]

Although Hammond had not publically declared a change in political orientation, privately he informed T. C. Power, "Our company will give the Republican party financial and moral support to a man." Acknowledging that Missoula County had "a natural Democratic majority" because of the number of Northern Pacific railroad workers, Hammond assured Power that "if you can bring the 'pinhead' politicians into line . . . our Committee should see that it [the election] is properly handled."[23]

However, Northern Pacific vice president Thomas Oakes had already decided, in reversal of his support of Carter the previous year, that it was in the railroad's interest to see Montana "go Democratic."[24] Martin Maginnis jumped to the task and began lining up the railroad workers. Maginnis, having gone to bat for Hammond and Bonner against Sparks and Lamar, felt particularly aggrieved at their defection and publicly denounced them. Meanwhile, Daly pressured Hauser "to do something to control Hammond," bring him into alignment with the Democratic Party, and dump Carter. Unbeknownst to Daly and Hauser, however, the Missoula Mercantile was supplying more than half of the territory's Republican campaign funds.[25]

In retrospect, Hammond's switch was hardly mystifying. Given his temperament, politics, profession, birthplace, and religion, it was quite natural that he should gravitate toward the Republicans. Descended from New England Protestants who were largely temperate, inclined toward hard work and little frivolity, Hammond was conservative, pro-business, and anti-labor in his economic outlook. Along with his New Brunswick brethren, Hammond soon formed the nucleus of Missoula's Republican Party.

With just a week to go before Montana's first statewide election, Daly swallowed his pride and journeyed to Missoula to plead with Bonner and McLeod to rejoin the Democrats. McLeod informed the copper magnate that while he regretted the loss of his friendship and business, "they had nominated a ticket" and fully intended to support it. Indicative of the importance of political affiliation in shaping a nineteenth-century businessman's identity, McLeod told Daly that abandoning the Republican Party "would simply be surrendering his manhood." Incensed, Daly rose to his feet and declared "he would make grass grow in the streets of Missoula." To top it off, he cancelled Anaconda's lumber contract with the Blackfoot Mill.[26]

While Daly could dominate the mining and smelting districts of Butte and Anaconda, he was poorly matched against the MMC in Missoula County. Bonner, Hammond, and McLeod were building their own political machine, one in which party affiliation was secondary to company loyalty. As the largest employer in the area, the Merc pressured employees and area residents into voting for its hand-picked candidates. While Hammond maintained a friendly correspondence with T. C. Power, who was running for governor, behind the scenes Hammond had McLeod instruct Merc employees to vote for Power's opponent, Joseph K. Toole, a Democrat, and for Carter, a Republican, for Congress. When the returns came in, both Toole and Carter won, as did five out of six of Hammond-supported legislators from Missoula County. In 1889, at least, it appeared that the Hammond-Bonner machine could outpoll the combined efforts of Anaconda and the Northern Pacific. Furthermore, Hammond's switch of parties sent reverberations clear to the U.S. Senate.[27]

Missoula's swing to the R column resulted in a state legislature equally divided. Accusations of voter fraud and bribery, combined with intense personal rivalries and political rancor, permeated the state following the election. When the legislature met the following January, both Republicans and Democrats refused to recognize the other as legitimate and convened separately. This resulted in Montana having two legislatures in 1890, each of which elected its own U.S. senators. The Democrats chose William A. Clark and Martin Maginnis, while the Republicans elected Wilbur Fisk Sanders and T. C. Power. In a bizarre maneuver, Montana sent four senators to Washington instead of the usual two, leaving it up the U.S. Senate to decide. Nationally, the Democrats badly needed Montana's two seats, but the slim Republican majority refused to seat Clark and Maginnis and sent them packing. The election results from Missoula County thus put two additional Republicans in the Senate.[28]

Furious at the results, Daly plotted to ruin Hammond and his associates and drive them out of Montana. Despite Daly's wealth and wrath, Hammond proved too entrenched. The competition between the two titans resulted in a massive building boom that transformed Missoula into the modern city that Hammond had long envisioned.

Daly's first step was to finance D. J. Hennessey's mercantile in Missoula to compete with the MMC. Intending to deplete Bonner and Hammond's cash reserves and credit, Daly severed his business ties with the partners. Resigning as president of the First National Bank of Missoula, Daly forced Bonner and Hammond to buy $102,000 worth of stock in the bank. He then backed out of a railroad construction contract that the three men had with the Northern Pacific, requiring Bonner and Hammond to come up with more cash. But without a doubt, the biggest hit to Hammond was Daly's cancellation of Anaconda's lumber contract. Not only did Hammond have to repay Daly $75,000, but he also lost his biggest customer.[29]

Although these calculated economic blows cost "a good many thousand dollars and a great deal of grief," Hammond appeared unperturbed by Daly's efforts. He simply succeeded Daly as bank president, and the Blackfoot Mill continued to produce lumber despite Anaconda's boycott. Throughout 1890–91, Daly and Hammond kept expanding their businesses, each seeking to outdo the other to gain control of Missoula. Hammond added a second story to the Merc and laid the foundation for a new First National Bank building across the street.

On the opposite corner, he began work on the four-story Hammond building.[30]

Meanwhile, Daly announced he would build a two-story building one block away to house the Hennessy mercantile as well as a new warehouse near the railroad depot. In February 1891, Daly announced the incorporation of a new bank in Missoula with $100,000 more capital than Hammond's bank. Furthermore, Daly announced plans for a railroad from Anaconda over the Sapphire Mountains to the Bitterroot Valley, where he was building a large sawmill to tap the region's vast timber reserves. Daly also began to plat Hamilton, a new town in the Bitterroot, to become the valley's commercial center.[31]

But then, in April 1891, Daly suddenly conceded the mercantile business in Missoula to the MMC, which purchased Hennessey's unsold merchandise. Daly also failed to establish his bank or railroad. Perhaps he decided the dispute was not worth the effort and cost. Perhaps his wife, who remained friends with Florence Hammond, convinced him to abandon the fight. Regardless, except for Hamilton, which surpassed Hammond's town of Grantsdale, Daly failed miserably in his attempt to crush Hammond on his own turf.[32]

Politically, Hammond held his own as well. With the Democrats fractured between Daly and Clark, Hammond could play the power broker among the Republicans. Patronage was still alive and well in Montana in the 1890s, and Hammond used his influence with both Power and Carter to reward his minions with federal appointments. Hammond's influence was so pervasive that one office seeker complained, "Mr. Carter has turned a deaf ear to everybody and everything that did not have A. B. Hammond's endorsement." Elected officials knew of their obligations to the Missoula Mercantile machine.[33]

Unlike many of his contemporaries, Hammond regarded politics solely as a business enterprise, signaling his transition to a modern businessman. Return on investment trumped party affiliation and the identity politics of religion and ethnicity. Daly, in contrast, saw himself as an Irish-Catholic Democrat; the three elements were inseparable.

While the infusion of business into politics was certainly nothing new in the late nineteenth century, Americans were becoming increasingly concerned over the concentration of economic and political power.

With the Missoula Mercantile, the First National Bank, the Florence Hotel, and the new Hammond building anchoring downtown Missoula, for many observers, the head of the beast arose as a tangible and domineering presence. Unabashedly partisan, newspapers across Montana condemned Hammond's growing hegemony. The Butte *Inter-mountain* referred to the MMC as "an octopus whose slimy tentacles reach out and envelop nearly every farm and ranch and herd and even the forest land of the richest county in the territory." The *Missoulian* personalized the attack, caricaturing Hammond as the "Missoula Octopus." The cartoon depicted Hammond's face in the center of a twenty-seven-tentacled octopus with each arm representing one of his concerns. Some of the octopus's arms were exaggerations and others, such as "Church" and "Fort Missoula" were farfetched. Nonetheless, Hammond did control the bank, the Merc, the Florence, the Blackfoot Mill and ninety thousand acres of timberland, the light and water company, the Missoula Street Railway, the *Missoula Gazette,* the Bitterroot railroad, real estate development in Missoula and the Bitterroot, the area's only flour mill, and even the cemetery. In truth, it was nearly impossible to live (or even die) in Missoula without doing business with A. B. Hammond.[34]

Hammond's economic domination of Missoula, however, paled beside Marcus Daly's Anaconda, Andrew Carnegie's Homestead, Pennsylvania, or George Pullman's Pullman, Illinois. But these were company towns intentionally built to be dependent upon a single corporation. In contrast, Missoula, like many western towns, was settled by independent-minded entrepreneurs who resented and resisted the rise of a single economic power. Many saw Hammond's economic and political domination as contrary to America's republican tradition.

The backlash began in 1891. The political pressures and personal animosities that had been bubbling under the surface came to a boil over replacing the bridge spanning the Clark Fork River in Missoula. Hammond wanted to retain the original south end of the bridge, while Frank Higgins—heir to the estate of his father, Cap Higgins, as well as to his enmity toward Hammond—lobbied for a new location. The issue became so contentious that the city held a special election on September 8, 1891, to decide the issue. Although the most logical site was the one Hammond favored, it lost, as the election had turned into a referendum on the popularity of Frank Higgins and the growing antipathy toward Hammond.[35]

Local resentment toward Hammond and the political activities of the MMC continued. Throughout the fall of 1891, Harrison Spaulding ran a series of editorials accusing "The Missoula Mercantile Monopoly" of taking advantage of Bitterroot farmers by blocking construction of a flourmill in Victor, forcing farmers to ship their wheat on Hammond's Bitterroot Railroad to his mill in Bonner. "Monopoly," in the parlance of the time, was a rather loose term that could refer to any large company that by virtue of sheer size could engage in unfair business practices.[36]

The acrimony spilled over into Missoula's next municipal election when Frank Higgins ran for mayor. Daly reinserted himself into Missoula politics and actively supported Higgins. The *Anaconda Standard,* which Daly owned, fired potshots at Hammond, hoping to sway the Missoula election. Accusing the Merc of total domination of Missoula's political landscape, the *Standard* called the MMC "a corporation which is famous for the fact that it is friendless and which has but one mission—that is to 'swipe' every man and everything in sight."[37] The obvious irony in the attacks of the *Standard* was that Daly had built the greatest company town in the West: Anaconda. Reading between the lines suggests that the editor of the *Standard* reflected popular misgivings about the influence of Daly and the pervasive power of the Anaconda Company. A week later, Higgins won the election. How much of the vote was a result of Daly's meddling and how much was due to an anti-Hammond electorate is impossible to gauge, but the result demonstrated that Hammond, losing two elections in a row, hardly had a stranglehold on Missoula politics; economic power did not automatically translate to political power after all.

In their attacks against the MMC, the editors of both the *Missoulian* and the *Standard* expressed a sentiment that was sweeping the country. During the summer of 1892, thousands of farmers and workers, fed up with corporate domination of politics, converged in Omaha in the first national convention of the Populist Party. The Populists called for such reforms as a graduated income tax, government ownership of railroads and telegraphs, an eight-hour workday, the secret ballot, popular election of U.S. senators, ballot initiatives, a subtreasury system to grant farm credit, and, perhaps most famously, the free and unlimited coinage of silver. Two years later, the election of 1894 marked the high point of voter engagement in the United States. The surge of the anticorporate outlook drew thousands into the Populist fold, as did the cry of free silver permeating the West.[38]

Similarly, by the 1890s, hostility toward Hammond and the Merc was palpable across the Treasure State. When Hammond received the nomination for delegate to the Republican National Convention, fellow Delegate George Irvin protested against the MMC's political control that was barely "short of slavery and abject servitude."[39] Similarly, the *Standard* noted that if anyone "whose defeat in a political convention should be a matter of easy accomplishment, that man is A. B. Hammond."[40] Electoral popularity and political control, however, did not always coincide, and the convention selected Hammond as one of twelve delegates to the national convention.[41]

Stopping briefly in Missoula on his way to the convention in Minneapolis, Hammond granted a rare interview with the *Missoula Gazette.* In a refrain that personified American business, Hammond essentially argued that what was good for A. B. Hammond was good for Missoula. Seeking to mollify residents against the rising tide of Populism and answer the charges of monopoly, Hammond detailed his accomplishments in developing Missoula through "pluck, perseverance and industry" by building "railroads, hotels, sawmills, bank buildings, brick and granite blocks and the grandest mercantile establishment in the northwest." He portrayed his accusers, who "howl 'monopoly,'" as profiting "through the industry and foresight of others." Shamelessly, Hammond called upon residents to "conduct their business . . . in the interest of honest government" and put aside "their personal ambitions and prejudices." Up to now, Hammond had avoided using the media as a forum to address the general public. As he became convinced of the righteousness of his position, Hammond grew increasingly adept at ignoring his own actions while spouting platitudes. Hammond concluded his soliloquy by informing his detractors, "whose stock in trade consists of spleen and malice," in no uncertain terms: "We are here to stay, and while we are ever ready to join with the people in all public enterprises, we are also equally prepared, if others do not join us, to go it alone."[42] For Hammond, like many other businessmen, "public enterprise" was synonymous with private profit.

Hammond's role in bringing the state university to Missoula illustrates how the pursuit of private profit, along with corrupt politics, could result in public gain. Since the Montana Constitution of 1889 postponed the location of the capital and other state institutions, in 1892 voters went to the polls to choose among seven cites vying for the capital. Missoula had removed itself from the contest so as to be in a position to bid for the university. Hammond struck a backroom deal with Hauser to support Helena's bid for the capital in exchange for Missoula getting the university, and throughout the winter of 1893, Hammond lobbied the state legislature. Although he was not interested in higher education, Hammond's Missoula Real Estate Association owned substantial acreage south of the river. Locating a university there would greatly inflate real estate values, pouring even more dollars into the Hammond coffers.[43]

Although Helena received the most votes for the capital, it fell well short of a majority, triggering a runoff election two years later. The battle pitched Democrats, Republicans, and Populists against each other and themselves, undermined party unity already fractured by the free silver issue, and this time aligned Hauser and Clark in a furious contest against Daly.

Marcus Daly desperately wished to see his personal creation—Anaconda—become the state capital. Hauser felt the same way about Helena. Clark, perhaps simply to oppose Daly, came out in favor of Helena and poured tens of thousands of dollars into the emotional and bitterly fought campaign. Coolly assessing the situation, Hammond recognized that if the MMC supported Anaconda, he could get anything he wanted from Daly, "but if we should elect to do otherwise we will have war such as we have had in the past."[44] Ultimately, he decided that while the location of the university remained unresolved, he would support Helena.

Money and booze flowed freely on the streets of Missoula that fall. The Merc dropped nearly ten thousand dollars on the campaign, but Daly outspent them ten to one. Perturbed, Hammond informed Hauser "Your committee sends Mr. McLeod $300 for the whole of Ravalli County, while they [Anaconda] give a common healer on the streets $500. . . . Besides suffering heavy losses in our business on account of the stand we have taken in favor of Helena, we have been assisting your committee here with money."[45] A few days later, however, Hammond believed the election would be decided on ethical grounds. Without a hint of hypocrisy he told Hauser, "Our men are working for glory and theirs for Boodle." Statewide, Daly admittedly paid out $450,000, while Clark figured he spent $1 million, but the actual amounts may have been much higher.[46] Even these conservative numbers added up to an absurd $30 per vote. With the adoption of the secret ballot in 1889, voters freely pocketed cash from both sides.

Ultimately, Helena won a narrow victory over Anaconda to become the capital. Concerned over corporate influence in politics, many voters worried about placing the state capital in a company town. In Missoula County, however, Anaconda came out on top. Thus, Daly and Hammond both suffered political defeat even though they were on opposing sides. Hammond's support of Helena cost the MMC many customers and thousands of dollars. Missoula, however, did receive the university, and in the long run Hammond profited by real estate sales, although less than he had hoped. It hardly seemed worth the effort.

Despite outright bribery, employer intimidation, and political chicanery and manipulation, none of the businessmen who dominated Montana politics could ever exert complete control and, in fact, were often frustrated in their ambitions by the vox populi. More than once, backlash against their efforts cost them victory. Clark lost every effort to become senator until 1901, and only occasionally did Daly's handpicked candidate win an election. While Hammond helped send Carter and Power to Washington, both were ousted in subsequent elections. Despite the accusations of economic and political domination in Missoula County, Hammond lost three political battles in a row. Not only had politics proved fickle and beyond his control, but it was also becoming bad for business. The fight over the state capital marked the end of Hammond's direct involvement in Montana politics. Besides, he had his eye on bigger things.

CHAPTER 9

Panic

Through a series of cascading events, the Panic of 1893 plunged the nation into its most severe depression yet. In February 1893, the Philadelphia and Reading Railroad declared bankruptcy, causing many investors to become jittery from the overextended and speculative nature of the economy. Stocks tumbled as European financiers sold off their railroad interests in exchange for gold-backed securities, which in turn depleted U.S. gold reserves. Then in June, Great Britain ceased the coinage of silver rupees in India, and within hours world silver prices dropped 20 percent, bringing chaos to the mining centers of the West. Dependent on mining, Montana had tied itself to the global economy, and its fortunes rose and fell in accordance with decisions made in distant financial capitals. As Montana journalist Joseph Kinsey Howard later put it, the state lay at "the end of a cracked whip."[1]

The Panic affected the business world much the way an understory fire acts upon a forest, leaving the big trees unscathed while clearing out smaller and marginal ones. Those with shallow roots and overextended financial commitments—Henry Villard, for example—toppled over, while men like John D. Rockefeller, Jay Gould, E. H. Harriman, and Andrew Carnegie profited from the reduced competition. The aftermath of the Panic also provided fertile ground for budding capitalists like A. B. Hammond.

As the epicenter of American business debt, railroad corporations imploded. One-third of the nation's trackage fell into receivership, including the Northern Pacific. Capitalizing on the situation, James J. Hill and J. P. Morgan bought the controlling shares in the floundering railroad.

The position of the Montana branch lines was even worse; with a total deficit of $525,000, all of these railroads went bankrupt as well.[2]

As the railroads came crashing down, so did the banks. Linking distant financial markets with local development projects, banks provided financing for everything from mines to mortgages and served as important bellwethers of economic conditions. However, with almost no federal oversight and minimal liquidity requirements, nineteenth-century banks depended upon the wisdom and frugality of their directors. Business practices could determine the difference between survival and failure in such dire times, as the story of two banks, one controlled by Samuel Hauser and the other by A. B. Hammond, reveals. Furthermore, without deposit insurance, if a bank closed, investors and depositors lost everything, making them understandably nervous during a depression. Financial panics were aptly named.

The decade prior to 1893 brought unprecedented prosperity to the nation, particularly the West. The years following the arrival of the Northern Pacific proved fruitful, financially and domestically, for Hammond. Seizing opportunities as they arose, in just ten years he advanced from the manager of a mercantile to one of the wealthiest men in Montana and turned the MMC into an economic powerhouse. Exploiting both the public domain and railroad grant lands, Hammond assembled the region's premier lumber company with retail outlets extending from eastern Montana to Washington State. Bonner Mill, the largest sawmill in the Intermountain West, supplied timbers and cordwood for the insatiable appetite of the Butte mines and smelters and shipped finished lumber as far as Salt Lake City. Hammond had quashed the major competition in the mercantile trade of western Montana and survived Marcus Daly's attempt to drive him out of the state. Not only had Hammond evaded federal charges of timber theft, but he even had his own U.S. senator—Thomas Carter—in his pocket.

Hammond was also keeping his young wife busy. Less than a year after giving birth to Edwina, in 1879 Florence had another daughter, named after herself. The following October, their first son was born—Richard, in honor of Richard Eddy. After finishing the construction of the Clark's Fork Division of the Northern Pacific Railroad, Hammond turned his sights to the house that his family shared with the Eddys. He added a second-floor sitting room and nursery, which was quickly filled in 1884 with the addition of another son, Leonard, named after Hammond's grandfather. Two more girls, Grace (1886) and Daisy (1888) rounded out the family. To help Florence with the six children, Hammond hired a cook and nanny, making for a rather crowded household.[3]

Although Hammond began looking for more spacious housing in Missoula, health issues compelled him to the more salutatory climate of the West Coast. Plagued by recurrent bouts of rheumatism, he sought treatment at various hot springs and health resorts. His condition became so bad that in 1889 the Hammond family began spending winters in Oakland, California, where the milder weather provided relief.

Hammond also found the social atmosphere of the Bay Area refreshing. Like many of the Gilded Age's nouveau riche who lacked a formal education, Hammond sought social recognition, and in April 1890 he joined San Francisco's prestigious Bohemian Club, whose honorary members included Samuel Clemens, Luther Burbank, and Jack London. Here, too, he could hobnob with elite capitalists like Charles Crocker of the Southern Pacific Railroad instead of battling it out with provincial power brokers like Daly and Hauser. In the prime of his life at forty-four, Hammond wanted desperately to move into the major leagues of industrial capitalists, and San Francisco—the West's social and financial hub—was the place to do it.[4]

Yet all of his businesses and social contacts remained in Missoula. Florence, at thirty-one, was the town's principal socialite, and nearly all of their children were of school age. Given these considerations, in the summer of 1891 they returned to Missoula and laid plans for a mansion on the south side of town. But before construction began, the economic bubble burst.

Hammond's trouble began in the spring of 1893. His lumber operations had been highly profitable the previous year, when the Big Blackfoot Milling Company netted $200,000. In anticipation of another big year, the company sent hundreds of men up the Blackfoot River to cut timber throughout the winter. With the arrival of the spring runoff, the men floated an astounding forty-four million board feet of timber down the river, enough to supply the Blackfoot Mill at full capacity for the entire year. Under the system of contract logging, Hammond needed to pay off these men all at once. While this substantially drew down his funds, Hammond counted on the mines to absorb all the lumber he could supply.[5]

But then silver prices began to tumble, sending ripples through Montana's economy. As the mines closed, they cancelled their lumber contracts and left thousands unemployed. Panicked, depositors descended on banks. Unable to cover the withdrawals, banks closed their doors, creating "runs" on other banks in a chain reaction. In Missoula, a series of quiet withdrawals on Friday, June 9, forced the Higgins Bank to close its doors, triggering a run on the nearby Western Montana Bank as customers pulled out $100,000 in deposits. Hammond's normally self-possessed, cool demeanor began cracking as his First National Bank of Missoula paid out $25,000. Nonetheless, it had avoided the massive run that hit other banks and managed to stay open. While Frank Higgins may have been more popular, Missoulians displayed more confidence in Hammond when it came to financial matters.[6]

Hammond believed his bank could stay open with checks coming in from the Northern Pacific Railroad, unaware that it was on the verge of collapse. But by the time the banks opened on Monday morning, Hammond grew desperate and circled his financial wagons. Well aware of Hauser's shortcomings as a businessman, Hammond penned a tense and authoritarian letter to the former governor, demanding that he send $20,000 to the Missoula bank, half of which was to come from the Northern Pacific. "*See that it comes without fail,*" Hammond insisted.[7] Uncharacteristically on the verge of panic himself, Hammond dashed off two or three letters to Hauser each day during the crisis. With all of his businesses intertwined, a collapse of the bank would bring everything crashing down.

By Wednesday, however, the scared depositors were bringing their money back to the bank and Hammond breathed a sigh of relief. But his renewed optimism quickly faded as the international economic situation deteriorated. President Cleveland (who had staged a comeback in 1892 to win re-election over incumbent Benjamin Harrison) called for a repeal of the Sherman Silver Purchase Act, which had guaranteed a market for silver as a component of the nation's monetary policy. Cleveland believed the act had caused the spate of bank failures around the country. In Montana, the nation's second-largest silver-producing state, just the opposite was true: the act propped up the mining industry, which kept the banks solvent.[8]

At the beginning of 1893, the First National Bank of Missoula could boast it had $700,000 on hand. By July reserves had plummeted to $400,000, and only the personal assets of the directors, including Hammond and Bonner, kept it afloat.[9] Underscoring the importance of the MMC, Hammond pulled $42,000 out of the Merc and "pledged all the assets of that company to the bank."[10] Hauser, still a major stockholder in the Missoula bank, was in no condition to help. With his own bank sinking into a financial abyss, Hauser beseeched Hammond for aid. Hammond replied in no uncertain terms, "In rustling for money to protect this bank we had a right to call on you . . . we were looking after your interest here at our own expense . . . [but] you are asking a good deal from us when you ask for *$30,000* in times like this." Nevertheless, he relented and sent Hauser the funds but upon the condition that "that under no circumstance" could he use the money other than to bolster his report showing that his bank was still solvent.[11]

After limping along for weeks, Hauser's First National Bank of Helena, the premier financial institution in the state and one of the largest in the Northwest, suspended operations, causing a run on the First National Bank of Missoula. Livid, Hammond blamed Hauser personally, writing, "Every stockholder outside of yourself has come in and put up money to keep the bank open and to do this we have exhausted our credit and resources East and West and still the run caused by your failure continues." Hammond then insisted Hauser make good on his debts to the Missoula bank so that it could remain open.[12]

And remain open it did. Although nearly every bank in the state closed, at least temporarily, during 1893, Hammond managed to keep the First National Bank of Missoula open by calling in notes from the Northern Pacific, drawing on the impeccable credit of the MMC and its financial reserves, and imploring his bank directors to pony up their personal savings. Above all, in all his financial dealings, Hammond thoroughly embodied the lessons of the boom-and-bust cycles of the New Brunswick timber industry. He kept impeccable records and carefully avoided the

Hauser-type boosterism. Even with the temptations of the boom years, Hammond carefully kept his expenses under control and avoided over-extending his finances. Despite the protestations of many stockholders, he resisted paying dividends until he had fulfilled all of his obligations.

National events, however, were far beyond Hammond's control. The economic situation continued to spiral downward. In October 1893, Congress complied with Cleveland's wishes and repealed the Silver Purchase Act. The effects in Montana and other mining-dependent economies were immediate and tragic. Virtually overnight, three thousand people fled the silver mines of the Philipsburg area. By the end of the year, one-third of Montana's twenty thousand workers were unemployed, many subsisting off the free lunches offered in saloons.[13]

Relations between Hammond and Hauser further deteriorated as Montana teetered on the brink of total economic collapse. The situation reached the breaking point when Hauser asked Hammond for yet another loan from the First National Bank of Missoula. By now, Hauser's financial empire had collapsed, turning Montana's leading citizen into a pathetic figure as he pleaded with Hammond. Hauser's mines sat shuttered and useless; his stock was worthless; he possessed no collateral whatsoever. His bank was bankrupt, in large part because of speculative personal loans to its directors—a point Hammond emphasized when he ruthlessly denied Hauser any more favors. Hammond cast himself as an unselfish benefactor to Hauser, conveniently forgetting that the older man was largely responsible for Hammond landing lucrative railroad contracts and had backed him in buying out Higgins's bank shares. Nonetheless, Hauser perhaps pushed his obligation too far. Hammond believed that Hauser had only himself to blame with his spurious and speculative activities. Hammond especially took umbrage at Hauser's appeal for a loan based on their friendship and closed his letter stating, "If friendship is to be held thus cheaply I want none of it."[14]

Prior to 1893, the net profits from Hammond's enterprises were running $300,000 to $400,000 per year, making him a multimillionaire by today's standards. Although Hammond suffered a 40 percent financial loss in the Panic, by the following year, the Merc and his bank were, remarkably, back in the black and his lumber business nearly so. Not only had the bank stayed open; by January 1894 it possessed $300,000 in accumulated undivided profits, and Hammond recommended paying a 20 percent dividend from then on. What makes Hammond's recovery in 1894 especially astounding is that the nation, and Montana, still lay locked in the grips of the depression. That same year eight thousand men fled thc state, seeking work elsewhere. Some six thousand unemployed remained wandering the streets of Butte, Helena, Great Falls, and Missoula looking for work.[15]

Politically, the Panic of 1893 provided an enormous boost to the Populist Party, which pointed to the depression as tangible proof of the failures of laissez-faire capitalism. In Montana, labor and farmer groups joined with mining moguls and businessmen to advocate for free silver, but to no avail. The continuing production of copper and the mines' unquenchable demand for wood products enabled some businessmen, such as Marcus Daly, William A. Clark, and A. B. Hammond, to endure and even come out ahead. But Samuel Hauser, with his heavy investments in railroads and silver mines, never fully recovered.

Although commonly portrayed as a hands-off approach to government intervention in the economy, laissez-faire was more akin "hot-house capitalism," in which government and big business partnered to stimulate artificial growth conditions. This "wink-and-nod" relationship encouraged wild speculation, spurious investments in questionable ventures, and a financial system without a solid foundation. Those who played this game, like Samuel Hauser and Henry Villard, soon found their empires little more than rubble.[16]

In contrast, Hammond's aversion to speculation—especially in mining schemes—his adroit handling of the banking crisis, and the solid foundation of the Missoula Mercantile helped him to weather the greatest crisis of his career relatively undamaged. Finding his finances, credit rating, and reputation intact, Hammond drew upon these to enlarge his empire.

CHAPTER 10

Montana's Pariah/Oregon's Messiah

In many ways, 1893 marked the end of the era of the pioneer entrepreneur and the beginning of the rise of the corporate businessman. The expanding railroad network that connected the source of raw materials with urban markets presaged the rise of modern corporations. Railroads provided the seedbed for national markets, technological innovations, the vertical integration of industry, and bureaucratic structures. Prior to 1893, American investments were primarily in railroads, but with their financial collapse, a new economic order emerged from the ashes.[1]

Although he retained many nineteenth-century practices in his businesses, Hammond readily adapted to the emerging industrial model. Samuel Hauser, with his wild speculations and old-boy network, failed to make the transition. Bonner, too, was confounded by new developments in the business world and soon retired. Having survived the Panic of 1893, Hammond saw the ensuing depression as an opportunity to move into the top tier of industrial capitalists. All the big names—James J. Hill, C. P. Huntington, Jay Gould—had made their fortunes off railroads, and Hammond began casting for similar opportunities. As Hammond shed old partnerships in Montana, he applied the new paradigm to the development of the Oregon coast, connecting its emerging lumber industry with national markets via railroads.

Wintering in California, Hammond often passed through Portland, the hub of the Pacific Northwest. As agricultural produce streamed down the Willamette Valley toward Portland and manufactured goods moved up the valley, the region appeared as a more established and grander version of Missoula and the Bitterroot Valley. As historian William Robbins noted, capital also flowed from one region to another, following investors' incessant quest for higher profits. "The interior Northwest paid tribute to Portland," he wrote, "just as Portland paid tribute to San Francisco."[2] Hammond realized that Montana was, and would remain, a hinterland, subordinate to the metropolises that controlled capital investment, and therefore, began eyeing the Pacific coast with great interest.

A single letter prompted a dramatic turn in Hammond's fortune. In the fall of 1894, Edwin Stone, a former employee of Eddy, Bonner and Co. who had moved to Corvallis, Oregon, wrote to Hammond, telling him that the Oregon Pacific Railroad was for sale at a bargain-basement price. Built in the 1870s by T. Egerton Hogg in a fit of opportunistic boosterism, the Oregon Pacific aimed to be Oregon's first transcontinental connection. Linking Yaquina Bay on the central Oregon coast with Boise, Idaho, Hogg's railroad would cut 225 miles off the trip from San Francisco to the inland Northwest by slicing through the Cascades. Bypassing Portland, it would also siphon traffic from the Willamette Valley. Boosters claimed that Newport, on the Yaquina Bay, would be the new metropolis. With the railroad bonded for $15 million, Hogg sank $5 million into building the line from Yaquina Bay through Corvallis and up the Santiam River to the base of the Cascades, where it halted. The Oregon Pacific opened in 1884, but the entrance to Yaquina Bay proved overly shallow, thwarting shipments. In 1890 the Oregon Pacific defaulted on its bonds with more than $1 million in debt. Subsequently, the line fell into receivership and disrepair. During the following four years, the county sheriff, acting on behalf of the receivers, made repeated attempts to unload the railroad to pay off the debts, each time lowering the required bid.[3]

In the fall of 1894, Hammond and Bonner traveled up the Willamette Valley to Corvallis to inspect the Oregon Pacific. Hammond's experience with boosters such as Hauser made him leery of enterprises that were, like the Oregon Pacific, long on vision and short on cash. Nevertheless, they bid on the railroad and the sheriff accepted their $100,000 offer. The railroad included 142 miles of finished rail and an unfinished road through the Cascades. Along with the train cars, they acquired sixteen locomotives, two tugboats, and four steamships. By paying off the debts on the ships and selling them, the partners quickly recouped their original investment plus $20,000.[4]

While waiting for the sale of the Oregon Pacific to go through, Bonner and Hammond headed to Astoria, Oregon, to investigate another struggling railroad proposition, the Astoria and Columbia River Railroad (A&CR), which was the unfinished portion of the Northern Pacific's transcontinental project. When the partners arrived in Astoria, they discovered two other promoters were already issuing proposals. In a behind-the-scenes deal, the competing capitalists, J. C. Stanton and H. I. Kimball, withdrew their offer when Hammond gave Stanton "several thousand dollars in cash," leaving Bonner and Hammond as the only ones available to build the road.[5]

Although they were just the latest in a long string of railroad promoters—all of whom had been disappointments—Hammond and Bonner stimulated renewed optimism in Astoria. Hammond's demeanor, poise, and professionalism impressed the leading citizens. For their part, town leaders regaled Hammond with stories of how Astoria was destined to be a great port city. But Astoria had yet to become the booming port at the mouth of the Columbia River that John Jacob Astor envisioned when he established the outpost in 1811. Having crossed the treacherous bar at the river's mouth, ships simply continued on to Portland. A railroad, local leaders believed, would transform Astoria into the entrepôt of the Northwest.

As early as 1853, Astoria began agitating for a railroad, but not until the coming of the transcontinental railroad thirty years later did Astoria seem assured of a rail connection to the rest of the nation. By the time the Northern Pacific reached Portland, however, Henry Villard had grossly overextended the railroad's finances, and he halted fifty-eight miles short of Astoria. Hoping to spur someone to finish the job, Astorians formed a railway company and offered a cash bonus. Railroad promoter William Reid took the bait and began grading a roadbed. Although railroad tycoon Collis P. Huntington backed Reid, the deal fell through.[6]

Astoria residents heaped subsidy upon subsidy, offering 1,000 acres of prime real estate plus $300,000 to lure a rail line. A new company formed and graded seventeen miles before construction suddenly halted and contractors disappeared. Astorians responded by making the subsidy even larger and advertising across the country for investors. There were five more attempts between 1892 and 1894, but all failed to secure enough capital to complete the job. The Union Pacific entered the picture briefly but withdrew because of the national depression in 1893. Desperate, Astoria increased the subsidy to 3,000 acres in Astoria and an additional 1,500 acres at Flavel, across the bay—which together had an estimated value of $1.8 million.[7]

While Bonner was reluctant to undertake building the A&CR, Hammond became swept up by the possibilities. The subsidy could attract investors and might become valuable real estate after the railroad's completion. Furthermore, Astoria indeed seemed an ideal location for a busy port. Hammond was usually calculating with his investments, but in this case, his optimism overflowed. He extolled the virtues of Astoria and nearby Seaside as tourist destinations while he escorted Northern Pacific officials on a promotional tour. Pointing out the projected railroad route, salmon canneries, and real estate potential, Hammond gushed to his guests: "Wait until we reach Seaside and I will show you one of the prettiest spots on the coast."[8]

Excited by his new prospects, Hammond journeyed to New York to secure financing. He confided in his friend C. H. McLeod, manager of the Missoula Mercantile, "The Astorians treated me in great shape and if things should go my way I will make quite a clean up. I don't propose to put up any money. I have the contract and subsidy and the other fellows must put up."[9] The city quickly accepted the Hammond-Bonner proposal and the Astoria newspaper assured its readers, "The financial ability of these two gentlemen is unquestionable. They have been connected with several very important financial enterprises in Montana and the Northwest and have a national reputation of not only possessing large personal fortunes but also controlling an unlimited amount of eastern capital."[10] Contrary to what the Astorians believed, neither of the men controlled unlimited capital. In fact, Hammond was having difficulty drumming up interest in New York. Despite the solid credit rating of the MMC and Bonner's financial contacts, he discovered that this was "the worst time that one could come to New York with a railroad proposition." By mid-January 1895, Bonner and "the other fellows" had all backed out.[11]

Since the Panic of '93, Hammond had grown increasingly frustrated with Bonner's inside deals, and a fissure between the old partners and friends began to widen. Citing Hammond's poor health, Bonner urged him to drop the Astoria project, but Hammond suspected his mentor was trying to squeeze him out and take control. Bonner, however, had displayed remarkable prescience, as Hammond contracted malaria a few weeks later. Still, relations between the two men would never be the same. Bonner was also reluctant to pour more money into the decrepit Oregon Pacific. Ever the booster, however, he assured Oregonians that he and his associates would repair and finish the road through the Cascades, thus providing transcontinental connections.[12]

Although confined to his hotel room in New York, Hammond continued to make plans for raising the capital to turn both Oregon railroad investments into profitable ventures. Far from being discouraged, Hammond informed McLeod, "I am sure that this is a grand scheme if it can only be properly floated, I propose to stay with it for a while yet." By late February, Hammond was even more optimistic, believing the nation was on the verge of "an era of great prosperity." Finding that the credit of the MMC was "gilt-edged," he enthused to McLeod, "We are in a position to take up most any ordinary enterprise in any part of the country, and carry it to a successful issue." Hammond had become so absorbed in his Oregon prospects that he turned the management of his Montana operations over to McLeod. Although unwilling to invest any more money in Missoula, he reassured McLeod that he would "not do anything to impair the credit" or divert the funds of his Montana enterprises.[13]

The Astoria contract required that Hammond begin construction by April 1, 1895, complete the line by October 30 of the following year, and spend at least $50,000 per month on construction. Although he still had not found financial backing, Hammond remained hopeful. As the April deadline approached, however, he became apprehensive about his ability to build the railroad. At the last minute, Hammond finally enlisted two of the nation's biggest railroad magnates—C. P. Huntington and Thomas Hubbard, both directors of the Southern Pacific—as financial backers. Although Huntington now owned 70 percent of the A&CR stock, he granted Hammond free rein in building the railroad. Hammond, in turn, idolized the railroad pioneer. Just as he once looked to Leonard Coombes and E. L. Bonner as models, he now found inspiration and guidance in Huntington. This marked the beginning of a long and fruitful relationship between Hammond and the Huntington family, as well as between Hammond and the Southern Pacific Railroad.[14]

Hammond returned to Oregon in April and, now armed with substantial financial backing, began to bargain from a position of power. Seeking to recoup the money he had given Stanton to back out and simultaneously prove his worth to Huntington, Hammond began renegotiating the contract with the Astoria committee. He asked for and received, much to everyone's surprise, an additional real estate subsidy, an agreement from the committee to cover $6,000 worth of expenses, remittance of back taxes, and a lower purchase option for the previously completed railroad from Astoria to Seaside.[15]

With the new contract, Hammond wished to complete the railroad as soon as possible and threw all his energy into building the Astoria line. By May, however, Astoria's boosters had neither obtained the right-of-way for the railroad nor turned over all the land titles listed in the subsidy. Hammond responded by laying off his engineers, stating he would incur no further expenses and waste no more time on the railroad until Astoria honored its contract. "Unless the right of way and subsidy matters are speedily closed up there will be no road as far as he is concerned," the *Daily Astorian* reported. Instead of chastising the extortion, the paper applauded Hammond as "a man of pre-eminent honor coupled with immense wealth . . . a man who expects and will have honorable treatment from others."[16]

Although the subsidy committee could sort out the land titles, obtaining the right-of-way proved sticky. The committee members had promised something that was not theirs to give: access through private lands they did not own. As a corporation and public carrier, the railroad could claim eminent domain, but Hammond wished to avoid lengthy condemnation proceedings. Finally prodded into action, the committee engaged in heavy-handed persuasion and deceit to obtain the right-of-way. Employing social pressure, the *Daily Astorian* reprimanded those who were blocking the railroad and placed those who did assign the right-of-way on a "roll of honor." The paper cited James Quinn, who donated a one-hundred-foot right-of-way across his ranch, as a "shining example of noble patriotism." Pleased at the action, Hammond wrote to McLeod, "I have succeeded both in subsidy and right of way matters beyond my expectations, and have all other matters in first class shape."[17]

With construction slated to begin in August, Astoria feted Hammond and his family in a massive kick-off celebration that included a parade, fireworks, and a twenty-one-gun salute by the National Guard. In contrast to Missoula, Astoria lionized Hammond. Merchants capitalized on his popularity by linking his name with brand products, imploring customers to buy "new, stylish HAMMOND SUITS."[18] On July 25, 1895, the Astoria newspaper evoked royal—or at least presidential—connotations, as it proclaimed across its banner, "Entire Community Celebrates Inauguration by the Noise of Cannon, Fireworks, Music and Speeches. Hammond Addresses the People." The newspaper declared, "The citizens one and all were a joyful lot of people because they realize that at last their city was to be connected with the rest of the world."[19]

That evening, Hammond addressed the people of Astoria from his hotel balcony. According to the local newspaper, he announced, "I did not come here to talk but to build the railroad (applause), besides for sometime past I have been doing the talking and the committee have been doing the work. Now I want to do the work and will let them do the talking." At that someone in the crowd shouted, "There will be plenty of it." Undeterred, Hammond continued, demonstrating that he could play the populist as well as anyone. To the crowd's applause, he stated, "We propose to give you value received when this railroad is built. It will be second to none on the coast. We propose to put the money into the road and not into our pockets. When the railroad is operated it will be operated for the interest of the people."[20] Despite the dashed hopes, false promises, and half-finished efforts of the past, Hammond apparently inspired confidence among townspeople who saw him as a man who got things done.

Hammond's name had also elicited cheers and celebration two days earlier in Corvallis, when the State Supreme Court announced it had approved the sale of the Oregon Pacific to Hammond and Bonner. The months Hammond had spent in New York finally paid off when he succeeded in getting financial backing from John Claflin, president of a leading wholesale and retail dry goods company. Here again the MMC figured prominently, as Bonner and Hammond built upon an established relationship with Claflin through the mercantile business. Despite having paid only $100,000 for the Oregon Pacific, Bonner and Hammond reincorporated the railroad as the Oregon Central and Eastern and capitalized it at $3 million but issued only $1,500,200 in stock. Claflin provided much of the capital and received two-thirds of the stock, with the rest split between Bonner, McLeod, and Hammond. Unlike the financing behind the Astoria and Columbia River Railroad, Hammond drew heavily on the Missoula Mercantile to provide the necessary cash to run the Oregon Central line, which badly needed repair.[21]

With no government oversight, many nineteenth-century entrepreneurs built their railroads as purely speculative ventures, which often led to shoddy construction. Such capitalists were more concerned with selling stocks and bonds based upon hopes and promises than they were with building a transportation system based upon wood and steel. The Oregon Central was a case in point. Its poor condition became evident when one of its bridges collapsed, smashing up a train, killing two men, breaking the legs of the conductor, and causing $25,000 in property damage. Hammond had been riding on the train that crashed but had disembarked at the station immediately before the accident. Rattled by his narrow escape, he dashed off an angry note, blaming his partners for their unwillingness to spend the necessary $80,000 to upgrade the railroad. "I would have made improvements before but have been kept from it by Bonner and Claflin," he wrote to McLeod. "I suppose [now] they will quit trying to run a railroad from New York."[22] This event, combined with his experience in building the Northern Pacific branch lines, cemented Hammond's commitment to building railroads that would withstand the demands of traffic and time.

The Astoria line demanded even more attention and capital than did the Oregon Central, and Hammond pressured the Astoria committee to secure title to the three thousand acres that it had pledged. Hammond promised that when he received the title, he would finish the line within a year. Losing no time in cashing in on the subsidy, he began "carefully and scientifically" platting a city on the peninsula across Youngs Bay from Astoria. Hammond announced he would complete a massive bridge across the bay linking the new city—New Astoria—with Astoria by December 1, 1895. The *Oregonian* believed New Astoria had the "greatest harbor on the Pacific coast north of San Francisco" and would soon be "a great city." Indeed, within a year, it housed six hundred people.[23]

Railroad construction began in August 1895. Although Astorians were banking on freight traffic, the tourist trade from Portland to the coastal resorts proved to be the mainstay of the line. Noting that "a great rush is now going on from this section of the country to the seaside," Hammond arranged for first-class steamer connections from Portland to New Astoria and upgraded the railroad from Astoria to the Seaside beach resorts.[24]

A year after beginning construction, in September 1896 Hammond laid off his construction crew. With the presidential election between William Jennings Bryan and William McKinley looming, Hammond, like many industrialists, feared Bryan's election would instigate social upheaval and economic chaos. Engaging in a business ploy common during the election, Hammond applied economic coercion. He announced that he and other capitalists were unwilling to invest in the railroad with the prospect of free silver on the horizon and that if Bryan was elected the road would never be built. In December, with McKinley's decisive triumph and the shift to a Republican Congress, work on the railroad resumed.[25]

Hammond promoted McKinley back in Missoula as well. Like other mining states, Montana supported free silver, but his employees put aside their own political inclinations and "did what Mr. Hammond wanted," campaigning hard for McKinley. Although Bryan carried Montana by a six-to-one margin, in the company town of Bonner, where Hammond's Big Blackfoot Milling Company was located, McKinley received 140 votes to Bryan's 10.[26]

With A&CR construction proceeding on schedule, Hammond proudly showed off his work to his primary stockholder, C. P. Huntington. In May 1897, Huntington, his wife, Arabella, and his nephew Henry joined Hammond and Florence on a steamer from Portland to Astoria. By now well acquainted with the ship's captain and eager to impress, Hammond took Huntington into the pilothouse and charted the progress of the railroad along the shore and on maps. The mayor of Astoria and prominent businessmen met the party at the docks and paraded them around the city for an hour before retiring to the newly opened, luxurious Hotel Flavel in New Astoria, where Hammond entertained his guests in grand style.[27] In addition to trying to impress the Huntingtons, Hammond hoped to attract investment in the Oregon Central. After traveling over the decrepit line, however, C. P. publicly declared, "From the time the ground was broken on that road, it cost not less than $28 million and not one man who put his money into it originally ever got a cent out of it." Twisting the knife in Hammond's project, he added that he would not pay $50,000 for the road.[28] Such a statement would only make Hammond that much more determined to prove the railroad's worth. Following C. P.'s death, in 1900, Henry Huntington inherited his uncle's stock in the A&CR and invested heavily in the Oregon Central as well.[29]

Astorians were so pleased with the progress on their railroad, even before its completion, that they created a public park on Tongue Point, just east of town, as a monument to Hammond "so that when Astoria reaches the greatness that is expected, the coming generations may have something to reveal the name of the man whose business foresight and energy made the upbuilding of Astoria possible." For several hours, town leaders escorted Hammond on a walking tour through the hills and forests of "Hammond Park," which included "some of the finest specimens of Oregon timber."[30] A lumberman like Hammond, of course, appreciated such trees more for their commodity value than for their aesthetics. In just a few years, Tongue Point would become the site of an immense Hammond Lumber Company sawmill and a shantytown for mill workers.

Finally, in May 1898 the subsidy committee boarded the first train from Astoria to Portland, meeting cheering crowds at each station along the way. At Clatskanie, where the last spike had been driven six weeks earlier, the committee adopted an impromptu resolution, congratulating and thanking Hammond. They noted with satisfaction that "no finer or more substantial piece of railroad building has ever been put up in this country."[31]

The A&CR became an immediate hit for Portlanders seeking escape from the summer heat, as it enabled them to travel to the coast for only two or three dollars round-trip. Portland families had long enjoyed summers at Seaside, but instead of taking two days by steamer, they could now make the journey in four hours. Railroad advertisements promoted Seaside as a place where "the pleasure seeker—impatient of limitations—can bathe in the surf, loiter on the sands, dig for the succulent razor clam, net the elusive crab, cast a fly for leaping mountain trout, hunt for bear, philosophize in sylvan groves or clamber over rugged heights just as the fancy seizes him."[32] The A&CR initiated a special weekend service in what became known as the "Daddy Train." Businessmen could now leave Portland on Saturday, spend the weekend with their families vacationing on the coast, and return to work Monday morning. The emergence of a middle class, with its increased leisure time, coincided with the new rail access to form the beginnings of a major tourism industry along Oregon's coast. During the following six years, the population of Seaside tripled.[33]

The success of the A&CR soon attracted attention from railroad magnates James J. Hill and E. H. Harriman. Hill, who had acquired control of the Northern Pacific, and Harriman, president of the Union Pacific, each believed the A&CR would be a nice addition to their transcontinental networks. But rather than simply purchasing the railroad, in typical robber-baron fashion they sought to undermine its value, drive it to the brink of bankruptcy, and then buy it. With his ownership of the Oregon Railway and Navigation Co. (OR&N), Harriman was particularly well poised to undercut Hammond's operation. Unbeknownst to both Harriman and Hill, however, Hammond was backed by Huntington and could fend off any attack.[34]

Although slower than the train, the OR&N's steamship line from Portland to Astoria directly competed with Hammond's A&CR. Noting that the steamships charged 40 percent less than the railroad did, in the summer of 1899 Hammond dropped his rate to match. The OR&N responded by cutting its fares. Hammond retaliated, slicing his fares to two bits for the Sunday excursion train, although the A&CR was able to recoup some of those losses by charging an additional seventy-five cents from Astoria to Seaside. It soon became a test of wills and financial resources as to who could operate at a loss for the longest time. Passengers, of course, delighted in the rate war, which lasted nearly two years.[35]

Harriman also used freight rates in his attempt to drive Hammond to the wall. Although the Union Pacific leveled a standard fare, known as a common point rate, for freight from Portland to the Midwest, Harriman refused to extend this rate to Astoria. This led the Astoria lumber mills to cancel their log-hauling contract with the A&CR. As a result, Hammond had to suspend freight service between Portland and Astoria for the winter and limit his business to passenger trains. C. P. Huntington, however, agreed to extend the common point rate on his Southern Pacific lines to Astoria and the Willamette Valley. He further reassured Hammond that he would back him in his fight against Harriman "until the Columbia River ran up towards the mountains, instead of to the sea."[36]

But Huntington died in August 1900, and the feud quickly ended. After suffering combined losses of $600,000, Hammond and the OR&N agreed to cease the rate war and set standard fares. As a result of the truce, the A&CR in the summer of 1901 posted an increase in net earnings of $18,000 over the previous year. Now with a proven track record and having successfully foiled attempts to bankrupt the railroad, Hammond and his investors agreed that "the time is near at hand when we ought to be able to sell the property."[37]

Hammond opened negotiations simultaneously with Harriman and Hill. Both discovered what a shrewd and uncompromising haggler Hammond could be. A. B. spent much of August 1902 in New York attempting to push Harriman's offer of $3 million for the A&CR up to $4 million. During their discussions, Harriman hinted that since he controlled the transcontinental traffic, Hammond was at his mercy. Refusing to be intimidated, the lumberman reminded Harriman of the recent rate war and suggested that Harriman would be wise not to engage in another fight. Seeing through Harriman's bluff and confident that he was fully supported by Henry Huntington and Thomas Hubbard, Hammond insisted he would accept no less than $4 million. Upon the railroad magnate's refusal, the negotiations reached an impasse—for the time being.[38]

Leaving Harriman to simmer, Hammond began to increase the value of both his Oregon railroads. Hoping to sell them as a package, Hammond made plans to build the A&CR from Seaside 138 miles south to Newport, where it would join his other line, the Corvallis and Eastern (formerly the Oregon Central). This connection would also provide access to the timber in the Nehalem drainage. The lumber hauls over the A&CR were steadily increasing, and Hammond looked forward to doubling the freight traffic in 1903. As profits increased, the value of the A&CR grew and Hammond's interest in selling it waned.[39]

Hammond, however, was at heart a lumberman, not a railroad magnate, and before long came to regard his Oregon railroads as a means to a greater end: accessing the vast timber resources of the Cascades and the Oregon Coast Range, previously "shut off by an impenetrable wall" of rugged mountains and twisting river valleys.[40] Shooting down the coast, the A&CR accessed the timber of the Necanicum River spruce belt, a watershed twenty-five miles long by ten miles wide that held a staggering eight million board feet per square mile. Adding to the original Astoria subsidy, Hammond bought timberlands in parcels ranging from as small as forty acres to as large as fifty thousand acres along his railroad lines. In 1901, the *Oregonian* recognized that he had "invested more millions in Oregon than any other man who has come into the state in the past 20 years." The newspaper closed its article on Hammond's purchases with a prophetic statement: "Much as we may regret to see the passing of the great forests which have made Oregon famous . . . [p]osterity may not bless the present generation for turning a forest into a field, but while the change is being made the lumber business will place in circulation an immense amount of money."[41]

Oregon's greatest concentration of timber lay in its four northwestern counties (Tillamook, Clatsop, Washington, and Columbia), which contained an estimated 56 billion bf of timber. Until the 1890s, loggers focused on timber that they could float down rivers, rarely venturing more than a mile or so inland. Technology and access limited the extent of logging; even so, timber was Oregon's leading industry. When Hammond completed the A&CR in 1896, Oregon had three hundred lumber establishments that provided $1.6 million in wages to 3,777 employees. Clatsop County alone produced 20 million bf per year.[42] These figures skyrocketed after Hammond's two railroads opened previously inaccessible timber lands. By the turn of the century, Oregon's timber industry was cutting 550 million bf per year, and boosters believed the state had enough trees to last six hundred years at that rate. With five hundred sawmills in 1906, Oregon doubled the previous year's lumber production to 2 billion bf, valued at $30 million. The *Oregonian* proclaimed, "While the timber of the Eastern States is rapidly becoming exhausted, that of Oregon stands almost intact, and this state is prepared to head the list in its lumber output for an indefinite number of years to come."[43] The newspaper failed to note the social and environmental costs that the East had incurred by exhausting its timber and that Oregon faced the same fate.

The A&CR certainly brought prosperity to Astoria. By 1910, the city had grown to fifteen thousand people and its lumber mills were running day and night, cranking out more than 263 million bf of lumber a year, almost all for export. Hammond's Tongue Point Lumber Mill was producing 250 thousand bf of lumber on every shift. Astoria also boasted twenty-seven salmon canneries, eleven of which belonged to Hammond's Columbia River Packers Association, an organization he had patched together in 1896 to consolidate the salmon industry.[44]

Meanwhile, neighboring Columbia County's population nearly doubled, from 6,237 in 1900 to 10,580 ten years later. As the *Oregonian* concluded in 1904, while "it was the timber resources . . . that made the Astoria road a profitable enterprise," its construction allowed for vastly increased passenger traffic. Indeed, the A&CR actively promoted the Necanicum River as a tourist and trout-fishing mecca, despite the yelps of loggers, trees crashing to the ground, and the rattling drone of lumber mills. While recreation and lumber production were hardly aesthetically compatible, financially they proved mutually reinforcing. From the A&CR's inception, tourism provided its profit margin during the summer months.[45]

The Oregon Central was another story, and profits from the line proved as elusive for Hammond as they had for previous owners. Other than providing a quick access from Albany (near Corvallis) to the coast, the Oregon Central and Eastern dead-ended at Detroit, Oregon, on the upper Santiam. East of Albany, 90 percent of the railroad's traffic came from Hammond's Curtiss Lumber Company in Mill City. With only half the locomotives in operating condition and the line in need of $96,000 in repairs, the chief value of the railroad lay in its potential.[46]

To inflate the worth of the railroad Hammond engaged in a corporate shell game, a standard practice among railroad capitalists. While Hammond sought total control over his companies, he also wished to conceal his ownership, possibly because he was still under federal indictment for timber poaching in Montana. Thus, in 1897, Hammond dissolved the Oregon Central and Eastern and reincorporated it as the Corvallis and Eastern. As he had done with the Big Blackfoot Mill in Montana, Hammond set up a "dummy" board of directors for which he tapped Edwin Stone, the former Missoulian who had first drawn his attention to the Oregon Pacific Railroad in 1894. J. K. Weatherford, who served as Hammond's attorney in Oregon, became president of the Corvallis and Eastern with one share of the company's stock. Hammond also employed

T. H. Curtis as chief engineer on the A&CR and made him treasurer of the Corvallis and Eastern. As Curtis later admitted, "The local people as stockholders were simply figureheads."[47]

In February 1898, Hammond mortgaged the Corvallis and Eastern, ostensibly to use the money to complete the line through the Cascades and then south to California. Curtis, acting as a placeholder for his uncle Thomas Hubbard, accepted $1 million in bonds in exchange for $200,268 in stock. Backed both by Hubbard and Henry Huntington, Hammond publically announced he would pump $10 million into finishing the railroad, pushing it through the Cascades and into eastern Oregon to connect with the Oregon Short Line. Although he had no intention of pouring more money into the Corvallis and Eastern, the ploy to promote the railroad's value succeeded, and in 1907 Hammond sold the railroad to Harriman for $1.4 million, a 1,400 percent gain on his original investment.[48]

Shortly after the sale, the murky finances of the Corvallis and Eastern attracted the attention of the Railroad Commission of Oregon, which began an investigation. Testifying under oath, two of the railroad's directors, Curtis and Weatherford, evaded questions and lied repeatedly to obstruct the inquiry. They denied any knowledge about the corporate organization of the Corvallis and Eastern or the identity of the principal stockholders—that is, Hammond, Hubbard, and Huntington. Demonstrating his loyalty, Curtis told the commission that Hammond "had very little stock," but under repeated questioning, Curtis acknowledged it was "common knowledge that nothing was ever paid for the stock that was issued or for the bonds; if money was ever realized from the sale of them it went to the Bonner and Hammond interests." As was the case of most railroads of the era, the Corvallis and Eastern's stock was grossly inflated so as to make the railroad appear more valuable. Failing to overcome the obstruction of the officers of the Corvallis and Eastern, the commission ended its investigation.[49]

For years Hammond had played Harriman and Hill against each other, as well as repeatedly announcing expansion plans that would threaten their domination of the Oregon market. Harriman's purchase of the Corvallis and Eastern made the A&CR even more valuable to Hill if he were to compete with the Union Pacific. Although a few years earlier

Hammond had offered the Astoria line to Harriman for $4 million, in 1907 he bumped the price up to $5 million, and this time Hill snatched it up. The sale of these two lines left Oregon with no independent railroads of any considerable length. The concentration of railroad ownership hardly concerned Hammond, however, as he used the infusion of cash to further his own attempts at consolidation in the timber industry.[50]

In the usual progression of western development, eastern capital financed railroads that linked urban markets to hinterland resources and provided consumer goods. In the wake of the Panic of 1893, however, Hammond's Oregon railroads reversed this process. In this case, it was Hammond's consolidation of the mercantile business in the hinterland of western Montana that enabled him to build (or rebuild, in the case of the Corvallis and Eastern) two railroads in the more developed region of western Oregon. The impeccable credit rating of the Missoula Mercantile and the potential growth of Astoria's real estate based on tourism and timber ultimately allowed Hammond to tap eastern capital—but initially his capital was both local and western.

Only after taking the initial risks did Hammond attract eastern investors, who served primarily as conduits of capital, channeling it west to increase their returns. While historians credit Huntington, Hubbard, Hill, and Harriman with being the great railroad industrialists, it was Hammond and numerous other regional entrepreneurs who actually built the railroads and transformed the western landscape. Extending the rail lines into the rugged Coast Range allowed Hammond to exploit Oregon's most valuable natural resource. The conversion of trees into capital began a process that would nearly eliminate the ancient forests of the Pacific Northwest over the next century and create a single-resource economy dependent upon a wildly fluctuating market—not unlike the timber colony of New Brunswick. But rather than being shipped overseas to Great Britain, much of Oregon's timber flowed south to California. As the financial capital of the Pacific Rim, San Francisco became the new London. Following the money, A. B. Hammond would relocate there by the turn of the century.

CHAPTER **11**

A. B. Hammond, Inc.

After leaving Missoula in 1894, Hammond spent four years living out of hotels, such as the Waldorf-Astoria, and shuttling between New York, Missoula, and Portland. During these years his family summered in Oregon and Montana but remained in the East during the school year, marking the beginning of an increasing physical separation between Hammond and his children. His sons, Richard and Leonard, preferred to spend their summers in Montana, camping and fishing. The girls, too, spent less and less time with their father. Edwina, the eldest, in particular enjoyed the sunshine and social life of Los Angeles for months on end.[1]

Like many West Coast lumbermen, Hammond discovered that an office in San Francisco was imperative if he wanted to join the major leagues of western businessmen. With a headquarters in San Francisco, Hammond could assess both domestic and foreign markets, secure sales, have a hand on the financial pulse of the western United States, and access capital more easily. Instead of settling in Portland, where he had purchased a fine house, Hammond relocated again, this time to San Francisco, where political, social, and financial elements were intertwined. Lumbermen, financiers, industrialists, and politicians not only shared office buildings but lived in the same neighborhood. In building a new mansion in Pacific Heights, Hammond joined other lumber barons—the Popes, Talbots, and Hoopers—within a six-block radius. Their offices, too, were clustered along lower California Street. Hammond, however, moved his office into the newly constructed Merchants Exchange at 465 Montgomery, a modern fourteen-story office building fashioned after Chicago's new skyscrapers. The ground floor opened into the great hall of the San Francisco Chamber of Commerce, "where captains of industry traded contracts for goods and speculated on future prices." The top seven floors housed the offices of the Southern Pacific. San Francisco's particularly tight geography would aid Hammond in forming a close social, political, and financial network.[2]

San Francisco was a multiple-day journey from both Portland and Missoula. With operations from western Montana to Northern California, Hammond stretched himself thin. His Montana operations were secure in the capable hands of C. H. McLeod, but with two railroads, three lumber mills and several more in the works, and the timber rush in full swing, Hammond desperately needed a faithful lieutenant to oversee his Oregon operations. Just as he had done in Montana, he tapped his loyal New Brunswick clan.

Twenty-year old George McLeod had arrived in Missoula in 1891 to join his cousin in the Hammond organization. George's first job was at the Blackfoot Mill as a janitor and stock boy. After selling the Blackfoot Mill to Marcus Daly in 1898, George stayed on as a clerk for the Anaconda Mining Company, but Hammond also employed him to look after accounts during the transition. George became disgusted with the new Anaconda manager, who was stealing goods from the company store. Taking a substantial pay cut, he returned to Hammond's employ at the Kalispell Mercantile Company. With the formation of the Hammond Lumber Company in 1900, Hammond needed someone trustworthy to oversee his Oregon land acquisitions, and McLeod's honesty and background as a civil engineer recommended him. His marriage to Hammond's niece didn't hurt, either. Eventually, George would become president of the Hammond Lumber Company and have the singular distinction of being its first as well as final employee. Throughout his career, George McLeod balanced Hammond's stubborn belligerence with a more reasoned approach.[3]

McLeod exemplified the rise of the professional managerial class and the decline of the owner-operator in the lumber industry. Similarly, the Weyerhaeuser Timber Company (also formed in 1900) employed George S. Long as its general manager. Hired in the same month to do the same job for two competing lumber companies, Long and McLeod would follow parallel career paths. Geography, however, provided a territorial separation between the two companies. Long based his operation out of Tacoma, while McLeod directed Hammond's Portland office.[4]

In 1900 A. B. Hammond joined the corporate revolution. All around him and across America, corporations were forming, reforming, absorbing, and consolidating, and Hammond wanted to be at the forefront of the gobbling. In moving from merchandising to railroads to manufacturing, Hammond followed the trajectory of American business in the Progressive Era.

Hammond came a generation behind the first wave of industrialists, which included Andrew Carnegie, Phillip Armour, James J. Hill, and John D. Rockefeller. Thus, Hammond could readily apply models they pioneered, such as Rockefeller's use of the trust and holding company. Hammond modified their methods to suit both his personality and the unique conditions of the lumber industry as he envisioned a lumber empire stretching from the Columbia River to Southern California fashioned along the lines of Carnegie Steel, Standard Oil, or Anaconda Copper. Such an undertaking would require substantial capital, changes in organizational structure, a transportation network, manufacturing facilities, a cheap labor supply, easy access to raw materials, and new technologies to increase productivity.

Until the 1870s, small, local, or, at most, regional businesses, primarily single-owner enterprises or partnerships like Eddy, Hammond and Co. dominated American commerce. As a store manager, Hammond personally knew his customers, distributors, wholesalers, and even the freighters who slept on the counters. Hammond chatted with customers as they shopped, and if someone had a complaint or sought a favor, he or she could walk into his office. Commercial exchange was a social as well as an economic transaction based on credit, and credit was based on a person's character. In agrarian America, customers and suppliers were often one and the same—farmers or ranchers who both bought goods from and sold their products to the local mercantile. Farmers and ranchers sold their goods only once a year, so merchants extended credit to their customers until they could pay off their debts.

The railroads changed everything. Railroads formed the first national market for industrial products such as steel, copper, and lumber and, in the process, catalyzed the profound transformation from a regional agrarian society to a national industrial one. The vast railroad network, in turn, fostered urbanization and urban markets. Perhaps even more important, railroads provided the model for the modern corporation with centralized bureaucracies. Hammond's experience managing his Oregon railroads taught him the value of a hierarchical organizational structure based upon separate divisions, much like a military command structure.[5]

Following the Panic of 1893, the financially overextended railroads fell out of favor with investors who, casting about for new prospects, hit upon manufacturing industries. At the turn of the century, finance capital joined industrial capital and spawned the corporate revolution with an explosion of new manufacturing enterprises, most of which had originated as proprietor-owned companies. In less than a decade, from 1895 to 1904, America moved from a nation of small-scale entrepreneurs to one of corporate giants.[6]

The corporate revolution divorced ownership from management. Stockholders, rather than company founders, now owned large corporations but had little control over day-to-day management. While increasing the number of stockholders allowed a company to raise more capital, it also diffused ownership. Stockholders, with few exceptions, were content to let a board of directors hire managers to run their companies, while those actually managing the operations usually had little or no ownership. While this arraignment seems normal in today's world, in 1900 most people were accustomed to a system in which owners and managers were the same. Although ensconced in the older system, Hammond needed capital to realize his imperial ambitions in the new order, and so he adjusted.

The lumber industry, however, differed from other industrial enterprises and retained many nineteenth-century characteristics, frustrating Hammond's aspirations. Most lumber companies were founded, owned, managed and operated by a single individual or, sometimes, a partnership. By the second generation, some had become family-run corporations. Few, if any, were publicly traded. Even the industry giant

Weyerhaeuser was a collection of smaller, family-owned companies rather than a single corporation—at least until 1900.

While Hammond considered his Oregon prospects in the late 1890s, Marcus Daly accelerated his Montana copper production and desperately needed to secure a supply of mining timbers as well as cord wood for his Anaconda smelter. Daly had bought sawmills in the Bitterroot Valley in the late 1880s but was running out of trees and mill capacity. He began to eye Hammond's Blackfoot Mill, the region's largest, with a production capacity of 125,000 bf per eleven-hour shift.[7]

For fifteen years Hammond had supplied Daly's timber needs, but the bitter fight over the location of the state capital in 1894 pitted the former partners against each other, and Daly resented his dependence upon Hammond. Thus, in 1898 Daly made a bid for the Big Blackfoot Milling Company. While Hammond wished to unload his Montana properties to invest in the more lucrative Pacific coast timber industry, he knew he could play Daly like a hooked trout.

Hammond set his price at $1 million, including the mill and timberlands containing five hundred million board feet, three-fourths of which were valuable ponderosa pine. Daly's agent, Mike Donohue, considered the price absurd. Hammond, however, knew Daly could afford any price he quoted and that the longer he held out, the more desirous Daly would become. When Donohue returned to Missoula in March ready to bargain, Hammond raised the price $50,000. Outraged, Donohue again departed for Anaconda only to revisit the following month. This time Hammond bumped up the price an additional $100,000. Donohue expressed his annoyance in a letter to Daly, stating, "Hammond is a very difficult man to deal with, but I succeeded in getting him to put a price on the business."[8] Hammond, however, was enjoying the game and added another $100,000 to keep Daly at bay while he traveled back to the West Coast. In the summer of 1898, Hammond returned to Missoula and resumed negotiations. Finally, in August, Hammond and Daly settled on a price of $1,479,179, nearly half a million more than the original asking price of a few months earlier.[9]

Flush with cash, Hammond assessed his prospects. Astoria would make an ideal location from which to base his operations. Not only was it surrounded by an immense forest; Astoria also accessed a national transportation network by both rail and sea. Indeed, the area enjoyed a long history of overseas lumber exports. Since transportation costs could double the price of lumber, low shipping rates were as crucial as access to rail lines and sea ports. Furthermore, Hammond believed he could compete with the Portland mills as he owned the railroad out of Astoria and had C. P. Huntington's assurance of low transcontinental shipping rates on the Southern Pacific Railroad. Low transcontinental rates, however, would evade Hammond until 1916, when James J. Hill, in his final business act before he died, guaranteed a common point rate for Astoria.[10]

Following the sale of the Big Blackfoot Milling Company, Hammond began laying plans for a "mammoth" industrial facility consisting of lumber, paper, and pulp mills near Astoria. The *Pacific Lumber Trade Journal* reported, with considerable hyperbole, "The purpose of the mill seems to be to supply about all the lumber needed in the universe."[11] After spending a week in Astoria trying to lure local capital, Hammond grew frustrated and looked toward the Cascades. Thick with old-growth stands of immense Douglas fir, the western front of the Cascades could be a bonanza if he could only access them. Draining the Cascade Range, the North Santiam River cut a narrow canyon before dropping into the wide Willamette Valley. Hammond's Corvallis and Eastern Railroad snaked up the river and, although it was intended to punch through the Cascades, the line proved indispensable in accessing the watershed's timber resources.

In May 1899, Hammond purchased three sawmills near the far end of his railroad line. All were in desperate need of improvements. Partnering with William Curtiss, a timberland purchaser, and J. K. Weatherford, a Corvallis attorney, Hammond formed the Curtiss Lumber Company at Mill City. While financing and actively running the company, Hammond kept his name off official records and once again set up a dummy board of directors who provided an interlocking directorship of his enterprises. Weatherford, for example, served as president of the Corvallis and Eastern and vice president of the Curtiss Lumber Company. Over the next few years, Hammond more than doubled the capacity of Mill City; he invested in logging machinery, timberlands, and lumber camps and brought in twenty carloads of logs a day via spurs of the Corvallis and Eastern.[12]

Still unwilling to give up on Astoria, Hammond proposed building a huge sawmill (with a capacity of 250,000 bf per day, twice that of the Blackfoot Mill) and using the electricity generated by burning the waste to power a flour mill. He had two conditions, however. First, Hammond wanted Astoria to furnish the site and guarantee that the city would buy at least five hundred barrels of flour per year from the mill, giving him exclusive control of the city's flour production and consumption. Second, he insisted that both the flour mill and sawmill stand together. Like Daly, Astoria learned that Hammond drove a hard bargain. In this case, however, Hammond's insistence on a monopoly over a basic food staple yielded too much control to an outsider, and Astoria's chamber of commerce refused the deal.[13]

Hammond again made front page news in 1903, when he announced that he would indeed build the giant sawmill, but this time across the bay in Flavel. He bought two thousand feet of waterfront property, giving him control of the entire west bay. In the interim, Hammond had acquired "immense holdings" of timber in the Nehalem and Necanicum watersheds just to the south of Astoria. Once again, town leaders believed their city stood poised to become the seat of a great lumber empire.[14]

Hammond needed substantial financial backing to bring his plans to fruition, and in 1900 he formed the Hammond Lumber Company with five others, each contributing $600,000. Hammond recruited his railroad partners, C. P. Huntington, Thomas Hubbard, and John Claflin, and brought in Francis Leggett, owner of New York's largest grocery firm, and Marcus Daly. Daly apparently held no ill will against Hammond for repeatedly raising the price of the Blackfoot Mill. Indeed, the two men had become "great friends" in the intervening two years. With this select group, all of whom were busy with other enterprises and happy to leave management of the lumber company to him, Hammond acquired the capital he needed while retaining the control he desired.[15]

Hammond, like many businessmen, sought total control over his companies and would not tolerate minority stockholders holding veto powers. Fortunately for Hammond, four years previously New Jersey had rescinded the "rule of unanimous consent," which had held that fundamental corporate changes, such as the sale of assets, required unanimous approval of stockholders. Nor did the state require directors to actually own stock in the companies they managed. New Jersey also allowed corporations to own stock in other companies. By sitting on more than one board, directors could control competition and coordinate financial activities. Taking advantage of such relaxed restrictions on corporate charters, Hammond incorporated both the Astoria Company and Hammond Lumber Company in New Jersey, even though they were Oregon firms. In doing so, he followed hundreds of others. New Jersey's (and later Delaware's) legalization of holding companies, such as Hammond's Astoria Company, spawned the Great Merger Movement from 1895 to 1904, during which 1,800 firms were absorbed by consolidations. By 1904 forty-two of these combinations controlled more than 70 percent of their respective industries.[16]

While the lumber industry was quick to accept technological innovations that increased production, it resisted corporate consolidation. By 1900, corporations produced less than half of the value of American lumber, and as late as 1919, individuals and partnerships still owned 85 percent of the nation's lumber companies. Even in the 1990s, lumbering ranked among the least-concentrated manufacturing industries, with the top fifty companies accounting for less than half of the total value produced. And yet the lumber industry was essential to America's industrialization. In 1910 it was the nation's third-largest industry, and even as late as 1937, lumber ranked third among manufacturing entities in the number of employees. Well into the twentieth century, lumber was the primary construction material for ships, railroad cars, and of course, houses. What, then, explains the persistence of small and independent lumber firms into the present day?[17]

Business historians such Alfred DuPont Chandler, Jr., have argued that increased efficiency explains the rise of corporations. Chandler tied the rise of the modern corporation to the expanding railroad network, which connected the source of raw materials with urban markets, and to the vertical integration of industry, which required hierarchical organization. Focusing on changes in organizational structure that increased production, Chandler contended that big business arose from the "managerial revolution" as the "the visible hand of management replaced the invisible hand of market forces." He posited that this new organizational structure was more efficient than the older proprietary capitalism and thus allowed companies to grow large and dominate their particular industry.[18]

Embracing the new business model, Hammond spent the remaining thirty-four years of his life engaged in a futile attempt to gain control of the West Coast lumber industry. Hammond used new technology to increase production; he reorganized his business on the corporate model and employed a managerial hierarchy, yet his need for personal control and oversight caused him to retain many nineteenth-century practices. The Hammond Lumber Company integrated both horizontally and vertically, controlling everything from raw materials to the retail sales of finished products. Eventually, Hammond would own timberlands, lumber mills, sash and door factories, railroads and shipping lines, and wholesale and retail outlets, and he would even assume construction contracts. While technological developments did not necessarily lead to the "managerial revolution" in the lumber industry, the Hammond Lumber Company did meet Chandler's definition of a "modern business enterprise" as having "many distinct operating units" and "a hierarchy of salaried executives."[19] Hammond's salaried executives, however, were part of an extended kinship network, suggesting that a corporation could adopt some modern properties while preserving aspects from the mercantile age.

If efficiency was necessary for corporate development, the lumber industry was notoriously inefficient. Unlike other industries, in lumbering, both labor and transportation costs went up as production increased. After cutting nearby forests, lumbermen needed to travel farther from the mill, building more miles of railroad to access ever-diminishing stands of timber. While small mills could move frequently to follow the timber, this mobility sacrificed the labor-saving machinery of larger mills. Furthermore, the migratory nature of the lumber industry favored decentralization [20]

Unlike many industries in which the acquisition of raw materials was hidden, the lumber industry was directly dependent upon natural ecosystems and individual trees, in particular. Ecological constraints impeded the development of large-scale economies in the lumber industry.

Topography, climate, soils, fire history, vegetative associations, and genetics determined which species of trees grew where and how large they grew. In addition, the unique quality of individual logs—varying by species, diameter, density, and knots—limited mill standardization.[21] As long as the lumber industry remained contingent upon a highly variable and widely dispersed natural resource, it would be impossible to establish a monopoly, no matter how hard men like A. B. Hammond tried.

CHAPTER 12

The Oregon Land Frauds

As the twentieth century dawned, A. B. Hammond began to implement his plan for a West Coast empire built upon wood. While he lorded over a fiefdom of sawmills and lumberyards, he wished to secure control over the supply of raw materials. Such vertical integration was a hallmark of the modern corporation, but it brought Hammond into yet another clash with the federal government.

Throughout the nineteenth century, sawmills purchased timber from independent loggers. While this worked well for small mills and operations, by the turn of the century rapid industrialization created an increasing global demand for lumber, prompting the increased mechanization of sawmills. The capital investments for new mills required consistent and reliable production, and lumbermen believed they needed to secure control over forests to ensure future supplies even if the mill continued to buy contract timber in the meantime. The need for raw materials, combined with federal policies that allowed for the acquisition of timberland at below-market value, set off a frenzy of timberland speculation in the Pacific Northwest. This led to flagrant abuses of the land laws, attracting the attention of Progressive conservationists, and ultimately provided the impetus for government control over the nation's forests.

During his tenure at Pope and Talbot in the late 1860s, Hammond witnessed how the firm used military scrip and school land grants to acquire thousands of acres of timberlands. In building the Northern Pacific and his Montana lumber business, Hammond simply logged both public lands and railroad sections. By the time he moved to Oregon, times had changed, and neither open access nor military scrip was an option. During the 1890s, spurred on by the Klondike gold rush, the development of the Pacific Northwest's railroad transportation and industrial infrastructure made lumbering the region's major industry.

Hammond arrived in Oregon at the very height of the timber rush and joined in the action, amassing timberlands as fast as possible. The problem for Hammond and other lumbermen was that the railroads or the federal or state governments owned nearly all timberlands. Although the days of truly free timber were over, *almost*-free timber was readily available.

The timber rush reinforced itself. As lumbermen snatched up land, they drove speculation. Locators and entrymen (those filing land claims, either for themselves or others), eager to cash in, filed claims and turned them over to lumbermen. Then, in 1900, lumber prices skyrocketed, advancing seven dollars per thousand board feet, adding fuel to the fire. The Astoria newspaper reported, "Cruisers are in every district and hundreds of men and women are scurrying to Oregon City to make filings, so as to sell at a small advance." Hammond raced against other lumbermen for not just acreage, but acreage at a low price. When Hammond snagged several thousand acres, Willamette Pulp and Paper bought even more. Nor was the rush limited to lumbermen. Agents for Rockefeller, E. H. Harriman, and Scottish syndicates took notice and began buying choice timberlands as well.[1]

Hammond was hardly the only lumberman to lust after the forests of the Pacific Northwest. The depletion of the forests of the Northeast and Great Lakes states, combined with the availability of cheap timberland and the arrival of the transcontinental railroad, propelled the nation's lumber industry to the West Coast. In the Midwest, Fredrick Weyerhaeuser had formed a tight group of investors that dominated the upper Mississippi lumber industry, but they were rapidly exhausting their timber supply. In 1891 Weyerhaeuser moved to St. Paul and next door to railroad magnate James J. Hill. The two men soon became close friends. Hill needed cash to pay off bonds following his takeover of the Northern Pacific Railroad. Weyerhaeuser needed timberlands, which the Northern Pacific had in abundance as a result of its land grant. The two tycoons struck a deal. On January 3, 1900, the Weyerhaeuser group bought nine hundred thousand acres of Northern Pacific grant land in western Washington for $5.4 million. The effort severely strained the companies' resources as "it took practically all the lumbermen on the upper Mississippi River to raise the money."[2] Indeed, the enormity of the purchase stunned the industry—after a momentary pause, it accelerated the speculative frenzy for Pacific Northwest timber.

Other lumbermen were less optimistic. The veritable Pope and Talbot considered selling out to Weyerhaeuser. Puget Mill manager Cyrus Walker—who had hired Hammond thirty years earlier—was becoming increasingly frustrated with the lumber industry and wanted to unload "this elephant of ours." However, the Weyerhaeuser group was not after sawmills but timberlands. To manage their holdings they formed the Weyerhaeuser Timber Company in March 1900.[3] The race for timberland was on.

Normally a solid booster, the *Astoria News* began to question the rapid consolidation of timberlands and its possible repercussions. In referring to Hammond's purchase near Astoria in 1899, the paper noted that the laxity of timberland laws had allowed "a single man, or a few men" to possess "hundreds of thousands of acres of timber." The paper lamented that "most of our coast timber is now owned by a comparatively few capitalists, most of whom do not live in this region and care nothing about it except for the wealth that it will yield them."[4] Although Hammond had given them a railroad, the town boosters wanted to see investments in manufacturing rather than speculation in timberland. Locals leveled this same charge at the Weyerhaeuser group when it made no move to buy or build mills to exploit its big purchase.

At first, Hammond focused on buying timberland that his railroads had made accessible, including twenty thousand acres of spruce timberland near Seaside. Then, Hammond's agents moved inland and scoured the state for available timberlands, acquiring parcels as small as forty acres for delinquent taxes. Many homesteaders had filed claims and either had not proved them up, had failed to pay taxes, or had simply abandoned them. But this tactic was time consuming and resulted in fragmented ownership.[5]

Hammond and other lumbermen also used less scrupulous means to amass timberlands. By hiring dummy entrymen to file claims under the Timber and Stone Act of 1878, western lumbermen acquired millions of acres of public land that were supposed to go to settlers. To encourage western development, the Timber and Stone Act allowed settlers to purchase 160 acres of government land for $2.50 per acre. Claimants had to sign an affidavit that the lands were for their personal use and settlement and not for speculation or sale. They further had to swear they had neither contracted with lumbermen nor intended to sell the lands. The discrepancy between the low purchase price and the market value led to widespread disregard for these legal provisions. Furthermore, in 1892 the Supreme Court loosened restrictions, ruling that the Timber and Stone Act forbade only prior *formal* agreements. A lumberman could make known his willingness to buy timberland above the government rate, and an entryman could buy land from the government and then transfer it without violating the law.[6]

Even with this legal latitude, lumbermen continued to engage in a suite of schemes to get around the law's requirements. Cash-poor settlers welcomed the chance to make a quick buck by simply filing a claim and turning it over to a lumberman or speculator. Under the Timber and Stone Act, a parcel of 160 acres actually worth $16,000 could be acquired for as little $415, yielding an astounding 4,000 percent profit. Timber speculation appeared so profitable and easy that it attracted many investors, including governmemt officials, the supposed guardians of the public domain.

In Montana, Hammond used the Timber and Stone Act to acquire 90,000 acres along the Blackfoot River. But after his battle with Land Commissioner William Sparks, Hammond kept his name off official records and instead employed others to make purchases for him. In Oregon Hammond still sought to distance himself, supplying the capital while employing others to file claims and hold the titles. To add another layer of bureaucracy and apparent legitimacy, Hammond had the entrymen mortgage their claims to one of his companies. After a few years, they then transferred the title over to his Astoria Company or the Hammond Lumber Company. In one such case, Hammond employed W. A. Geer, who filed thirty-one claims from 1899 to 1903. Geer sold many of these to the Astoria Company even before they were patented, in clear violation of the law. In October 1899, the Astoria Company paid Geer $6,345 for 4,394 acres on the Nehalem River. This payment, coming to about $1.45 per acre, likely represented Geer's profit, as Hammond had probably advanced Geer the expenses of filing fees and the purchase price of $1.25 to $2.50 per acre. Hammond saw himself as a refined businessman and, while he approved each sale, preferred to remain in the background unsullied by disreputable dealings. More important, this also kept him below the legal radar while others took the heat.[7]

Although the Timber and Stone Act was poor public policy, some analysts maintain that fraudulent use of the act provided the only means for lumbermen to acquire tracts large enough to make logging operations economically feasible.[8] While they certainly made use of the act, the big lumber companies of the West Coast, including the Hammond Lumber Company, purchased the majority of their timberland directly from the railroads. Out of western Oregon's 223 billion bf of timber, the Southern Pacific owned 18 percent, Weyerhaeuser owned 5 percent, and the next three largest holders, including Hammond and Charles A. Smith, owned a combined 12 percent. Eighty percent of Weyerhaeuser's 1.5 million acres originated from the Northern Pacific land grant.[9]

Likewise, Hammond acquired his biggest units from the Southern Pacific. In December 1900, he bought 14,500 acres east of Albany from the railroad. The next year, in conjunction with Charles Winton of Wisconsin, Hammond made the largest purchase of unbroken timberland in Oregon to that date. For half a million dollars, the Southern Pacific sold the pair 50,000 acres in the Tualatin and Trask River drainages. Combined with nearly 20,000 acres in Clatsop and Tillamook counties, the purchase made Hammond one of the largest owners of timberland in the Oregon Coast Range. In 1911 Hammond and Winston found an even greater bargain, buying another 46,000 acres from the Southern Pacific for $321,807.[10] These large purchases at $7 to $10 per acre indicate that lumbermen were willing to pay such prices without having to resort to the Timber and Stone Act.

The sale of large parcels of railroad grant land, however, was still fraudulent. Congress had intended for the railroads to sell off their grants to individual settlers and even made this an explicit condition in case of the Oregon and California grant (which was subsumed by the Southern Pacific). Ultimately, the collision between two ideologies—that of the Jeffersonian farmer and industrial capitalism—created the public policy failure of timberland management.

For more than thirty years officials had beseeched Congress for a solution that would allow for a controlled and rational means of supplying timber. Finally, in 1897 the Forest Management Act authorized the sale of timber within the forest reserves. Despite the guarantee of federal timber, filings under the Timber and Stone Act exploded in Oregon following the act's passage, increasing from 18 (covering 2,110 acres) in 1897 to 4,209 (covering 645,578 acres) six years later. The rampant fraud under the Timber and Stone Act led every secretary of interior since the act's passage to urge its modification or repeal. Typical was Ethan Allen Hitchcock's warning to Congress in 1902 that "the timber and stone act, will, if not repealed or radically amended, result ultimately in the complete destruction of the timber on the unappropriated and unreserved public lands."[11] Lumbermen fraudulently acquired land for $1.25 an acre, which they sold a few years later for more than $100 per acre. For example, in 1903, Hammond flipped a 160-acre homestead in southern Washington to the Cowlitz Lumber Company for $20,000, or $125 per acre. By 1909 the Hammond Lumber Company was paying up to $50 per acre for timberlands. Lumbermen clearly used the Timber and Stone Act more to ensure enormous profits in land speculation than to achieve economies of scale.

Lumberman Charles A. Smith admitted as much, stating it was "a mistaken belief that the manufacture of lumber is a profitable business. Instead, Smith insisted that it was the increased value of timberlands that made lumbermen wealthy. The smaller parcels attained by fraud had another purpose. They strategically isolated smaller landowners, who then had little choice but to sell their holdings or the timber thereon. In this manner, Hammond and other timber barons could effectively control large blocks of land they did not own.[12]

By the turn of the century, uncontrolled and illegal harvests had stripped entire mountainsides, drastically increasing erosion. For years Americans had called for the protection of mountain watersheds from timber depredation. Downstream water users joined with forest advocates to forge a conservation coalition, resulting in the Forest Management Act of 1897. In supporting the designation of the Cascade Forest Reserve, one Oregon letter writer summed up the connection while also invoking the importance of the commons: "The timber lands of the Cascade Reserve, Mr. President, should be held inviolate for all time to come. They are the common heritage of the people. These forests act as reservoirs to hold in storage the rains and snowfalls which go to make up our navigable streams and water courses, and are of vital importance to the people who inhabit this county."[13]

The Cascade Forest Reserve encompassed the high, icy peaks of the Cascades, while the lower slopes, both inside and outside the reserve, harbored stands of immense Douglas fir, red cedar, and hemlock, trees so large and straight that they would make a lumberman drool in a Pavlovian response. With mills in the foothills and a railroad piercing the heart of the Cascades, Hammond was well situated to take advantage of what was about to unfold.

Ironically, the establishment of the forest reserve system provided yet another avenue for privatization of timberlands. The Forest Management Act contained a provision that allowed claimants within the reserves to exchange their tracts for vacant land elsewhere. While seemingly well intentioned, this invited even more land fraud. In the time lag between the designation of the reserves on paper and their actual survey, speculators and locaters swarmed into new reserves to claim homesteads on inhospitable and inaccessible land that they could swap for valuable federal timberlands elsewhere. Often speculators, having located forest reserve claims, simply sold their exchange rights as scrip to lumbermen. Hammond, too, sent his agents to buy lands in the newly expanded Cascade Forest Reserve as quickly as possible, paying $4 an acre for land that he could exchange for timberland worth $100 an acre. "We are in the greatest kind of hurry for this land," noted his assistant, George McLeod.[14]

Problems stemming from the Forest Management Act's "lieu-land provision" were readily apparent, and Interior Secretary Cornelius Bliss recommended modifications to the law. The abuses were so flagrant and so obviously favored railroads and large lumber companies that western congressmen, usually those most amenable to government land disposal, led the efforts to repeal or change the act. Senator Thomas Carter of Montana, who still enjoyed Hammond's support, proved a notable exception in opposing the repeal of the lieu-land provision.[15]

In 1900, Congress limited lieu selections to surveyed lands, substantially restricting selections. Upon the insistence of Senator Carter, however, Congress delayed implementation for four months, giving lumbermen ample time to file for exchanges. Hammond seized the opening and sent men into to the Cascade Forest Reserve, the Priest River Forest Reserve in northern Idaho, and the Sierra Forest Reserve in California to file on lands that he could turn in for scrip, which could then be used to block up valuable timberlands elsewhere. In this manner, he acquired several thousand more acres in the Oregon Coast Range and in Northern California.[16]

Weyerhaeuser also capitalized on lieu-land selections. In Oregon, Weyerhaeuser acquired nearly half of its 380,599 acres by buying Northern Pacific lieu-land scrip. The most infamous of the lieu selections accompanied the designation of Mount Rainier National Park in 1899. Congress allowed the Northern Pacific to select timberlands outside the park to replace its grant lands within, much of which lay under glaciers. In 1900, Weyerhaeuser bought 40,000 acres of this scrip from the Northern Pacific and used it to file timber claims in Idaho.[17]

These lieu-land selections engendered much hostility in the West toward the forest reserves. In Oregon, locals supported the actual creation of the reserves as a means to protect their water supply, but they resented the lieu-land provision, which primarily benefited railroads and large lumber companies. Finally, in 1905, Congress repealed the lieu-land provision, albeit after the transfer of three million acres of ice, rock, and cutover lands in exchange for prime timber. Ironically, the lieu-land selections, while protecting the land and watersheds inside the reserves, exacerbated the destruction of those timberlands outside the reserve boundaries.[18]

With the Oregon timber rush in full swing, Hammond discovered his chief rival was Minnesota lumberman Charles A. Smith. In scoping out timberlands, Smith traveled incognito, conducted his negotiations with great secrecy, and relied heavily upon the intrigue of independent agent Stephen Puter, who described the lumberman as "one of the greatest criminals that ever went unwhipped of justice." On the other hand, Puter had nothing but praise for the savvy of A. B. Hammond, who also engaged Puter's services. Based out of Portland, Hammond held the home-team advantage as well as employee loyalty and kinship ties.[19]

A large part of the race between Smith and Hammond was to see who could more quickly defraud the federal government in acquiring timberlands. These Oregon land frauds nearly proved Hammond's undoing, but not only did he escape unscathed; he also acquired thousands more acres and bested Smith in the bargain. It all hinged on Puter.

Working the timberlands of the Oregon Coast Range, Puter served as the agent for the timber barons. As townships opened to settlement following land surveys, he hired entrymen to file homesteads or claims under the Timber and Stone Act. Theoretically, settlers were supposed to reside on these lands to obtain title. Financed by lumber capitalists, Puter paid the entrymen up to $750 for each 160-acre claim, which they then signed over to the lumbermen. Sometimes Puter and his accomplices secured a number of applicants for no more than a glass of beer.[20]

Early in 1900, Puter entered into an agreement with Smith to acquire 9,120 acres of timberland, with Smith advancing the money needed to hire entrymen and file claims on the South Fork of the Santiam River. In a rather labyrinthine process, Puter paid all expenses, including land office fees, the cost of the land ($2.50 per acre), and $100 to each entryman upon transferring the deeds. Entrymen then mortgaged their claims to Fredrick Kribs, another of Smith's agents, for $600 each. Kribs, in turn, paid Franklin Pierce Mays, the U.S. district attorney for Oregon, $50 per claim to approve the patents and arranged for Oregon's U.S. senator John Mitchell to receive $25 for each patent issued. Smith promised Puter $5.50 an acre for his work.[21]

In a similar scheme, Hammond employed several locaters to hire entrymen, who would then mortgage their titles to his Astoria Company or one of its employees. In 1899, Hammond employed W. E. Burke, a former Oregon legislator, to conspire with William Gosslin, secretary of the Astoria Company, to hire twenty unemployed men from Portland's seedy North End to file on timberlands in the Cascades, paying them two dollars apiece. Meanwhile, Hammond was acquiring lieu-land scrip to "plaster on the land." Unfortunately, the Northern Pacific was after the same tracts and protested the fraudulent entries.[22]

On October 19, 1899, Burke, Gosslin and the twenty entrymen were indicted on charges of defrauding the federal government. In an astute combination of politics and business, Hammond hired Oregon state senator Charles Fulton of Astoria to defend Gosslin, while the Northern Pacific employed Franklin Mays, the U.S. district attorney, to represent its interests. Behind the scenes, the two lawyers readily reached a compromise; the Northern Pacific received half the claims and Hammond got the other half. The charges were dropped, and neither the railroad nor Hammond was ever investigated. When Fulton advanced to the U.S. Senate in 1902, Hammond paid him ten dollars to expedite each new land patent, less than half of what Smith was paying Mitchell. Hammond could add now another senator to his account. With Carter and Fulton, the lumberman enjoyed as much representation in the upper chamber as did an entire state.[23]

Discovering that lands in the upper Santiam River were to be included in the new Cascade Forest Reserve, Puter rounded up ten people to file homestead claims under the pretense of being longtime settlers. The "homesteaders" had to make affidavits that they had been living continuously on their claims in order to claim lieu-land scrip. Puter simply had each "homesteader" act as witness for another while his partner bribed the deputy county clerk $100 for each claim. The lands filed on were high in the Cascade Range; "in fact, not a soul lives nearer than 30 miles," Puter later admitted.[24]

Receiving complaints of fraud, the General Land Office began examining the claims. To head off the investigation, Puter bribed both the agent for the local land office and the superintendent of the Cascade Forest Reserve. Still fretting that the deal would go sour, Puter traveled to Washington, D.C., to see Senator John Mitchell and Land Commissioner Binger Hermann. Anxious for action, Puter pulled out two $1,000 bills and laid them on Mitchell's desk. A few days later all twelve patents were issued.[25]

Back in Oregon, Franklin Mays was furious that Puter had spent so much. Mays thought Puter could have bribed the pair for $500, and feared that Mitchell and Hermann would expect such sums in the future. Puter then sold the claims to Smith for $10,080. Thinking he could cheat a thief, Smith paid Puter only $1,000 and sent him back to his old stomping grounds in Humboldt County, California, to scare up more timberlands. Smith's duplicity would cost him dearly and profit Hammond most generously.[26]

Both men wanted to buy a redwood lumber company, and one of California's largest and most established lumber firms—the John Vance Mill and Lumber Company—was for sale for $1 million. The company's assets included 9,600 acres of redwood timberland, the largest redwood lumber mill in existence, a secondary mill, 3,900 acres of grazing and cutover land, an eighteen-mile-long railroad with five locomotives, two bay steamers, and four lumber schooners. In short, here was everything Hammond needed to build a West Coast lumber empire, if only he could get to it before Smith.[27]

When Puter arrived in Eureka in June 1900 on the SS *Orizaba* to negotiate a deal for Smith, one of his fellow passengers was George Fenwick, former secretary of the Big Blackfoot Milling Company and Hammond's brother-in-law. Upon investigation of the Vance properties, both Puter and Fenwick sent back favorable reports to their respective bosses. Within weeks, Edgar Vance, who had inherited the company from his father, gave Smith a thirty-day purchase option.[28]

Not content to stand by, Hammond stepped up his efforts. Vance had recently made large investments in building a new mill and was apparently less interested in selling out than in acquiring capital to fend off cutthroat competition from other producers. Hoping to impress Vance, Hammond brought him to Missoula to show off the Blackfoot Mill and the vast economic engine of the Missoula Mercantile Company. Hammond made Vance an offer: he would form a new company capitalized at $2 million to assume the assets of the Vance mill, issue Vance $700,000 in preferred stock, pump massive capital into the firm, and institute a plan to gain control of the redwood lumber industry.[29]

When Puter discovered that Hammond was eyeing the Vance property, he hurriedly called Smith to tell him that Vance would extend the option if Smith would just come to Eureka. But Smith declined and continued to string Vance along, hoping to get a lower price. According to Puter,

Smith believed "he was the only man in the country who could swing such an amount." Smith however, had failed to account for Hammond, who "was quick on the trigger when he caught sight of a good thing."[30]

Smith still owed Puter more than $9,000 for the Oregon land deal and was reneging on payment. Disgusted, Puter went to Vance and exposed Smith's duplicity. Meanwhile, Hammond had not only sent George Fenwick to Eureka but also dispatched his brother Henry, former manager of the Blackfoot Mill. Regardless, as soon as Smith's option expired, Vance terminated negotiations with him and accepted Hammond's offer with a $100,000 deposit. The Vance purchase catapulted Hammond into the top tier of West Coast lumbermen and provided the foundation of his empire.[31]

Returning to Oregon in May 1901, Puter decided to pull another scam in the Cascade Forest Reserve. With the complicity of Marie Ware, the U.S. land commissioner at Eugene, Puter dispensed with the formality of securing actual "settlers" and simply used fictitious names. Ware provided Puter with papers already filled out; all Puter needed to do was insert the location and have someone sign the forms. In exchange, Puter paid her $100 for each claim. Franklin Mays wanted in on the racket and demanded half the profits for smoothing the legal road.[32]

On the job just over a year, George McLeod, treasurer of Hammond's Astoria Company, received deeds from Puter, notarized the forgeries, and forwarded the deeds to Ware to be executed and returned. In exchange, the Astoria Company received claims for the cut-rate price of five dollars per acre.[33]

The next summer, the federal government opened timberlands in the Deschutes River country near Bend to public entry following official surveys. Loading up their buckboards, buggies, and wagons, men and women came for miles around and lined up for blocks at the land office to claim lands. Puter was on hand, lining up dummy entrymen to acquire 17,280 acres.[34]

Instead of Smith, this time Puter turned to Hammond, who agreed to go in on the deal and advanced him the necessary cash. District Attorney Mays arranged for Senator Mitchell to receive $25 for each claim he escorted through the land office. McLeod, however, pointed out that Hammond was paying Senator Fulton only $10 to get patents approved and thought he could get a job-lot rate from Fulton. It might seem absurd to have two U.S. senators bidding against each other for such paltry bribes, but in 1900 a senator's salary was only $5,000, much of which was chewed up in expenses. With hundreds of claims passing across his desk every year, an unscrupulous senator could double his income.[35]

To thwart possible government investigation, Hammond engaged in complex maneuvering to provide a cloak of legitimacy. He had his secretary, William Gosslin, print two sets of blank mortgages: one was for the entrymen to execute to Hammond for $450 (the amount he advanced for final proof and payment); the second went to Puter for $160 in location fees. To further insulate Hammond, McLeod would sign for the entrymen's payments to the land office.[36]

Entrymen were well aware of the illegality of their actions. Worried about repercussions, they refused to sign the mortgages unless assured that all their future legal costs would be covered. One of them wrote to McLeod, "You have got us in trouble about the deal we made with you in regard to our homestead. . . . You assured us that night in the hotel Parlor before witness that we would not get in no trouble. And I hope you will prove to be a *man* of your word. That we will have no trouble in regards to deal [*sic*]."[37] McLeod then tried to purchase forest reserve scrip as there were only ten days remaining until the unclaimed land inside the reserves reverted back to the government. All of Hammond's and McLeod's maneuvering was to no avail, and the Deschutes deal fell through. Had any of these endeavors succeeded, it would have gotten Hammond into even more trouble with the federal government and might well have spelled an early end to his career.[38]

By the turn of the century, land fraud had become a time-honored tradition. Then, in 1901, an assassin's bullet and McKinley's death propelled Theodore Roosevelt into the White House. Suddenly forester Gifford Pinchot was not-so-quietly whispering in the president's ear about the egregious theft and exploitation of the nation's timberlands. The government's attitude toward graft and corruption shifted from a wink and nod to active purging. With the Progressive reformers now firmly in control of the administration, they set their sights on the timber industry in much the same way as had the Mugwumps of the previous generation. However, whereas William Sparks had targeted the lumber companies for fraud and depredations fifteen years earlier, Progressives saw corrupt public officials as the problem while largely ignoring lumbermen. Despite his rhetoric of "trust busting," Roosevelt and his administration were less concerned with corporate malfeasance than with ferreting out government corruption.[39]

Interior Secretary Ethan Allen Hitchcock joined Roosevelt and Pinchot in their Progressive campaign. Hitchcock, with the president's blessing, began to clean house beginning with Land Commissioner Binger Hermann. For the previous two years Hermann had resented Hitchcock's active interest in the land office. Hermann insisted, "The General Land Office is fully competent to deal with this subject [timber fraud]."[40] But in 1902, Hermann blocked investigation of a land ring in Arizona, possibly fearing exposure of his own involvement in the Oregon land frauds. Angered by Hermann's suppression of the report, Hitchcock asked for his resignation. Before leaving office, however, the land commissioner destroyed numerous files and letters that he later declared were of a personal nature. Hermann returned to Oregon, and six months later voters elected him as one of the state's two U.S. representatives. He attributed his victory to Oregonians' displeasure over reformers like Hitchcock.[41]

With Hermann out of the Land Office, Hitchcock began a relentless investigation of land frauds with help from Attorney General Philander Knox. Suspecting that Oregon U.S. district attorney John Hall's investigation would be half-hearted, Knox appointed the incorruptible and fiery Francis Heney as Hall's assistant. Heney would pursue the case with bulldog determination wherever it led, even if it probed the highest levels of government. Hall and Heney's investigation soon led to Puter, and, given his connection to the timber barons, it looked as if Hammond was about to become embroiled in another legal battle. Unlike the timber civil suits in Montana, these were serious criminal charges that had the full backing of the Roosevelt administration.[42]

Politics and business, however, continued to dominate the judicial process. Much to Heney's surprise, Hall chose to prosecute the case that implicated Hammond and McLeod, instead of the much stronger one that involved Puter's bribing Senator Mitchell and former commissioner

Hermann. Hall, however, faced reappointment and, needing the recommendation of Senator Mitchell, was reluctant to drag him into the quagmire.[43]

Frustrated with Hall's apparent complicity in the case, Heney traveled to Washington, D.C., to meet with Knox and Hitchcock. The three men suspected that the entire Oregon congressional delegation, as well as Hall, might be involved in the land frauds. Knox therefore appointed Heney as special prosecutor to take charge of the investigation. Heney then substituted the stronger case, which implicated Mitchell, for the weaker one, which involved Hammond. Not surprisingly, both Senators Mitchell and Fulton protested Heney's appointment.[44]

Hammond once again proved adroit at manipulating the legal system, but this time he did so with more finesse. The skill in cultivating political relationships that he had learned long ago from Bonner proved quite useful. Although Hammond had insulated himself through his employees, he would not let them hang. Parallel to the criminal investigation, the United States brought suit against the Astoria Company to cancel the land patents it had received from Puter. McLeod testified that he was an innocent purchaser of the deeds. Hammond again retained Senator Fulton to clear up the matter, and McLeod escaped further investigation. In the criminal proceedings, McLeod acknowledged that he had arranged to purchase claims from Puter at five dollars an acre, yet both he and Hammond remained unblemished. Hammond's unflagging support of McLeod during this episode cemented the younger man's loyalty for the rest of his life.[45]

Puter, on the other hand, was quickly convicted and decided to tell Heney everything, even offering to testify against Franklin Mays and Mitchell. Following Puter's confession, Heney's grand jury returned twenty-six indictments against one hundred individuals in Oregon. On the day that Heney announced his indictments against Senator Mitchell and Representative Hermann, President Roosevelt fired District Attorney John Hall, who was eventually convicted but later pardoned by President Taft. Other convictions in Oregon included the former mayor of Albany, two state senators, and two surveyor generals. Franklin Mays, also convicted, was sentenced to four months in jail and fined $10,000.[46]

Nationally, the land fraud investigation of 1903–1905 indicted 1,021 people in twenty-two states and resulted in 126 convictions. In terms of the number of people involved, the land frauds were larger than all the government scandals of the nineteenth century combined. Ironically, the timber barons who ended up with the land were rarely investigated and simply walked away with the titles to thousands of acres. Although Heney indicted several lumbermen, including Robert A. Booth, an Oregon state senator and vice president of Booth-Kelly Lumber Company, none of them ever faced prosecution. A. B. Hammond and Charles Smith were notably absent from any prosecution, indictment, or investigation.[47]

The Oregon land fraud trials signaled a sea change in attitudes regarding the nation's forests and an end to the Great Barbecue. When Heney subpoenaed Mitchell to appear before the grand jury, the senator publicly proclaimed, "I am as innocent as a babe unborn of any complicity in any land frauds." First elected in 1872, Mitchell came from a generation and mind-set that viewed the land laws as a type of spoils system in which the whole point was for the government to divest its property. He thus saw no impropriety in freely doling out favors.[48]

Mitchell attempted to cast himself as the victim of the rift between the conservative and progressive wings of the Republican Party. Indignant, Mitchell insisted that the prosecution was "a most damnable and cowardly conspiracy in which Secretary Hitchcock and this man Heney are the chief conspirators, their motives being partly revenge and partly politics." The Oregon legislature backed its senator and adopted a resolution endorsing Mitchell's character, innocence, and twenty-two years of faithful service. Fulton joined in, referring to Hitchcock as "hostile" toward Oregon.[49]

The land frauds underscored a fundamental divide within the Republican Party. Hitchcock and Heney saw themselves as reformers dedicated to ridding the nation and party of corruption while instituting moral virtue and efficiency. Hitchcock believed that the convictions would "have a wholesome moral effect."[50] Oblivious to political fallout, Hitchcock insisted that prosecution of land fraud was based upon "the high principle of honest government."[51] The *Oregonian* portrayed the land fraud trials as a battle between the youthful, energetic Heney and Senator Mitchell, the aging but dignified patriarch. The juxtaposition of the two images conveyed Heney as the modern Progressive with a belief in law and the federal government, replacing Mitchell's nineteenth-century old-boy network of patronage and laissez-faire capitalism.[52]

On January 17, 1905, Mitchell addressed the full U.S. Senate, proclaiming his innocence. Six months later, he was convicted, but while his case was pending appeal, Mitchell died from a diabetic coma. By 1907, Hitchcock's dragnet had pulled down three-quarters of Oregon's congressional delegation, and it probably would have included Senator Fulton had the statute of limitations not expired. Oregon representative John Williamson received a ten-month prison sentence for his part. Although he and Hermann appealed their convictions, they both declined to run for reelection in 1906. The attorney general finally dropped their cases in 1910.[53]

In 1907, the ailing and worn-out Hitchcock resigned, having served longer than any previous interior secretary. Soon thereafter, lumberman Thomas Walker was charged with fifty counts of fraud, but the new land commissioner, Richard Ballinger, declared he would not investigate, signaling that the land fraud prosecutions had ended. By the time Ballinger assumed the helm of the Interior Department under Taft, the administration had lost interest in timber fraud.[54]

Hammond and his associates continued to acquire lands through extralegal maneuvers, although they grew much more circumspect. Urging a focus on small parcels, Hammond's attorney J. K. Weatherford noted, "The announced policy of the Government is not to attack any sales made to individuals where the amount purchased is 160 acres or less."[55] Since such relatively small parcels (40–160 acres) were often sold for speculation or absorbed into larger holdings, it is impossible to know precisely how much land Hammond obtained through fraud. Under the Timber and Stone Act, Hammond and other lumbermen alienated nearly 3.8 million acres from Oregon's public domain, far more than in any other state.[56]

Nationwide, the General Land Office reported that by 1909, the Timber and Stone Act had resulted in the sale of twelve million acres of timberlands, ten million of which "were transferred to corporate or individual timberland investors by the entrymen." The sale of these lands brought in $30 million to the government, yet the lands were worth more than $240 million. The report estimated that less than 1 percent of lands purchased under the act were logged by those who made the entries.[57]

The land frauds had long-term national repercussions in motivating significant changes in public policy. By exposing illegal activity and wild speculation, the land fraud trials generated public support for stronger federal control over the public lands. Both open access and privatization had failed as viable policies to alleviate the environmental degradation of natural resources. It was now time to try active government regulation.[58]

Defending his designation of new national forests (created from the forest reserves in 1905), Theodore Roosevelt cited the abuses of the lumber barons as a primary factor. He noted, "Without such action it would have been impossible to stop the activity of the land thieves."[59] He further maintained that the only ones who were adversely affected by the designation of national forests were the "owners of great lumber companies" whose purpose was "by illegal, fraudulent or unfair methods" to gain "possession of the valuable timber of the public domain, to skin the land, and to abandon it when impoverished well nigh to the point of worthlessness."[60]

While the timber frauds catalyzed the conservation policy of the Roosevelt administration, the new policy predictably spawned a counterreaction. In Oregon, the Commercial Club of Eugene protested the expansion of the Cascade Forest Reserve. In Congress, none other than Senators Fulton and Carter, the two men Hammond had helped put in office, led the anticonservation movement. Fulton expressed his distaste for the newly created Forest Service, stating before Congress, "While these chiefs of the Bureau of Forestry sit within their marble halls and theorize and dream of waters conserved, forests and streams protected and preserved throughout the ages and ages, the lowly pioneer is climbing the mountain side where he will erect his humble cabin, and within the shadow of the whispering pines and the lofty firs of the western forest engage in the laborious work of carving out for himself and his loved ones a home and a dwelling-place."[61]

Despite his flowery rhetoric, Fulton was fully aware that his "lowly pioneer" was far more interested in making a few bucks by turning over "the whispering pines and lofty firs" to the lumber companies than in hewing a house out of the forest. Not only did the creation of the national forests eliminate the kickbacks Fulton had received for assisting the patenting of land claims; it also signified the rise of increased government controls over private industry.

In a bizarre concourse of political logic, Fulton and Carter accused Hitchcock of responsibility for the abuses under the lieu-land provision of the 1897 Forest Management Act. In addition, Fulton introduced an amendment to the 1907 appropriations bill that prohibited the creation of any more national forests in Oregon, Washington, Idaho, Montana, Colorado, and Wyoming except by an act of Congress. Carter thought this did not go far enough and wanted to insert a provision to rescind some of the reserves. Congress approved the amendment without Carter's provision. With the bill sitting on his desk, Roosevelt designated twenty-one new national forests, then signed it.[62]

Despite the backlash, forest conservation was widely accepted, even in the lumber industry. George Cornwall, editor of the *Timberman*, regarded the establishment of the forest reserves as "commendable," since he believed it would help stabilize the industry. Elected in the midst of the land fraud trials, Harry Lane, mayor of Portland, stated his support for "the Government to have absolute control over the forests of this country." Like Pinchot, he noted that such a system in Europe had allowed "people of the present age" to saw "lumber from trees that were planted by their ancestors one hundred years previously."[63]

The land fraud trials not only promoted the reform of land laws but also helped create a base of popular support for forest conservation. Indeed, Gifford Pinchot attributed Roosevelt's support of the national forests as a response to the land frauds that had consolidated timberlands in the hands of a few lumbermen. For the Progressives, it was just a small step from the prosecution of graft to government regulation. The trials thus indicated a shift in government policy away from the disposal of public lands and toward permanent federal control.[64]

Although Hammond and other lumbermen accumulated great fortunes in timberlands through their fraudulent activities, they soon confronted a number of unintended consequences. The land frauds created a timber baron phobia in Oregon such that a few years later when lumbermen sought state appropriations for fire protection, lingering hostility defeated their proposal. The public felt that since lumber barons had stolen the timber, they should have to protect it at their own expense.[65]

The larger problem that cheap timberlands produced for lumbermen was intense competition to liquidate their holdings. The end of land frauds stabilized the market value of timberlands, but taxes and carrying charges soon exceeded the value of the timber. Lumbermen were left holding substantial assets on which they either had to pay taxes or liquidate. Most chose liquidation in the form of accelerated logging, which, in turn, promoted overproduction—the lumber industry's chronic nemesis—as well as environmental destruction.[66]

Hammond, drawing upon his capital resources and his long experience in the timber industry, turned this situation toward his advantage. In the decade from 1895 to 1905 he had amassed a fifty-year supply of raw materials that would serve as the foundation for his lumber empire. He was now ready to turn his attention to production.

CHAPTER **13**

California Dreaming

In late February 1854, the steamship *Santa Clara* edged across a hazardous sand bar into Humboldt Bay in Northern California. On board was a thirty-five-year-old ship's carpenter from the Maritime Provinces who had joined thousands of others in the California gold rush. Like many of his fellow argonauts, John Vance found little gold and fell back on his former occupation. Building up a head of steam, the *Santa Clara* aimed straight for the small town of Eureka and rammed into the shore. The intentional grounding allowed a quick career change for the old steamer as Vance and the crew converted the ship into Humboldt County's first substantial lumber mill. Two years after landing, Vance purchased his own sawmill and began assembling a timber fiefdom. Vance built more sawmills, acquired thousands of acres of timberlands, and punched a railroad up the Mad River to access groves of immense redwoods. By 1881 Vance employed 150 men and accounted for a quarter of all of Humboldt County's lumber production.[1]

From Monterey Bay to Puget Sound, enormous trees flanked the Pacific coast, but sheer cliffs and rocky shorelines discouraged landings. Surrounded by a seemingly endless expanse of redwoods, the world's tallest trees, Humboldt Bay attracted the attention of lumbermen hoping to supply the booming California market. As the largest harbor between San Francisco and the Columbia River, Humboldt seemed an ideal location to launch a timber industry. Unfortunately, only twin sandy spits like crab pinchers protected the harbor from the ocean storms, and a shifting sand bar across the mouth of the bay often proved treacherous. But with lumber prices in California jumping from $80 to $200 per mbf, it was worth losing a few ships. Before long, Humboldt dominated California's lumber industry, and Eureka grew to become the largest city on the West Coast between San Francisco and Portland.[2]

With a flowing white beard and pince-nez spectacles perched on a large bulbous nose, Vance personified the nineteenth-century lumber baron. Known as "Uncle John," each year he transported Eureka residents by ferry and railroad to his Big Bonanza Mill on the Mad River for a giant Fourth of July picnic. When Vance died in January 1892, during his second term as mayor of Eureka, the *Humboldt Times* produced a two-page supplement detailing his accomplishments and contributions to the development of the region. Mourners lined the street and packed the Pioneer Hall for his funeral. The paper noted, "In his death the working men of the county have lost a friend who was using all his means on their behalf." The paper described Vance as gruff, eccentric, forceful, and self-reliant as well as intelligent and "a great reader," all characteristics shared by A. B. Hammond, who, after buying Vance's lumber company, would impose on it a vastly different business model.[3]

Like many industrialists of the time, John Vance and Humboldt's other pioneer mill owners developed a paternalistic relationship with their workers. These men were their friends and neighbors and, like themselves, were Anglo-Saxons from New Brunswick, Nova Scotia, and Maine. Owners believed in a reciprocal obligation between employer and worker. For example, during the 1887 depression, Vance kept on logging so his men would not be out of work. Similarly, Noah Falk, owner of the Elk River Mill, provided a bonus to all of his employees after negotiating a major sales contract.[4]

Humboldt mill owners were not always so magnanimous. Faced with a severe economic depression in 1877, they all agreed to shut down until lumber prices rose, throwing three hundred men out of work. When they resumed operations the following year, the owners cut wages by 50 percent or more. Lumbermen also sought to influence their workers politically. Like many businessmen, mill owner William Carson feared the populism of Democrat William Jennings Bryan, and in the 1896 presidential election he promised his employees a raise if Republican William McKinley won.[5]

Lumber barons such as Vance, Carson, and Falk formed the center of Humboldt County's economic, social, and political landscape. As pioneer settlers, they had spent the better part of their lives both in the woods and mills. Such a life, combined with Humboldt's geographical isolation, fostered a deep sense of community within these men that translated into a commitment to the welfare of their employees. [6] This paternalism also stemmed from the mill owners' desire for "a steady, reliable workforce and [a] prominent position in local society."[7]

Although Humboldt's lumber barons occasionally engaged in fractious trade practices, they also abided by informal rules of reciprocity and shared the challenges of new technologies, finances, marketing, labor, and transportation. Just as it required a team of men to cut down a redwood, it took men working together to develop an industry, and no one man could dominate. Or so the pioneers believed. There were more than enough trees to go around, and with demand from Australia, South America, Hawaii, and California, the market boomed. Although rivals, Humboldt's pioneer lumbermen regarded each other as equals and formed both business and social relationships. Humboldt's cooperative ventures, social ties, labor relations, workforce composition, and proliferation of locally owned lumber companies all changed radically with the arrival of A. B. Hammond.[8]

While Hammond had also maintained a paternalistic relationship with his employees in Montana, his entry into the redwood industry in 1900 represented an entirely new way of doing business spawned by the corporate revolution. Many Americans, however, did not universally embrace the move toward the corporate structure. Historian Martin Sklar suggested that the rise of corporate America is best understood not as a "natural process" or the "outcome of economic evolution" but, rather, as a social movement, like populism, unionism, or feminism. Hammond's appearance in Humboldt County and his subsequent conflict with rival lumberman William Carson illustrates the resistance of many businessmen to the corporate model.[9]

As an example of old-school proprietary capitalism, Carson looked the part of the paternalistic mill owner. Over six feet tall, stately, with a well-trimmed beard, he appeared as a "venerable and august patriarch drawn from Biblical scenes."[10] Carson, like Vance, had emigrated from New Brunswick bound for the California gold fields. After two years as a logging contractor, Carson bought a small mill in Eureka and then, in 1863, formed a partnership with John Dolbeer. Eight years later, Dolbeer and Carson were tied with Vance as the second-largest lumber producers in Humboldt County.[11]

In 1885 Carson built himself a lavish Victorian mansion on a bluff overlooking his Eureka mill and waterfront operations. At sixteen thousand square feet, the house still sports an imposing tower and lavish trim in a fusion of Queen Anne and Italian Baroque. Architectural historian James Buckley noted that the Carson mansion "calls up the image of a feudal manor house that marks the center of power on a medieval estate," reinforcing his contention that paternalistic mill owners "often constructed architectural landmarks that they hoped would announce their benevolence to the whole town and remind workers of their obligation to the lumber company."[12]

Carson solidified his reputation among his workers in 1890, when he cut his mill's working day from twelve to ten hours with no reduction in pay. Vance and Evans quickly followed suit. Nearly a thousand surprised workers accompanied a band to their bosses' mansions and gave them a rousing reception. While the local press heralded the ten-hour day as supremely magnanimous, historian Daniel Cornford attributed the mill owners' actions to two factors, neither of which was entirely benevolent. Employers and supervisors themselves wanted to work fewer hours, and the economic rebound in the lumber market provided an opportunity. Furthermore, the ten-hour day had already come into standard practice in the Great Lakes states and the Pacific Northwest, and the Humboldt mill owners likely regarded the move as means of defusing impending labor agitation. Cornford argued that paternalism, while alleviating social conflict, did not necessarily mean subservience on the part of labor. Once they had achieved a ten-hour day, nothing would induce mill workers to give it up.[13]

Carson, however, appeared genuinely altruistic, often donating to many causes and individuals. Not only did he provide Christmas bonuses, but Carson and his wife hand-delivered Christmas turkeys to employees, driving their buckboard from house to house. Perhaps more meaningful, Carson also paid half to full wages to employees incapacitated by injury or illness, in contrast to other lumbermen, who would dock their employees' pay. In tacit recognition of Labor Day, on May 1, 1899, Carson raised wages across the board by five dollars per month. Dolbeer and Carson adopted the nine-hour day, but then a boom in lumber prompted longer hours. Walking through the lumberyard, Carson noticed a group of men gathered around the bulletin board unhappily discussing the notice that read, "Effective July 1, the mill will resume a ten hour day." Carson reportedly "took a blue lumber crayon out of his pocket and in big letters, he wrote 'eight hours straight time, two hours over time.'" And in his final gesture, Carson provided for thirty-three of his longtime employees in his will.[14]

Although far less philanthropic than Carson and notoriously tightfisted, Hammond, during his years in Missoula, had also hosted community picnics, gave Christmas bonuses to his employees, and provided them with credit and loans. Even after relocating to the West Coast, Hammond regarded Missoula as his hometown, and he continued to support both the MMC employees and the community. But sometimes Hammond's generosity required the goading of C. H. McLeod, who beseeched his friend to make contributions to the YMCA and other charities. Also at the urging of McLeod, Hammond permitted an employee profit-sharing plan for the Merc whereby longtime employees could purchase company stock.[15]

Occasionally Hammond received a letter from someone in need, and he usually agreed to help. In granting a job to a young man as a favor to his parents, Hammond let C. H. McLeod know that "he must be given to understand that we are not running a benevolent institution in Missoula, but that you expect him to earn every dollar he receives."[16] Hammond was considerably more sympathetic toward an old man laid up with rheumatism who lived alone in a shack in the mountains west of Missoula. Upon receiving a news of the fellow's plight, Hammond responded by telling McLeod to issue provisions for the old miner. All of these actions stood in sharp contrast to his operations on the West Coast.[17]

Ultimately it was the respective business models of Hammond and Carson that determined the relationship each had with his host community. Carson did business the old way: owning and operating his company and building it from within by his own reinvestments. Although Hammond started out in such a manner with the Blackfoot Mill, his capital now came from his stockholders. The difference was profound. The corporate structure was a new way of organizing business and ultimately society. But at what cost?[18]

During the first half of the twentieth century, the redwood industry displayed a continuum of business models all operating successfully and simultaneously. At one end was William Carson's proprietary capitalism. At the other was Hammond's arch rival, the Pacific Lumber Company, which typified the corporate model of multiple stockholders divorced from immediate management. Hammond, who owned and managed his organization through kinship ties but also tapped eastern capital, embodied an intermediate approach. With all three methods in full swing in Humboldt County, which would prevail remained undetermined. Access to power and wealth, social conditions, geography, legal structures, and even personality factored into both industry consolidation and resistance to corporate concentration.

Lacking a railroad connection, Humboldt County's only link to the rest of the country was by sea. Its isolation helped maintain a system of proprietary capitalism. However, the extraordinary profits emanating from redwood country could not escape detection. With the depletion of white pine in the Northeast and Midwest, redwood became highly desirable. Resistance to rot, fire, and insects made redwood a premier choice for railroad ties and construction lumber, and it was in demand across the globe, especially in the tropics, as nations embarked upon industrialization. Furthermore, a single two-thousand-year-old redwood could produce as much lumber as an entire stand of white pine.

Hammond was not the first to attempt a takeover of the redwood industry. As early as the 1880s, the redwoods had attracted the attention of a Scottish syndicate, resulting in the infamous California Redwood Company. As with the Oregon land frauds that followed, speculators hired dummy entrymen to file claims in Humboldt County under the Timber and Stone Act and bribed government officials. Although Land Commissioner William Sparks canceled nearly two hundred patents, the California Redwood Company fraudulently acquired fifty-seven thousand acres containing $11 million worth of timber. Daniel Cornford attributed Humboldt County's eventual embrace of populism and labor movements to the animosity stemming from the land frauds.[19]

Like much of rural America in the 1890s, Humboldt County proudly espoused the ideology of a democratic republic of independent farmers, artisans, and small-scale entrepreneurs and feared the rise of "monopolies." In 1890 the *Humboldt Times* warned of eastern capitalists cruising the coast looking for timberlands. The paper clearly favored locally owned businesses, stating, "The large land captures by syndicates of foreigners for future sale of lumber markets are not nearly so desirable to the State as more modest ventures." Desirable or not, Hammond's purchase of the John Vance Mill and Lumber Company marked a new era, and conflict was inevitable.[20]

In 1900, five months after forming the Hammond Lumber Company, A. B. made headlines up and down the West Coast when he outmaneuvered Charles A. Smith and acquired the Vance Lumber Company. The *Arcata Union* wondered what this foretold, but within a week a letter to a local newspaper presciently declared, "This is but the beginning of the end" as "lumber lords and barons" consolidated their power. The lumber industry trade journal *Timberman,* noting Hammond's connection with the Southern Pacific, speculated that the Vance sale would lead to an extension of Hammond's Astoria and Columbia River Railroad down the coast to California, eventually connecting with the Southern Pacific. Despite the formidable terrain, this was a reasonable assumption given the close ties among the railroad, Hammond, and the timber industry.[21]

The month after he bought the Vance Lumber Company, Hammond began buying as much prime redwood timberland as possible, most of it from the California Redwood Company. Having transferred their timber assets to the American Lumber Company to avoid government prosecution for timber fraud, the investors in the California Redwood Company were willing sellers. Although fully aware of the ill-gotten lands, Hammond knew he would be immune from prosecution as an "innocent purchaser" and paid $400,000 for 12,760 acres. Wishing to avoid the problems he was encountering in Oregon, Hammond also placed ads in the local newspaper announcing that his company was buying timberland but "title must be clear."[22]

Then in May 1902, Hammond, in combination with the Merrill and Ring Lumber Company of Michigan, bought 36,314 acres of timberlands on Prairie and Redwood Creeks for $600,000, also from the American Lumber Company. Containing 2.5 billion bf of timber, this was the second-largest purchase of timberland thus far on the West Coast, exceeded only by Weyerhaeuser's deal with the Northern Pacific.[23] In just two years, Hammond had become one of the largest owners of redwood timberland, with outright ownership of 30,000 acres and partial interest in another 40,000. By 1911 the Hammond Lumber Company owned 94,000 acres of redwoods and had attracted the attention of the Bureau of Corporations.[24]

In 1902 Theodore Roosevelt had created the bureau as an arm of the Department of Commerce and Labor. Its ostensible purpose was to investigate corporate consolidation and charges of monopoly in America's essential industries. Historian Gabriel Kolko, however, characterized the establishment of the bureau as a conservative move to assuage the Populists and avert their more radical proposals for the nationalization of these industries. Kolko argued that the purpose of the bureau was "not to enforce the anti-trust laws" but to gather information and provide "a shield behind which business might seek protection." Upon investigating U.S. Steel, Standard Oil, International Harvester, and the meatpacking and lumber industries, the bureau's primary conclusion was the "absurdity of the [Sherman] antitrust act." Instead, the bureau called for a standardized national incorporation law. Not surprisingly, most business leaders supported this typical Progressive proposal.[25]

In undertaking a detailed analysis of the lumber industry's timberland concentration in 1912, the bureau's chief worry was that such concentration would allow a few companies to set prices and control the industry. Concerned that concentration of land ownership would lead to "serious and lasting effects upon the economic, social, and political life of the country," the bureau noted that ownership was most concentrated in the redwood industry, in which the largest corporations held 93 percent of the timber and the top six companies owned 40 percent of the land.[26] Whereas the lumber industry as a whole was notoriously decentralized, the redwood industry was dominated by large corporations, in large part due to the timberland concentration. What the Bureau of Corporations failed to realize, however, was that large, heavily capitalized firms required extensive tracts of land to guarantee a long-term return on their investment. While the bureau suggested individual greed as the cause of timberland concentration, it was more a product of the emergence of corporate capitalism.

Especially offensive to the Progressive reformers at the bureau was the discovery that people profited from speculation "without rendering any corresponding public service." The report decried, "That great profits should be made by artificially withholding a commodity from use is economically wrong; but such a withholding of the timber, for the very purpose of raising prices and increasing their own gains, is precisely the policy that the great timber owners are proposing."[27] Yet for the lumbermen and the land itself, overproduction was the great evil to be avoided. The bureau thus contradicted another branch of the federal government, as Gifford Pinchot's Forest Service promoted the creation of more national forests, in part, to take timberland out of production and thereby reduce the glut on the market.

Overproduction plagued the lumber industry, as it simultaneously drove down prices and encouraged increased cutting to meet ever-more-marginal profits. Unlike petroleum, copper, or iron ore, timber was widely available and easily accessible. It required minimal capital to set up a sawmill, and small producers could supply local or regional needs competitively. With little invested, small producers could shut down during a slow market. In describing nineteenth-century mills in Oregon and Washington, Norman Clark wrote, "There were hundreds of mill owners around the two states to whom even simple bookkeeping was an esoteric art. Their manufacturing plants were simply a series of rusty saws in leaky shacks, powered by belts from steam engines, locked in a dreary rhythm of overproduction and close-down."[28]

New technologies, however, simultaneously increased production and required greater capital investment. In shifting from proprietary to corporate ownership, the new, larger firms issued bonds to raise capital to purchase the latest machinery. Companies now had fixed costs they needed to meet, which fed overproduction. As Andrew Carnegie pointed out, it was better to lose one dollar by running at full production than to lose two by shutting down and defaulting on payments.[29]

While Hammond engaged in land speculation, he also invested heavily in lumber mills. Although well aware of the dangers of overproduction, Hammond thought he could outproduce his competitors to gain control of the redwood industry. In 1902 he spent $200,000 installing new machinery and increasing the capacity of his new flagship mill in Samoa, California, from twenty-five to eighty million board feet per year. This required shutting down the mill for six weeks, but to avoid a slowdown Hammond leased the Bayside Mill in Eureka and used his Samoa mill workers to run the night shift.[30]

Hammond's production practices—running his mills day and night and logging continuously through the year to ensure a steady stream of lumber—caught other mill owners by surprise. Traditionally, West Coast lumbermen closed their mills for repairs in the fall and ceased logging during the rainy winter months. Their pause in production for repairs and winter weather cost them dearly, for upon restarting they discovered they had no new orders waiting for them because Hammond had captured their business. Just as he had done in Montana, he took orders regardless of his supply, ramping up production or contracting out if necessary. Justifying his winter closure, the manager of the Elk River Mill, Irwin Harpster, declared that with the "lumber market in such an unpromising condition. . . . We do not feel that it would be prudent to begin manufacturing lumber until we have orders to begin."[31] When the Elk River Mill started back up in mid-April, Harpster worried that Hammond's continuous production would lead the other mills to do the same, leaving the industry "in a demoralized condition" as a result of overproduction.[32] Indeed, by 1905 Hammond's Vance Redwood Company (formerly Vance Lumber Company) had eighteen million board feet of lumber on hand in its Samoa yard.[33]

Hammond's year-round logging operation also took its toll on his workers and did little to ease rising labor tensions. The spring rains made the hillsides so slippery that the logs hit the landings out of control, worsening already-dangerous working conditions. Moreover, the nearly continual rain kept the men in poor sprits. Nor did the millworkers have it any easier. In stark contrast to Carson, Hammond worked his men twelve to thirteen hours a day for over a month. Finally, recalling the customary ten-hour day, the men simply stopped working after ten hours, citing exhaustion and illness.[34]

Following the upgrade in 1903, Hammond had a fully modern and mechanized mill at Samoa. Equipped with two double band saws, the mill cranked out 250,000 bf per day, the largest output in the county, and employed 875 men. With a mechanized system of rollers, chains, and sixteen trimming saws, humans did not even handle the wood until it was spit out for them to separate the grades for shipment. The dozen drying kilns could handle 100,000 bf per day, while the expanded docks at Samoa allowed for eight ships to be loaded at a time. Sixteen steam boilers fueled by a sawdust- and wood-powered 750-horsepower engine drove the machinery, while an electric plant furnished lighting. Hammond estimated the improvements would increase production to 100,000,000 bf annually.[35]

Hammond began to integrate vertically from raw materials to finished product. In late 1900 he built a sash and door factory next to the Samoa mill that produced a thousand redwood doors per day. When the door factory burned down two years later, Hammond simply rebuilt it and doubled its capacity, making it the largest on the West Coast.[36]

After an unprecedented run of eighteen months of continuous operations, Hammond had to close the Samoa mill for repairs and lay off the mill employees for a month. He took this opportunity to update equipment, doubling his lumber output capacity. While new technologies increased mill capacity, they also raised fixed production costs, which in turn created a demand for more trees. Thus Hammond needed to expand his logging operations by opening new camps and building logging railroads deeper into the forests, decreasing the ultimate source of his wealth, the trees themselves.[37]

While he ramped up production, Hammond plotted to take over the redwood industry. In June 1901, Hammond and Hugh Bellas, an international lumber tycoon, announced the formation of a combination that would consolidate and control the industry. Capitalized between $15 and 20 million, the combination planned to acquire more than a million acres of redwoods in Humboldt, Mendocino, and Del Norte Counties. Hammond boasted that he "practically" controlled "the lumber business of California" and would "acquire all new companies that may be started."[38] Unfortunately for Hammond, the plans with Bellas disintegrated, forcing him to seek financing elsewhere.

Hammond's rapid ascent in the redwood industry instigated a rash of acquisitions by outside capitalists. Sensing profits, in 1901 the Santa Fe Railway became "very anxious" to invest in the industry. Proposing a giant railroad/lumber combination, Hammond approached both the Santa Fe and the Southern Pacific. In his negotiations, Hammond suggested the combination acquire vast acreages of timberlands in Humboldt County, along with all the area's railroads, and connect these with the main lines to the south.[39]

The roadblock to Hammond's plan was his primary competitor, the Pacific Lumber Company. After a fire destroyed Pacific's mill in Scotia, in 1895 the company rebuilt, constructing the world's largest redwood mill, only to be surpassed by Hammond's upgraded Samoa mill a few years later. Scattered investors held Pacific's stock until 1900, when Charles Nelson, who already owned stock in four other Humboldt mills, and Detroit lumberman Simon Jones Murphy consolidated ownership. With management divorced from ownership, Pacific Lumber epitomized the corporate lumber company. For the next half century, the Hammond and Pacific lumber companies vacillated between rivalry and collusion and mirrored the dance between their respective allies, the Southern Pacific and the Santa Fe. But at the turn of the century all was in a state of flux.[40]

After the misadventure with Bellas, Hammond attempted another industry coup d'état. With financing from E. H. Harriman, president of the Southern Pacific, Hammond dispatched his secretary William Gosslin to buy as much Pacific Lumber stock as possible. By November 1902, Hammond had acquired 41 percent, enough to install himself and Gosslin as directors of Pacific Lumber—but short of a majority interest, a fact that would soon be critical. Shortly thereafter, Gosslin began acting "rather strangely," and suspecting treachery, Hammond accepted his resignation.[41]

Hammond's suspicions appeared well founded when his negotiations with the Santa Fe proved fruitless and the railroad instead opted to buy the majority stock in Pacific Lumber that Gosslin had failed to acquire. Backed by the Santa Fe, agents of Pacific Lumber bought timberlands and railroads in the southern end of Humboldt County. Turning to Harriman, Hammond implored the railroad executive to move quickly to take control of the redwood industry. Harriman, however, refused to have anything to do with his major competitor, the Santa Fe.[42]

Unable to acquire the majority interest of Pacific Lumber, in January 1903 Hammond sold his Pacific stock to the Santa Fe Railway for $250,000. He made a tidy profit but spent the next three decades regretting his decision to sell. Meanwhile, Harriman continued to dither. Frustrated, Hammond noted, "His delay has enabled our competitor to accomplish what we had expected to do." Giving up on Harriman, Hammond resumed his timberland-buying spree despite the advancing prices.[43]

Abandoning his effort to form a giant combination, Hammond attempted another tactic: colluding with Pacific Lumber to set prices and restrict production. Such efforts met with a mixed response. The *Los Angeles Times* attacked the "Redwood Lumber Trust" for pushing prices "25% higher than what they should be," while the *Humboldt Times* defended redwood producers as having to charge higher prices to overcome past depression and high labor costs. The *Humboldt Times* noted, "For nearly 10 years our mills have been running without a profit" because of cutthroat competition and "the Democrats."[44] The political jab notwithstanding, cutthroat competition was precisely what the industry sought to overcome, but whether this would occur through trusts, giant corporations, trade associations, or government regulation remained an open question.

The corporation (or trust) question lay at the center of the nation's political and economic debate from the late nineteenth to the early twentieth century. Would America continue to be a republic of independent proprietors engaged in an open market, thus upholding the democratic tradition, or were property and the market to be reorganized along the lines of the unfolding corporate model, which consolidated economic and political power? Even the Supreme Court split over the issue, seesawing back and forth for more than twenty years in a series of decisions regarding the Sherman Antitrust Act of 1890.[45]

As Hammond tried to take over Pacific Lumber, others also sought to consolidate the redwood industry. In 1903 lumberman Henry Jackson and Charles Nelson, a steamship magnate interested in guaranteeing cargos for his ships, formed the Northern Redwood Lumber Company, and bought the Korbel and Riverside mills. Northern Redwood instantly became Humboldt's third-largest producer behind Pacific Lumber and Hammond's Vance Redwood Company. Wasting no time, Hammond lined up the top redwood producers in a cooperative venture to control redwood prices and production.[46]

With both the Northern Redwood and Pacific Lumber on board, the next largest firm in Humboldt was Dolbeer and Carson, so Hammond invited the venerable William Carson to join his project. On their first meeting, Hammond attempted to beguile Carson with his accomplishments and "sharp business dealings." To demonstrate his business connections and assert his place among America's elite, Hammond presented Carson with a favorable letter he had received from C. P. Huntington. Clearly, Hammond wished to convey that he was the lumber equivalent of the railroad tycoon. However, Hammond's egotistical and pushy manner offended Carson, who was appalled at the idea of collaborating with someone he viewed as unscrupulous. Furthermore, Carson favored proprietary capitalism and was suspicious of Hammond's new corporate approach. The pious Carson thus declined Hammond's offer to become one of the "big four" redwood producers.[47]

Carson's rejection marked the beginning of a deteriorating relationship that put the two men at "sword's points" for the rest of their lives. John Hanify, the San Francisco partner of the Elk River Mill, worried that "should a fight begin [between Hammond and Carson], it is hard to say where it will end."[48] Hanify had good reason to worry, as a battle between the giant lumber barons could easily crush the smaller Elk River Mill.

Although Carson refused to join Hammond and subsume control of his own company to that of another, he was open to cooperative agreements among lumber companies. As early as 1854, Humboldt mill owners had banded together to regulate prices and production. Realizing that ruthless competition threatened their profits, trade associations attempted to infuse stability and order into wildly fluctuating markets. But as membership was voluntary, these trade associations waxed and waned with market conditions. Attempts to extend trade associations across the industry repeatedly failed. Sooner or later one or more of the producers would bolt and undersell the others, and the agreements would crumple. After years of depressed prices and overproduction, the incentive for cooperation vanished. Mills engaged in unrestricted competition when the wholesale price of lumber increased and demand exceeded supply.[49]

Both Carson and Hammond recognized the perils of unrestrained competition, but they responded with two very different strategies based on two different ideologies. Carson sought cooperation by independent producers through trade associations—in this case, the Humboldt Lumber Manufactures Association (HLMA). Hammond, in contrast, saw corporate control of the industry as the solution. Stymied for the time being on outright ownership, Hammond attempted another tack to gain control of the redwood industry through the HLMA, applying both the force of his personality and economic might. The HLMA, by negotiating export sales and allocating quotas to each producer, enabled smaller operators to stay in business. As Hammond would demonstrate, however, large corporations could manipulate the trade associations to their advantage and drive out the smaller producers.[50]

Hammond's first step was to increase his "former small quota" that the HLMA had assigned him. Well aware of Hammond's intentions, Carson noted that "his object is to run business on the Bay" and feared "he will be a difficult man to control." Indeed, before long, Hammond had pressured the association into granting him one-third of its export orders, allowed him to ship lumber on consignment, and charged him less than others for port towing fees.[51] Growing increasingly frustrated, Carson repeatedly complained that the HLMA "has almost been turned over to Hammond" and threatened to withdraw unless the directors took "steps to bring Hammond in line."[52]

Dolbeer and Carson was Humboldt's fourth-largest producer, so its withdrawal from the HLMA would have had serious repercussions. The company was large enough to operate independently, negotiate its own export contracts, and undercut competitors.[53] Furthermore, Carson was widely respected in Humboldt County and in the industry in general, so he could align his own forces against Hammond's hegemony. On the other hand, membership in the HLMA provided stable prices and a guaranteed slice of the export market. Ultimately, Carson was committed both to the Association and to abiding by its ground rules. Hammond, however, faced no such constraints. He used the association to get export contracts and then undermined it by selling below the set price.

While this reduced his profit per unit, he made up the difference by boosting production, which ultimately increased his business. Since the HLMA lacked any legal standing, Carson could do little but complain. Indeed, price fixing and combinations such as the HLMA were of dubious legality under the Sherman Antitrust Act.[54]

Irwin Harpster of the Elk River Mill noted the emerging discrepancy between producers. He wrote to his partner, Hanify, "It has looked very strange to us for some time that some of the mills should have an abundance of orders and other mills very little business." Harpster recognized Hammond's de facto control of the HLMA and that Elk River needed to negotiate with him for a percentage of the quota.[55]

Small mills like Elk River faced a conundrum. They were hard pressed to produce and deliver on large orders, preventing them from taking on more business. "The amount we have to furnish staggers us," noted Harpster. Having to decline orders caused small mills to lose business and eventually shut down for lack of customers. The quota system provided a solution whereby large orders, usually from foreign railroads, were divvied up among the mills. In contrast, the size and financial resources of the Vance Redwood Company allowed Hammond to produce continuously, stockpile lumber, build greater mill capacity, or even purchase lumber from other producers if necessary.[56]

As the market began to slow in 1904, Harpster hoped to avert a collapse in the redwood industry. He noted "considerable uneasiness among the manufacturers" and regretted that they had "so little faith in each other." This was precisely the problem with trade associations: they were based entirely upon the good faith of all producers to hold prices steady. Indeed, anticipating a lull in the market, Hammond sold a considerable amount of lumber well below the association's agreed-upon wholesale price. Worried that such action would have "a demoralizing effect on the business," Harpster lamented, "unfortunately the innocent have to suffer with the guilty."[57]

Just as Hammond had strategically blocked up timberland to isolate small owners so they had little choice but to sell their timber to him, he used a similar strategy to isolate smaller mill owners to go along with his marketing schemes. Carson, however, proved more difficult. Even into his eighties, Carson stubbornly, perhaps heroically, resisted Hammond's attempts to control Humboldt's redwood industry.

In 1908 Carson headed off an attempt by Hammond, Northern Redwood, and Pacific Lumber to move the office of HLMA to San Francisco. Fearing they would lose what influence in the association they still had, Carson and Noah Falk again threatened to withdraw, and the "big three" relented. A year later, however, they paid for their intransigence when the "big three" cornered the export railroad tie business and denied other HLMA members their quota. Carson correctly predicted, however, that demand would exceed supply and that Hammond would have to contract with Dolbeer and Carson to supply some of their tie orders. Previously, Carson had wryly admitted that receiving such contracts "tends to soften our feeling toward him." Eventually, Hammond's attempts to control the HLMA unraveled as the "big three" got into a fight over exports to Australia—in large part due to personality rather than business conflicts.[58]

Most lumbermen wished to avoid overproduction, but the need to meet bonded indebtedness to cover large capital investments in new machinery and railroads forced them to continue to crank out more boards. In addition, many financial lenders were leery of the volatile lumber industry, which put producers on even shakier ground. But for Hammond, increased production was a means to another end: the elimination of competition. Because of his financial prudence, Hammond rarely became overextended and had little need to issue bonds. Furthermore, unlike other lumbermen, Hammond could draw upon the Missoula Mercantile Company and its "gilded" credit rating for credit and cash, and he often did so to pay off short-term loans or to buy more timberlands. Thus, Hammond's substantial financial resources allowed him to undercut prices while his ramped-up production could eliminate smaller producers, giving him an even greater share of the market.

The conflict between Hammond and Carson mirrored the dispute between proprietors and corporations that dominated the national discourse from 1890 to 1920. The resolution of this debate, historian Martin Sklar maintained, was "the single most important issue in the nation's politics" during this time. Indeed, these three decades marked a furious deliberation within the Supreme Court over the Sherman Antitrust Act and the role of the corporation in American life. In his voluminous history of the American lumber industry, Michael Williams wrote that corporate ownership, holding companies, mergers, consolidations, and vertical integration succeeded where trade associations failed. However, as the dispute between Carson and Hammond demonstrates, the failure of associations and the success of industry consolidation were as much a product of personal ambition, cooperation and coercion, political influence, and power as they were of institutional efficiency.[59]

Dolbeer and Carson's continued existence through the first half of the twentieth century underscores the viability of proprietary capitalism. Dolbeer and Carson readily adopted new technologies, notably with Dolbeer's inventions of the steam donkey and logging locomotive, which revolutionized the industry. Yet the firm maintained its moderate size, avoided mergers, and continued to operate in a manner that embraced paternalism, cooperation, and mutual aid. American business, apparently, could incorporate values other than the bottom line. Rather than an inevitable "natural" outcome based upon efficiency, the rise of the modern corporation proceeded along multiple pathways. Specific circumstances and individuals shaped institutions to guide the process of industrialization, often tailoring the process to fit their own needs and desires. In doing so, they also shaped the world around them.

In articulating capitalism as an economic model, the Scottish economist Adam Smith suggested that individuals pursuing their own economic interests and engaging in free competition would create a stable economic order. While Smith pinned his argument on a basic human quality—greed—he failed to take into account another fundamental human trait: that all humans are, by nature, social beings who interact, cooperate, bargain, and act, both as individuals and as groups. When unrestrained competition began to reduce profits, the free market quickly yielded to collusion and cooperation among producers, prompting the government to intervene to regulate and restrain a "free market," undermining Smith's basic premise. Was monopolistic control of an industry the natural outcome of free market capitalism? A. B. Hammond seemed to think so.

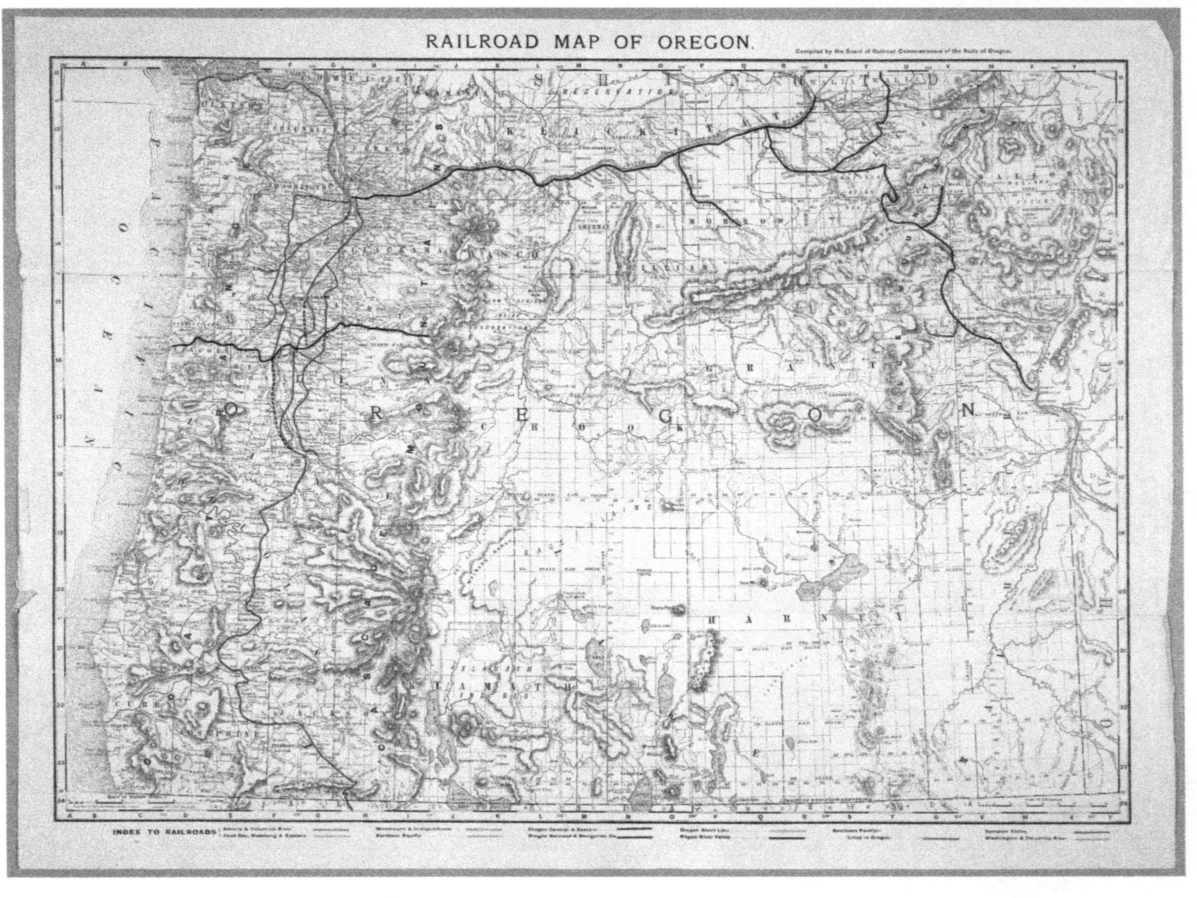

Railroad map of Oregon, circa 1897. Hammond's Astoria and Columbia River Railroad runs from the upper left to Portland while his Oregon Central and Eastern is the line running from the coast to the Cascades. (Huntington Library, empMPOR_007)

After a half-century wait for a rail connection, Astoria celebrated the inauguration of the Astoria and Columbia River Railroad in May 1898. In building the railroad, Hammond envisioned it primarily as a freight line, but passenger traffic proved a mainstay. The weekend service from Portland to Seaside, known as the "Daddy Train," became a hit with the emerging middle class. Here, both passengers and lumber await transport to Portland. (Oregon Historical Society, ba008328)

During the first decade of the twentieth century, lumbermen frantically competed to acquire as much Oregon timberlands as possible. In this timber rush Hammond amassed over 120,000 acres and soon began logging to supply his mills in Oregon and California. Posed here are a foreman (or owner) and two loggers, circa 1905, in the process of felling an old-growth Douglas fir on the lower Columbia. Standing on springboards to avoid the swollen base, the choppers worked in tandem (one right-handed and one left-handed). In this photo a logger reclines inside the cut while the other two men pose on springboards. (OrHi 37858, Oregon Historical Society)

Early-twentieth-century Hammond Lumber Company woods crew standing in a redwood undercut, Humboldt County, California. The world's tallest trees, redwoods can exceed 350 feet in height and contain enough wood to build twenty-two houses. Such trees were relatively common when Hammond arrived in Humboldt in 1900, but by the 1980s less than 2 percent of the redwoods remained, nearly all in state or national parks. (OrHi 35060, Oregon Historical Society)

Hammond log train on Humboldt Bay Flats on the way to the Samoa mill. Note the size of the logs—one per flatbed. (Ericson Collection, #1999.02.0233, Humboldt State University)

The SS *Necanicum.* Hammond owned scores of such lumber schooners to transport lumber and raw logs from the forests of the Northwest to his giant mill in Los Angeles. Specially designed for the shallow harbors of the West Coast, the low-cut gunwales allowed forty-foot logs to be stacked on deck. Hammond would eventually acquire some seventy-two ships, and the "Hammond Navy" would be one of the largest lumber shipping lines (and the only nonunion one) on the West Coast. (Author's collection #1)

A cigar log raft on the Columbia River, ready for towing. An empty cradle to build another is on the far right. Hammond pioneered the use of such cigar rafts on the West Coast. Towed by his lumber schooners, these rafts allowed him to greatly multiply each shipment. Bound together with anchor chains, these rafts ran up to 700 feet long and 55 feet wide and contained 5 to 10 million bf of timber. (OrHi 11436, Oregon Historical Society)

This Brawley, California, lumberyard (circa 1908) was one of the many Hammond Lumber Company retail yards in Southern California in the early twentieth century. (Author's collection, #2)

The HLC Tongue Point Lumber Mill in Astoria, Oregon, shortly before it was destroyed by fire in 1922. (OrHi bb009693, Oregon Historical Society).

Horse logging in the redwoods. Logging by horse and ox teams caused extensive but localized ecological damage. Use of the skid road (logs laid crosswise) inadvertently minimized soil compaction. Note the bucket used to grease the skid road in the foreground. (Ericson Collection #1999.02.0269, Humboldt State University).

By the early twentieth century, new technology, increased mill capacity, and capital investments encouraged clear-cutting. This photograph depicts a typical early-twentieth-century Hammond Lumber Company operation in Humboldt County. (Swanlund-Baker Collection #1999.01.0128, Humboldt State University)

Steam donkey logging in Humboldt County. Mounted on a wooden platform, the Dolbeer Steam Donkey could winch itself up a mountainside, allowing lumbermen to log previously inaccessible areas. Using a rope (and later, a steel cable), the donkey could haul in logs from any direction. The adoption of steam donkeys rapidly accelerated the pace and scale of logging operations in the early twentieth century. (Swanlund-Baker Collection #1999.01.0196, Humboldt State University)

High-lead logging operation with spar tree and steam donkey. Rather than towing logs across the ground, the use of a spar tree and cable rigging allowed the donkey to haul logs through the air. Incoming logs became "enormous battering rams," uprooting stumps, saplings, rocks, and anything else in their path, including workers. Such scenes led observers to describe this as a "factory without a roof" (see chapter 15). (Palmquist/Yale #2012.02.0186, Humboldt State University)

A typical bunkhouse in an Oregon logging camp, 1894. Such overcrowded conditions gave rise to the Industrial Workers of the World (IWW) and precipitated the great strike of 1907. (OrHi 35971, Oregon Historical Society)

A Hammond Lumber Company beer party in Stella, Washington, 1916. The original caption read "John Welke (German Spy) center back row." More likely he was a labor organizer that the HLC wished to cast as a dangerous subversive. (Author's collection #3)

Billed as the world's largest lumberyard in the 1920s, the Hammond Lumber Company distribution facility at San Pedro Harbor, Los Angeles, processed raw logs from Hammond's operations in Washington, Oregon, and California and shipped finished lumber around the world. (Author's collection #4)

A. B. Hammond at eighty-two in 1930. (73–0003, Archives and Special Collections, Mansfield Library, The University of Montana)

Stand of timber on Hammond Lumber Company (HLC) lands in the Tillamook forest, Oregon Coast Range, circa 1920. This old-growth stand of Douglas fir and hemlock is representative of Oregon coastal forests before logging. Note man's hat at base of large tree in foreground. (OrHi 106035, Oregon Historical Society)

Logging in the Tillamook, 1920s. The World War I demand for Sitka spruce had opened up the rugged Tillamook forest, while high-lead logging allowed for rapid cutting on steep mountainsides. (Dan Strite photo, author's collection #5)

As a result of poor logging practices, the Tillamook forest went up in flames in 1933, 1939, 1945, and 1951. All told, the fires destroyed 355,000 acres of timber and wiped out the HLC's Tillamook operation. (P029:Acc 87:6i, Oregon State University)

The HLC main facility at Samoa in 1947. In 1936 Leonard Hammond became the third president of the HLC. Within ten years he had retired the company's debt, paid all missed dividends, and amassed a $5 million cash reserve. Unfortunately, he died of leukemia in 1946. (Shuster, 2001.02.0151, Humboldt State University)

CHAPTER **14**

Assembling an Empire

Late on Saturday night, April 20, 1901, a crew of men surreptitiously gathered on the Eureka docks. Speaking only in whispers, they quietly unloaded steel rails, a small locomotive, and construction equipment. At one minute past midnight, the men—employees of the California and Northern Railway (C&N)—began laying track along the waterfront. The hammering of steel upon steel awakened the town. It was now Sunday morning and the crew no longer had any reason for secrecy, as A. B. Hammond would be unable to stop them. Any other day of the week, Hammond could have sought an injunction, but by the time court opened on Monday morning, it would be too late as the track would already be in place.[1]

Crowded with warehouses, mills, and wharves, Eureka's waterfront had room for one only rail line. Both the C&N and Hammond's Eureka and Klamath River Railroad (E&KR) desperately sought access, and the two competing railroads were furiously laying track in a race to be first to reach the city. Hammond believed he had the upper hand, as the neighboring town of Arcata had already granted the E&KR a franchise and he had an agreement pending before the Eureka City Council. But then, in early April, the city unexpectedly awarded the contract to the C&N, owned by the Pacific Lumber Company and lumbermen Charles Nelson and Francis Korbel—Hammond's rivals. When word reached him in New York, Hammond was livid and immediately wired the mayor and city council. He then hurried back to San Francisco to seek legal action to stop the C&N, but the track was already laid.[2]

Nevertheless, Hammond persuaded the city to grant him a franchise as well, whether through force of reason, legal arguments, or some other, less legal, inducement. With both railroads laying track into Eureka, an impasse was imminent, and the two companies began negotiations. In a series of meetings over the next few months, Hammond appeared particularly acrimonious and hostile, repeatedly requesting proof that the C&N's lawyer had authority to execute a contract. Finally, after much wrangling, the two sides forged an agreement. Each railroad would give the other one-half interest in its track already laid through town and a right-of-way along each other's track.[3]

Hammond's fears, however, proved well founded. The next year, the C&N denied that it ever authorized an agreement with the E&KR. The renewed dispute landed in court, which, despite Hammond's terse and contradictory testimony, ruled in his favor and compelled the C&N to comply fully with the contract and fulfill its obligations. In the end, the C&N never got off the ground—or under steam, as it were—and leased their four miles of track to another railroad.[4]

Such midnight plotting and legal maneuvering underscored the vital importance of railroads in the development of the American West. Just the promise of a railroad could generate capital. A railroad with a city franchise and a legal right-of-way was often enough to attract investors to fund construction of a line. The situation was especially pronounced in Humboldt County, which lacked inland waterways and a rail connection to the rest of the country. Blocked by rugged terrain, steep canyons and swollen rivers, the redwood region was effectively isolated from the rest of the world except by sea. Railroad capitalists eyed the region's vast natural resources and anticipated fat profits if they could only get there.

Hammond realized that Humboldt County would be the final railroad frontier in the continental United States. This was where the great railroad powers—the Santa Fe and the Southern Pacific—would play out their game for control of California's transcontinental traffic. Gaining access to Humboldt County, the state's largest timber producer, would be a coup for a railroad with imperial ambitions. And so the lumber barons jockeyed for position, intending to sell their paper and half-built

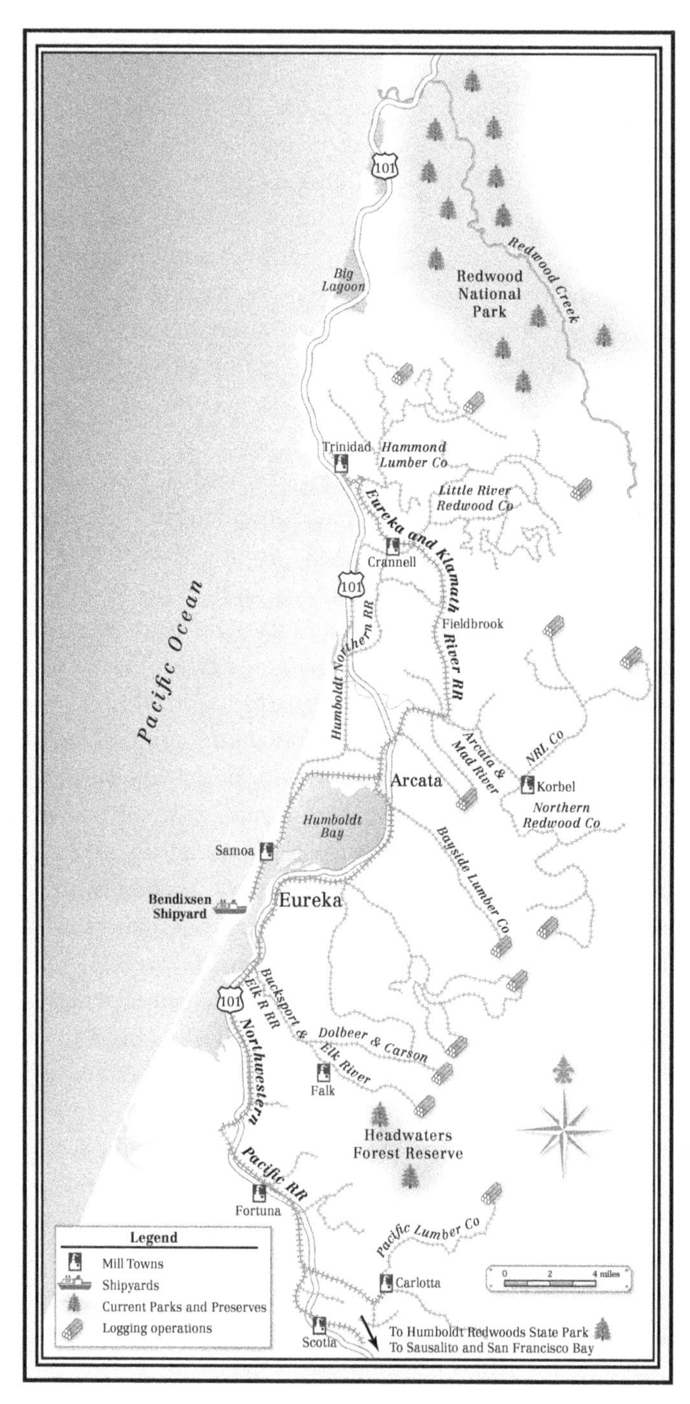

Railroads of the Redwood Empire. Map drawn by Gerry Krieg. Copyright © 2014 by the University of Oklahoma Press.

lines to the highest bidder. In the meantime, the roads could be used to haul timber from the forest to the mill.

When Hammond purchased the Vance Lumber Company in 1900, it included the 18-mile-long E&KR railroad, and he found himself enmeshed in a confusing web of railroad ownerships that merged social, political, business, and family interests. Once again, Hammond turned chaos to his advantage, drawing upon his forceful personality, legal maneuvering, business connections, superior bargaining ability, and knack for playing powerful interests off each other.

With the E&KR as a base, in March 1901 Hammond incorporated the Humboldt Railroad Company, capitalized at $2 million. This was an outrageous sum for a logging railroad and suggests Hammond had "watered" his stock—overvalued the company's assets to lure potential investors—a common Gilded Age practice. Hammond used his charter to propose a line from Dyerville on the Eel River 130 miles to Crescent City as justification for the inflated stock value, essentially banking on a paper railroad. Nonetheless, with spurs up the Eel River, Van Duzen River, Mad River, Little River, and Redwood Creek drainages, it appeared as if Hammond's rails would lace the entire region.[5]

Given Hammond's ownership of two railroads in Oregon and his close ties with the Southern Pacific, industry observers drew the logical conclusion that he had both the intent and wherewithal to run a line clear up the coast, connecting Eureka with his Corvallis and Eastern Railroad in Oregon, despite the formidable topographic challenges. Confirming such suspicions, on May 7, 1903, Hammond announced he would build a railroad from Eureka to San Francisco.[6] It was not only the public Hammond hoped to impress; he also wanted to attract the attention of the nation's two largest railroads: the Southern Pacific (SP) and the Atchison, Topeka and Santa Fe Railway, commonly known as the Santa Fe. In doing so, Hammond had a dual purpose. While he endeavored to sell his Eureka railroad at a handsome profit, he also wanted to gain a rail connection from Humboldt to the outside world so as to reduce shipping costs and better access the national market.

A week after Hammond's announcement of his grandiose plan to connect Eureka to San Francisco, the Santa Fe, the SP's arch rival, began to cobble together a railroad between the two cities. The Santa Fe bought the Eel River and Eureka Railroad for $700,000 and then purchased the C&N and several smaller lines. Brandishing the E&KR as the key that would give the Santa Fe full control of Humboldt County, Hammond put a $1 million price tag on his railroad. Meanwhile, he was secretly bargaining with E. H. Harriman, the new president of the Southern Pacific.[7]

Behind the scenes, Hammond and his partner, Henry Huntington, worked out a strategy. Pretending to drop an insider tip, Huntington told Harriman that Hammond and the Santa Fe were on the verge of an agreement. The ploy worked. With Santa Fe president E. P. Ripley on the way to Eureka to examine the E&KR, Harriman took the bait and bought Hammond's railroad for $1,150,000, sight unseen. Furthermore, Harriman offered cash, while Ripley proposed railroad bonds. Hammond had successfully played the railroad rivals against each other. He not only sold E&KR for a hefty profit but also was granted a twenty-year lease of the tracks by the Southern Pacific so he could continue to operate his logging railroad, which he reincorporated as the Oregon and Eureka.[8]

It appeared as if the Southern Pacific "octopus" was growing a new tentacle to supply its huge demand for railroad ties and construction lumber. Indeed, Hammond sold two of his railroads to the SP for far more than he paid and purchased thousands of acres of Oregon timberlands from the SP for a fraction of their value. Hammond's financial and personal relationship with the Southern Pacific, however, was diluted by C. P. Huntington's sudden death in 1900. As one of the original incorporators of the Hammond Lumber Company (HLC), Huntington owned one-sixth of the HLC stock and actively supported Hammond's enterprises. Although Huntington's widow, Arabella, and his nephew, Henry, inherited the railroad empire, Henry failed in his bid to become the SP president. After Henry and Arabella married, they sold most of their railroad interest to E. H. Harriman. They retained most of their HLC stock, however. Thus, the relationship between the HLC and the railroad was one more of mutual interest than of control. Purchasing railroads from a company in which they owned stock made financial sense to Harriman and Huntington, as did selling railroad grant lands to Hammond so he could exploit the timber resources.[9]

The HLC was hardly a subsidiary of the Southern Pacific, although Henry Huntington sat on the board of directors of both. By 1906 Huntington held less than 13 percent of the HLC stock and Harriman had even less. Hammond insisted on maintaining complete control over his business and disliked having major stockholders, other than himself, on any of his boards. When John Claflin attempted to increase his share of the Astoria Company, Hammond borrowed money from his Missoula enterprises to ensure that he would remain the majority stockholder and retain the capacity to handpick his directors.[10]

Although they operated independently, Hammond and Henry Huntington both invested heavily in the development of Southern California. Huntington's enterprises, such as streetcars and housing developments, helped create the modern Los Angeles. Hammond, in turn, flooded the region with cheap lumber, assumed construction contracts, and invested in hotels and apartment buildings. With such similar interests, both Hammond and Huntington were members of the antiunion Citizen's Alliance, held to a policy of "no concession to labor," fired any union employees, and worked hard to keep organized labor out of Los Angeles. Both men were part of an economic oligarchy dominated by white, Protestant males with shared values and objectives—the nation's "power elite"—who formed a tight social network and presented a united front against labor and any opposition to their supremacy.[11]

Just as railroads restructured the business world, they redefined the timber industry, allowing it to overcome the constraints of nature. With railroads, logging became more dependent on the flow of capital than on the flow of rivers. Before the advent of steam power, the cheapest and easiest way to get timber to the mill was via water; loggers simply piled cut logs into a river and floated them downstream until a boom of logs chained together caught the timbers. For nearly two centuries, lumbermen from New Brunswick to the Great Lakes had used this method with great efficiency. But unlike the large, placid waters of the East, the rivers of the Pacific Northwest tended to be steep, rocky, and choked with debris. Furthermore, the winter storms could quickly turn a creek into a torrent filled with thousands of logs capable of tearing out bridges and booms and washing out to sea, wasting an entire season of logging. And in a dry year, the logs often failed to make it down the river altogether.[12]

By the 1870s the redwood industry had stripped Humboldt's tidewater and river corridors of their trees. As lumbermen pushed farther up the valleys, they built crude railroads—nothing more than horse-drawn wagons and rough two-by-fours for rails. Then, in 1874, John Vance punched a railroad five miles up the Mad River to his new mill. Other lumbermen followed suit, extending lines up the river valleys to reach each timber "show" or logging operation. But the redwood region presented topographical challenges: these logging railroads had to bridge steep gullies, cross tidal flats, or snake up steep mountainsides. Consequently they required substantial capital investments in bridges, trestles, and pilings. Construction costs typically ran from $30,000 to $60,000 per mile.[13]

Nevertheless, the combination of capital investment, potential profits, increased mill capacity, and timberland concentration allowed for the construction of standard-gauge lines capable of hauling huge loads by steam locomotives. Railroads also permitted mills to move away from the bay and toward the source of timber, leading to the rapid expansion of the lumber industry.[14]

Building a permanent main line required greater infrastructure and thus multiplied the costs. In conducting surveys, both the Santa Fe and Southern Pacific discovered that because of the rough terrain it would cost $150,000 per mile, or $15 million, to build a railroad into Humboldt County from Willits, California. While the redwood region could produce significant freight traffic, it would not be enough to justify two competing railroads. Setting aside their rivalry, the Santa Fe and Southern Pacific declared a truce in November 1906. The railroad giants combined their resources to form the Northwestern Pacific (NWP) to serve the Northern California coast. No longer in a race to pick California's last plum, the NWP bided its time and carefully assessed expenses, but the lack of competition delayed an outside rail connection to Eureka until 1914.[15]

The NWP marked a significant shift from cutthroat, Gilded Age competition to twentieth-century cooperation and industry consolidation. Long dependent upon railroads to transport their logs to the mills, Humboldt's lumbermen were now at the mercy of two of the world's largest corporations working in harmony—the Southern Pacific and the Santa Fe. Under such circumstances, Hammond sought to consolidate his hold on the redwood industry by controlling the railroad transportation corridor within Humboldt County.

In 1896 Edgar H. Vance reincorporated his father's old Mad River Railroad as the E&KR and signed a ten-year contract with Dolbeer and Carson to haul logs from their logging camp near Fieldbrook to Humboldt Bay. But when Hammond took over the E&KR, he refused to renew the contract, forcing Dolbeer and Carson to close its logging camp and throwing 120 men out of work. At great expense, Dolbeer and Carson had to build its own line (the Humboldt Northern) paralleling the E&KR. Hammond piled on an additional injury by refusing to grant Carson's railroad a crossing so that it could reach Humboldt Bay. Carson, in turn, brought a condemnation suit against Hammond's railroad. The animosity between the two even carried over to their employees as recurring brawls broke out between the rival railroad workers.[16]

Although Hammond had sold the E&KR to the Northwestern Pacific, he maintained de facto control over the tracks through his lease, a situation that frustrated his competitors. By October 1906, the Little River Redwood Company had acquired timberlands not far from Hammond's extensive holdings and wanted to begin logging but lacked a means of transport. Less than a mile away, the E&KR line presented an option. Although forewarned of Hammond's business practices, Levi Crannell, an officer of the Little River Redwood Company (LRRC), met him in Eureka to negotiate an agreement to haul timber from its logging camps.[17]

Six months later, Crannell was still without a formal contract, but Hammond kept putting him off, citing the inability of the directors of the NWP to obtain a quorum. By May 1907, an impatient Crannell complained the delay was causing "serious inconvenience." Although Hammond continued to haul the LRRC timber on his railroad, the lack of a long-term formal contract precluded the company from investing in building a mill and logging camps. While Crannell threatened to withhold his business, Hammond continued to string him along, assuring him that the LRRC had a "good lease" and needed only Harriman's approval.[18]

Despite his conciliatory tone, Hammond's delaying tactics were readily transparent. The LRRC possessed the best timberlands in Humboldt County, and Hammond hoped to leverage his control of the transportation corridor to keep the LRRC from developing its timber before he could buy those lands. But Hammond's obstructionism directed toward Crannell and Carson backfired. When Hammond informed the LRRC that the NWP had finally approved the agreement, it was too late; Crannell had already contracted with Carson's new Humboldt Northern to haul lumber for eighty cents per thousand board feet, 35 percent less than what Hammond charged. Apparently Crannell found the proprietary capitalism of William Carson much more efficient than the corporate bureaucracy of the "managerial revolution" that A. B. Hammond embodied.[19]

Although he was willing to do it to others, Hammond had no intention of being held hostage over transportation rates. Building his own railroads enabled him to get the trees out of the woods and to the mills, but Humboldt still lacked a rail connection that could transport finished lumber to the outside world. Furthermore, shipping costs factored heavily in the cost of lumber. At the turn of the century it cost about $7 per thousand board feet (mbf) to turn a redwood tree into lumber, then about $4.50 per mbf to ship it to San Francisco by sea, plus additional rail costs to the eastern market. Redwood's selling price of $13 per mbf made for a very tight profit margin that could be justified only by volume, thus favoring larger companies. While lumber prices fluctuated on the open market, shipping costs remained steady, despite improved technologies. Clearly, Hammond needed to address transportation.[20]

Given the lack of outside rail connections to the region, West Coast lumbermen used ships to get their product to market. While Hammond, primarily a lumberman, moved into shipping, others, like Charles Nelson, used their lumber mills to provide cargo for their shipping empires. Still others, like William Carson and the Weyerhaeuser Timber Company, contracted out their shipping needs. More than anyone else at the time, Hammond integrated transportation and marketing as part of the lumber production process and soon controlled everything from raw materials to retail sales and construction contracts.

With his purchase of the Vance Lumber Company in 1900, Hammond acquired four sailing schooners, but within the year he sold them to build a fleet of steam schooners. Although most steamers of the time were coal-fired, Hammond, taking advantage of California's low oil prices, favored oil-burning ships. The first of these was the *Arctic,* which had a lumber capacity of 325,000 bf. Hammond would eventually acquire some seventy-two ships (not all were owned simultaneously), and the "Hammond Navy" became one of the largest lumber-shipping lines (and the only nonunion one) on the West Coast.[21]

Shortly after purchasing the *Arctic,* Hammond contracted with Henry Huntington's shipyard in Newport News, Virginia, to build a giant lumber carrier. With a capacity of 1.5 million bf of lumber, the *Francis H. Leggett* would be the largest ship to enter the Pacific lumber trade at the time. His fellow lumbermen referred to it as "Hammond's Folly," as it could enter only deep harbors, which were in short supply on the West Coast. But Hammond had bigger plans for his flagship.[22]

The *Francis H. Leggett* arrived on the West Coast in 1903 and began to carry wheat from Portland as well as lumber from Hammond's mills. In 1905 Hammond's two ships netted $62,000, greater than the combined profits of his Curtiss Lumber Company in Oregon and his operations in Southern California. Clearly shipping was a highly profitable enterprise, and Hammond ordered more vessels.[23]

Hammond's timing was impeccable: the disastrous fire that followed the San Francisco earthquake of 1906 resulted in a massive rebuilding effort, and the increased demand for construction materials pressed every ship on the West Coast into service. All along the West Coast the shipbuilding industry took off as shipbuilders recognized that Douglas fir—strong, lightweight, and available in great lengths—was just as good as East Coast oak for ship construction. Hammond leased the shipyards adjacent to his Samoa mill and entered the shipbuilding industry himself. Over the next ten years he built thirteen ships there.[24]

With the *Francis H. Leggett,* Hammond initiated a long line of ships named after his business partners, including the *General Hubbard,* the *Edgar H. Vance,* and the *George W. Fenwick.* In 1910, beginning with the *Nehalem* and *Necanicum,* Hammond started naming ships after places from which he had extracted significant wealth. This trend continued with the *Astoria, Eureka, Samoa, Santiam,* and, of course, the *Missoula,* the largest of them all. Although he continued to build wooden ships in Samoa until 1918, his larger ships required steel hulls and were built elsewhere. All flew the Hammond flag: a blue *H* set in a white diamond, an insignia that also appeared on the smokestacks.[25]

All of these ships were specially designed lumber schooners peculiar to the Pacific coast. Modeled on the old sailing schooners, with low-cut gunwales that allowed forty-foot logs to be stacked on deck, these ships were often so heavily laden that the water nearly reached the rails. Their broad, flat hulls provided a larger load capacity and allowed them to enter the shallow harbors of the West Coast. Many of Hammond's steam schooners also boasted twin booms, permitting easy loading and unloading.

As Hammond increased his mill production to provide both domestic markets and exports to China, Australia, and South Africa, he boosted his shipping capacity. In 1912 Hammond's fleet earned more than $200,000. The following year he launched the largest lumber carrier on the West Coast to date; costing $300,000, the *Edgar H. Vance* was 308 feet long and could carry 2.5 million bf of lumber. Even with the lumber industry in a slump in 1916, Hammond noted, "We have passed through the worst year that we ever experienced in the lumber business; still the results are better than we expected. This is due to the operation of our steamships." So much for "Hammond's Folly."[26]

Although ships allowed Hammond to transport lumber at a low cost, he found an even cheaper way of moving his product to market. On the St. John River in his native New Brunswick, lumbermen had long constructed log rafts composed of several "joints" of up to one hundred logs chained together, and they floated these downstream to port. In 1884, Hugh Robertson, an established sea captain, attempted to tow one of these rafts from Nova Scotia to Boston. Running low on coal, his tug cast loose the raft and headed to the nearest port. A few months later, the raft was discovered intact off the coast of Norway, demonstrating the viability of the concept. After moving to the Pacific coast, Robertson experienced several more false starts before linking up with A. B. Hammond.[27]

Robertson, who was as determined and stubborn as Hammond, continued to improve the design of the log raft. Using a floating cradle, Robertson fashioned rafts into a cigar shape, with a rounded bottom and tapered ends, thus increasing the draft so it could withstand the ocean's lashing. In 1898 Robertson convinced Hammond to give his new design a try. Hammond recognized the economic potential: log rafts could greatly reduce shipping costs. In 1901 he purchased a half interest in the Robertson Raft Company, renaming it the Oregon Rafting Company.[28]

For years Hammond had steadily acquired timberlands on both sides of the Columbia River. Timber flowed off his lands in the Oregon Coast Range toward Astoria, where his mills processed it and from which his Astoria and Columbia River Railroad could transport it east. But on the Washington side, Hammond's timberlands were more fragmented and often interspersed with Weyerhaeuser holdings. Furthermore, since James J. Hill controlled the railroads in Washington, Hammond found it easier and cheaper to tow the logs to California, where he could sell them as piling. Rapid worldwide industrialization had created a market for pilings to build trestles, docks, and port facilities up and down the West Coast as well as overseas. These pilings required long logs, up to one hundred feet, with little to no taper, requirements best met by the stately Douglas fir of the Pacific Northwest.

In a slough off the Columbia River near Stella, Washington, Hammond and Robertson set up a facility to build cigar rafts. Bound together with anchor chains, these rafts ran up to 700 feet long and 55 feet wide, containing 5 to 10 million bf of timber. Using his biggest ships, primarily the *Francis H. Leggett* and *Edgar H. Vance,* to tow the rafts, Hammond could greatly multiply each shipment of raw logs from his operations on the Columbia to Southern California.[29]

Most lumbermen were unwilling to chance such technology, and Robertson and Hammond enjoyed a monopoly for five years, until Simon Benson began assembling rafts across the Columbia River from Hammond's operation. In response, Hammond and Robertson built the largest raft of all time. At 60 feet wide and 835 feet long, it contained 11 million bf, enough to build a thousand houses. In just six days, the *Francis H. Leggett,* laden with her own 1.5 million bf of lumber, towed the raft to San Francisco. But after several successful years of rafting, in 1910 one of Hammond's rafts broke up off Point Reyes, and a few months later he lost another off Astoria. Valued at $60,000 each, these represented substantial losses to the company. Furthermore, ocean storms scattered the logs up and down the coast, creating a significant menace to shipping. Thus in 1912 Congress introduced legislation banning ocean rafting. The bill failed, however, and Hammond continued the practice until 1922. Despite occasional losses, in the long term rafting saved thousands of dollars in shipping costs. Over twenty years, Hammond shipped fifty-three rafts from the Columbia River to San Francisco and lost only three.[30]

After securing transportation, the next steps toward achieving a vertically integrated lumber company involved distribution and retail sales. Until the twentieth century, lumber sold itself, since it was needed for everything from building bridges, houses, and railroads to manufacturing railroad cars, wagons, barrels, and even cigar boxes. Nearly every lumber producer on the West Coast simply dumped the product on the wholesale market in San Francisco. In the national market, however, consumers still favored white pine, so to promote their product, redwood lumbermen began an aggressive marketing campaign linked with a national distribution system. In 1903, California's largest mills, including the Pacific Lumber Company, Union Lumber, Northern Redwood, and Hammond's Vance Redwood, formed the Redwood Manufacturers' Company. Notably absent from this venture was the old proprietary firm of Dolbeer and Carson. To coordinate lumber shipments, the Redwood Manufacturers' Company purchased a large storage area in Pittsburg, California, just east of Oakland. Located where the Sacramento River entered the San Francisco Bay, the site offered deepwater access for ships and transcontinental rail connections to both the Southern Pacific and the Santa Fe. Furthermore, Pittsburg was beyond the influence of San Francisco's powerful labor unions.[31]

Lacking the leverage of large corporations, San Francisco's firms, which competed in a tight market, largely acquiesced to union demands. By 1900 San Francisco's Building Trades Council represented fifteen thousand construction workers who refused to handle nonunion building materials. As if it suffered from a contagious disease, Hammond told his friend, C. H. McLeod, "This City is still in the throes of Labor Unionism. The business men seem to have surrendered unconditionally to the labor unions."[32] Indeed, the City Front Federation strike of 1901 had nearly paralyzed San Francisco in a battle between the Employer's Association and sailors, teamsters, and longshoremen. It was no surprise, then, that lumbermen, who were ardently anti-labor, began to look beyond San Francisco to establish their shipping yards and wholesale facilities. In 1903 Hammond leased thirty-nine acres at San Pedro harbor from the City of Los Angeles and Southern Pacific Railroad to construct a massive distribution center.[33]

Hammond envisioned much more than a warehouse at San Pedro. His rival, William Carson, picked up on Hammond's intentions, observing that his "mill at Samoa manufactures lumber pretty fast which forces him to make an outlet for it."[34] Carson astutely noted that large-scale production required large-scale consumption. As both Hammond and Carson were well aware, the lumber industry suffered from chronic overproduction, which decreased prices. Carson and other manufacturers attempted to address the problem by curtailing production industry-wide; Hammond's solution was to move into retail, and the booming Southern California market provided an ideal opportunity.

While other lumbermen—notably A. M. Simpson—had previously established retail yards, Hammond was among the first to establish an aggressive sales and marketing arm as an integral part of his operation. Just as he had redefined lumber manufacturing, so did he modernize lumber marketing. Hammond's emphasis upon retail sales resulted largely from his long association with the Missoula Mercantile Company, which he continued to rely upon for cash loans and its "gilt-edged" credit rating with New York financiers. Moreover, merchandising lumber was nothing new to Hammond. When Marcus Daly pulled his lumber contract in 1889 in retaliation for Hammond's political activities, Hammond changed his operating plan. Instead of cutting mine timbers, he began making building materials, such as two-by-fours, sheathing, and roof boards. In addition to setting up retail yards along the Northern Pacific route from Idaho to North Dakota, Hammond assembled a fleet of salesmen to market his lumber, an unusual move for a nineteenth-century lumberman in the far West. Retail sales proved similarly indispensable to the success of the HLC in the new century.[35]

In Montana, Hammond made a fortune off the combination of lumber and real estate development, and he recognized the potential for the same in Southern California. As the real estate market there began one of its many booms, in the beginning of the twentieth century Hammond purchased twenty acres in Los Angeles from Henry Huntington for a lumberyard. Meanwhile, Perry Whiting, a thirty-two-year-old former logger sporting a crew cut, jug-handle ears, and an "innocent hayseed look," had already established a retail lumberyard in the city and found his business growing rapidly. As he expanded his operations into the Imperial Valley, Whiting needed to keep a large amount of lumber on hand, which he bought on credit from the Caspar Lumber Company. When the head salesman from the HLC discovered this, he struck a deal with Whiting to make HLC his exclusive supplier in exchange for a $100,000 line of credit. In a single stroke, not only did Hammond glean 6 percent interest off Whiting's credit, but he also eliminated competition and found an outlet for his product. Furthermore, if Whiting should default, Hammond would own his retail yards outright. Within two years, Whiting had racked up $70,000 in debt, and Hammond showed up in person to collect.[36]

Impeccably dressed, as always, Hammond strode into Whiting's lumberyard, frowned at the stock on hand, and asked him how much he owed. When Whiting told him $100,000, Hammond said, "Well, you may not realize it, but you are broke." Whiting disagreed, stating he had that much in assets. Hammond replied, "You owe me $70,000, and if I should demand that money you would have to close your doors, as you could not pay." Whiting reminded Hammond that he could not demand payment until the notes came due and when they were, he would have the money to pay them. Failing to intimidate Whiting, Hammond glared at him with "his piercing gray eyes for at least five seconds, then said, 'You are a smart young man, and I hope you come out all right; as long as you pay your notes when they are due, we will give you credit.'"[37]

But the next year, Whiting was $125,000 in debt, and Hammond returned, this time with a proposal: Hammond would pay off Whiting's debts, buy his lumberyards, form the National Lumber Company, and allow Whiting to buy one-fifth interest in the new organization and become manager for a salary of $500 a month. Backed into a financial corner, Whiting assented, and Hammond wrote him a check for $265,000, out of which Whiting paid off his debts and invested $50,000 in the new firm.[38]

Hammond's accomplishments during the early twentieth century were remarkable considering that for seven months, from October 1905 to May 1906, he suffered from severe rheumatism and was largely confined to his bedroom. During this time he visited several hot springs resorts in California seeking alleviation, but with little result. Despite the nagging and often acute pain, he conducted business hobbling around between his bedroom and office on the third floor of his four-story mansion in San Francisco. In the spring of 1906, his condition worsened to the point that he scarcely left his bed.[39]

Then, just before dawn on April 18, a terrific rumble shook Hammond awake. Temporarily forgetting his condition, he ran into the street in his nightclothes, joining thousands of others as buildings began to crumble, the result of San Francisco's infamous earthquake. As downtown burst into flames, George McLeod rushed to the Merchants Exchange Building to save the company records. Weaving in and around those killed by the collapsing wreckage, he retrieved the most valuable records before the building caught fire that evening. For the next two years, the HLC operated out of the third floor of Hammond's house while the Merchants Exchange was rebuilt.[40]

The earthquake and fire shook Hammond to his very core. Although his mansion in Pacific Heights survived the disaster, a few blocks away, Hammond witnessed three hundred thousand homeless residents of the city cooking their food "on fires built in the street." Unlocking his compassion, Hammond opened his house as a safe haven for friends and associates, allowing some forty people, mostly women, to sleep on the floor. The head of the Red Cross stopped by to inform Hammond that hundreds of thousands would need food and clothing for the next six months. Sympathetic to their plight, Hammond dispatched his largest ship at the time, the *Francis H. Leggett,* to bring two hundred tons of fresh water and food from Los Angeles.[41]

Recovering his business sense, Hammond recognized the tremendous need for building materials that would arise. He observed that a majority of the leading citizens "are dazed and still fail to comprehend what a tremendous task it will be to rebuild the City." Hammond complained that landlords refused to repair damages, "claiming that it is the business of the tenant to do so," and that "the insurance companies are seemingly in no hurry to pay losses . . . delaying matters in every way they can." Such actions greatly reduced the money in circulation; San Francisco's banks began issuing scrip in lieu of cash and threatened to summon their call loans—loans repayable on demand that businessmen often relied upon for short-term funds.[42]

Should the banks call in their loans, it would throw Hammond into a financial crisis. For the first time in his business career, he was overextended. Hammond had just spent $165,000 on timberland purchases; he had taken out loans to upgrade the Samoa mill, build ships, and stitch together his lumber empire. In doing so, he had suddenly, and with uncharacteristic carelessness, racked up a debt of $1.4 million, with $540,000 owed to San Francisco banks alone. He thus decided to offer $1 million in HLC stock to current stockholders. This, however, would allow John Claflin to acquire the majority stock in the Astoria Company, something Hammond desperately wanted to prevent. Normally, he would have turned to his old standby, the Missoula Mercantile, to bail him out, but it was also more than $200,000 in debt and unable to advance the funds Hammond needed to keep Claflin at bay and retain control of the HLC.[43]

To Hammond's great relief, the San Francisco banks declined to call in their loans, and the HLC sold its treasury bonds to pay off some debts. In addition, Hammond had sold his Missoula Light and Water Company to William A. Clark for $900,000 a few months earlier, and the first payment would soon put $150,000 into his pocket. Hammond also realized that First National Bank of Eureka, of which he was president, was in solid financial condition and could easily underwrite a loan. But by far the largest factor in erasing Hammond's debt was the rebuilding of San Francisco.[44]

Although the earthquake and fire destroyed much of the city, causing financial ruin for hundreds of businesses and pushing others to the brink, the disaster proved an unexpected windfall for lumbermen and gave the West Coast lumber industry one of its best years ever. Once reconstruction began, demand for lumber skyrocketed. Pilings to rebuild the docks shot from eleven cents per foot to fifty cents. Redwood lumbermen pointed out that many of the buildings that survived the fire were built of fire-resistant redwood, thereby increasing the demand for redwood products.[45]

Suddenly Hammond was flush with cash, and the HLC posted its highest net profits to date—$825,573—while the Missoula Mercantile earned an additional $250,000. Hammond poured the profits back into his company, purchasing a half interest in thirty-one thousand acres of redwood timberlands in conjunction with Weyerhaeuser for $1.5 million. With another forty-six thousand acres of redwoods owned outright, plus mills, lumberyards, and steamships, HLC held about $12.5 million in real assets with "practically no liabilities" except for $200,000 owed for a new ship under construction.[46]

Then, in the midst of the boom, the Panic of 1907 "descended upon Los Angeles like a thunderstorm from a clear sky. Every creditor demanded their money," recalled Perry Whiting. To make matters worse, the wholesale price of lumber dropped from $26 to $11.50 per mbf in thirty days. The rebuilding of San Francisco, however, allowed Hammond to ride out the depression and even profit from it. As he had during the Panic of 1893, Hammond found opportunity in the financial crisis.[47]

Perry Whiting was not so fortunate. Speculative irrigation ventures in the Imperial Valley had encouraged an overoptimistic Whiting to sell lumber on credit to area farmers. In 1904 an ill-conceived water project attempted to irrigate the below-sea-level valley with a canal from the Colorado River. Not surprisingly, the spring flood burst through the cut and flooded the valley, destroying farms and entire towns and creating the Salton Sea. In the meantime, Whiting had accrued $20,000 in bad debts in the valley, and Hammond, who had always disliked the practice of extending merchandise credit to farmers, insisted that as manager of the National Lumber Company, Whiting was responsible. With Whiting deep in debt, Hammond took direct charge of the Los Angeles lumberyard and sent Harry McLeod (George's brother) to oversee operations. Hammond demanded a full accounting of Whiting's every move. Whiting stated, "I became just a figurehead with no authority. I was unhappy, dissatisfied and a nervous wreck, trying to please them, so I resigned and made a settlement with Mr. Hammond," and took a loss of several thousand dollars.[48]

Whiting was not the only retailer who found himself pulled into the Hammond orbit. Across Southern California and Arizona, numerous other lumberyards fell so far in debt to Hammond that his salesmen could obligate them to accept a set wholesale price or sell out. In this manner, the HLC expanded across the booming Sun Belt.[49]

The Panic of 1907 resulted in a substantial drop in the price of lumber, and Hammond admitted that the retail yards of Southern California "will probably show a considerable loss." Nonetheless, redwood, because of its widely recognized qualities, retained its value, especially in overseas markets. Hammond noted, "Our business is fairly good and we expect to show a profit of over half a million dollars."[50]

Having simultaneously reduced his debt and recapitalized the HLC, Hammond surged ahead. In 1908 he turned back toward Astoria and bought the newly renovated Tongue Point Lumber Mill for $1.25 million, finally providing a manufacturing facility for his extensive timberlands in the Oregon Coast Range. The next year, Hammond purchased controlling interest in the F. A. Hihn Lumber Company of Santa Cruz, which included redwood timberlands, two sawmills, and three lumberyards. Capitalizing on the renewed growth in Southern California, Hammond opened retail yards in Pomona, Pasadena, and Riverside, bought existing outlets in the Imperial Valley, and eventually expanded to twenty-six lumberyards spread from L.A. to Arizona. The retail yards proved so profitable that Hammond even bought lumber from other manufacturers to sell.[51]

By 1907 Hammond had everything in place. He was one of the largest owners of timberland on the West Coast, giving him a nearly unlimited supply of raw materials; his production facilities included some of the nation's largest lumber mills; he held a vast transportation network to move his product to market; and he owned retail and wholesale outlets in America's fastest-growing region. Hammond told the National Bank of Commerce, "We are now in a position to control our output from the stump to the consumer." Company profits netted $750,000 a year at a time when the average annual wage was only $500. By avoiding indebtedness and fully integrating his company to control production from raw materials to retail sales, Hammond prepared his businesses not only to profit from but also endure the seemingly endless cycle of boom and bust that characterized the extractive economy of the American West.[52]

CHAPTER 15

Oxen, Horses, and Donkeys

When Andrew Hammond entered the woods as a teenager in 1864, nature constrained the timber industry. Rivers and mountains determined the location of lumber mills and logging operations. Trees could be cut and hauled out of the woods only by human and animal muscle, and water powered the mills. Unlike other industries, tree genetics and the physical environment—climate, soils, and topography—influenced where and how lumbering would occur.

By the turn of the century, however, capital intensive technologies, consolidation of land ownership, increased market access from railway connections, and changes in business organization had radically restructured the industry. In New Brunswick this process unfolded across three generations of Hammonds, but in the Pacific Northwest the transformation was compressed into just three decades. Hammond and other lumbermen believed that innovations would allow them to escape the dictates of nature. This belief would soon prove an illusion.

In terms of biomass per acre, the world's most productive forests lay in a relatively narrow band stretching from Alaska to San Francisco. A rare combination of geology, genetics, climate, and environmental conditions came together to produce the biggest and tallest trees on the planet. As Pleistocene ice sheets inched south one hundred thousand years ago, the north-south coastal mountains isolated the forests on the west side, where the moderating effect of the Pacific Ocean provided an ice-free refuge for hundreds of plant species. This high vegetative density encouraged shade-tolerant species and multilayered canopies.

Reproductively isolated in a relatively stable environment, West Coast conifers developed an intense genetic selection for height.[1] More efficient than hardwoods at photosynthesizing at lower temperatures, and resistant to summer drought, conifers tolerate the low nutrients and highly variable pH of soil in the Coast Ranges.

Contemporary climate patterns feature warm, moisture-laden air moving east from the Pacific, while high-pressure cells often create relatively dry conditions in the summer. The Coast Ranges and the Cascades pull moisture from the clouds, producing between 60 and 140 inches of rain per year, depending on elevation. The ocean currents moderate temperatures so that Astoria, Oregon, for example, averages forty-three degrees Fahrenheit in the winter and sixty-one in the summer. Farther south, Eureka, California, experiences one of the smallest fluctuations in daily temperatures in the United States, varying from only fifty-two to sixty-two degrees in July.[2]

Fires also helped forge the forests of the Pacific Northwest. The cool and wet conditions that developed about 3,000 years ago led to infrequent, but intense, stand-replacing fires with a return interval of 275 to 300 years. Trees of the Pacific Northwest, therefore, display a wide variety of adaptations, from the highly fire-resistant Douglas fir and redwood, with their extraordinarily thick bark, to the Sitka spruce, which survives only in the wettest locations. However, changes in the fire regime—from large, stand-replacing fires to low-intensity fires—resulted in a wider diversity of forest structures, ranging from young or even-aged stands to old-growth stands.[3]

In addition to genetics, climate, and fire, a unique suite of historical events converged to create these particularly large trees. About five hundred years ago a catastrophic fire swept through the Coast Range and Cascades. In the wake of this monumental disturbance, alder—which has the ability to fix nitrogen in the soil—sprang up in the newly created clearings. Young Douglas fir seedlings profited from the increased nitrogen in the soil, but because of the dense brush, only a few seedlings managed to overtop the alder. Those that did drew a perfect combination of nitrogen, abundant sunlight, and virtually no competition from other trees. Such low stand densities permitted the trees to grow rapidly during their first thirty to forty years and subsequently produced large-diameter trees.[4]

These Douglas firs could top three hundred feet in height and thirteen feet in diameter—so large that seven men with their arms outstretched could not encircle one. Not surprisingly, these forests inspired awe. One turn-of-the-century traveler in the Oregon Coast Range encountered "trees straight as a lance . . . their smooth trunks unbroken by limbs for a hundred feet." This visitor found himself "in a mysterious semi-twilight" where "dim cathedral like aisles radiate in all directions."[5]

Such forests impressed both lumbermen and foresters. In 1903 Henry Gannett of the U.S. Geological Survey estimated that Oregon's original forests could yield 75,000 bf per acre, enough to build five 2,000-square-foot houses. Others reported yields of more than 100,000 bf per acre, twenty times that of eastern forests. More recent scientific surveys confirmed these claims, with old-growth Douglas fir running from 50,000 to 100,000 bf per acre and hemlock stands containing up to 200,000. Gannett also concluded that Oregon was the state with the most timber—more than twenty million acres containing 225 billion bf. Ten years later, the Bureau of Corporations raised the estimate to 546 billion bf. This increase was largely due to surveyors including hemlock (previously disparaged as a timber species) and areas once deemed inaccessible, such as the mountainous Coast Ranges.[6]

Until large-scale logging and fire suppression, wildfires were the primary disturbance in most western forests. The intensity and frequency of fire varied widely across the Pacific Northwest, from recurrent low-intensity fires in dry ponderosa pine forests to rare but catastrophic fires along the coast. Over millennia, each species evolved a different response to fire. The ecological role of fire determined forest composition, with important implications to the timber industry.

Although ecologists consider western hemlock the climax species in the lower elevation forests of the Cascades and the Coast Range, Douglas fir, an early successional species, dominated the region because of the presence of fire. Douglas fir/hemlock forests averaged 230 years between fires, well within the lifespan of an individual Douglas fir. Therefore, the less fire-resistant hemlock rarely succeeded in reaching the climax state and overtopping the Douglas fir before a fire returned. In areas with a longer fire interval—300 to 600 years—hemlock and Douglas fir became codominant, and if fire was absent for more than 700 years, Douglas fir dropped out of the mix altogether. When a fire did occur in a hemlock forest, it could be a high-intensity and stand-replacing one. This allowed for reestablishment of Douglas fir once again, provided the fire did not destroy the organic mat, which would have encouraged red alder as the pioneer species.[7]

Lengthy fire intervals allowed for the development of old-growth forests that covered most of the Coast Range, while the mix of stand-replacing and low-intensity fires generated a range of forest structures and patch sizes. This variable fire interval, which ecologist James Agee characterized as "episodic" rather than displaying a regular cycle, created a highly diverse forest, including both old-growth hemlock groves and thick stands of hundred-year-old Douglas fir.[8]

East of the Cascades, the light-loving ponderosa pine, with its thick bark and high crown, experienced a fire on average every four to seven years, primarily at the hands of American Indians, who for centuries deliberately set fires to increase game forage and berry production. At the other extreme, on the coast, Sitka spruce forests rarely burned, perhaps once every 1,200 years. In the 1840s, however, fires in the Oregon Coast Range burned into the Sitka spruce zone and all the way to the sea, even driving Indians in the area into the ocean for safety. While significantly less extensive than the fires of the fifteenth century, the fires of 1845–49 nonetheless burned more than eight hundred thousand acres between the Siuslaw and Siletz Rivers, creating conditions for extensive Douglas fir regeneration and thereby providing the raw materials for Oregon's timber industry in the following century. Indeed, Douglas fir became the "money tree" largely responsible for the development of the Pacific Northwest.[9]

Hammond and other western lumbermen built their fortunes largely off three species—ponderosa pine, Douglas fir, and redwood—all of which displayed remarkable adaptations to fire. In sculpting the forest, fire promoted these species over all others. Ironically, however, lumbermen failed to recognize the importance of fire to their industry and attempted to reverse natural fire cycles by suppressing fire in dry areas while burning extensively in wetter sites to reduce their slash piles.[10]

Southward down the Pacific coast, slightly different conditions resulted in a different forest, although fire continued to play an important part in determining species composition. Once widely distributed throughout the northern hemisphere, the range of redwoods (*Sequoia* spp.) began to contract about thirty million years ago. The uplift of the Cascades, Sierra Nevada, and Coast Ranges three to five million years ago further confined these trees to a strip along the West Coast. The availability of moisture also restricted the range of coast redwoods. Although the summers were generally dry, prevailing winds along the coast created an upwelling of subsurface water, resulting in the formation of clouds and fog. Fog was so vital to redwood survival that they grew only along a narrow belt stretching 450 miles from Big Sur into southernmost Oregon, where fog draped the coast. Moving inland just a few miles, Douglas fir replaced redwoods as the dominant tree species.[11]

For trees, fog moderates the summer aridity and heat, restricts water loss, and, most important, provides a unique water source. The towering redwoods actually strip the fog of its moisture. Recent studies indicate that each redwood can contribute 2.5 inches of moisture a day to the soil through fog drip, contributing a quarter to a half of the ecosystem's annual water budget. When redwoods are cut down, water input declines by 30 percent. Some botanists think that fog may explain why redwoods are so tall. By absorbing fog directly through their leaves, redwoods can overcome "the hydraulic limitations to height that are faced by species growing in other climates."[12]

While redwood thrives in a cool, moist habitat, it can also survive frequent wildfires. With bark up to forty inches thick, redwood can withstand heat that otherwise kills its competitors, such as Sitka spruce and hemlock. In Northern California, where conditions are drier than those in Oregon's Coast Range, Yurok and other tribes regularly burned the redwood forest to encourage tan oak and a rich understory of herbaceous plants and berries. Although burning by American Indians affected understory species composition, large redwood trees were unharmed by these low-intensity fires. These relatively frequent fires also benefited redwood regeneration by killing soil pathogens, removing duff, and preventing the establishment of grand fir, hemlock, and Douglas fir, which becomes fire resistant only when mature. Larger, stand-replacing fires, however, favored Douglas fir, which thrives in open, sunny conditions, compared with the shade-tolerant redwood. In many respects the fire-return interval played a large part in determining whether Douglas fir or redwood dominated a site. The fire-return interval in the redwood region varied from five hundred years near the coast to thirty-three years farther inland. Studies indicate a fire interval as short as nine years in some areas. This wide range led to a variable forest in terms of species composition and physical structure.[13]

Although fire resistant, redwood, unlike Douglas fir, is not fire dependent. Redwood can grow on mineral soil in sunny glades or on logs in deep, shaded forests, but it favors alluvial terraces that provide deep soils, mild temperatures, and a year-round supply of water. These areas are prone to flooding, but here, too, redwood has readily adapted. Although intolerant of prolonged flooding, when high sediment loads deprive the roots of oxygen, redwood, when toppled by floods, wind, fire, or logging, has the unusual ability (for a conifer) to resprout from the base of a tree, earning its scientific name, *Sequoia sempervirens* (ever living). Furthermore, young redwoods can grow incredibly quickly, up to six feet in one year, pushing their flammable needles and crown above the reach of ground fires or establishing roots in a flood zone.[14]

The term "sempervirens" can also apply to the apparent immortality of redwoods. While most of the few remaining old-growth redwoods sprouted during the medieval era, some predate Christianity. Much of this longevity is due to the tree's ability to repel fungi, insects, and disease, a feature that adds to redwood's commercial value, especially in tropical climates. Redwood's immunity stems from tannins and phenolics in the heartwood that make it difficult for microbes and insects to digest and provide its deep, rich color. While 320 species of fungus species associate with redwood and 54 insect species feed on the tree, none can kill it. Additionally, as redwood grows older it increases the volume of decay-resistant heartwood, making lumber from old growth more valuable than that from younger, second-growth trees.[15]

The redwood invites a plethora of superlatives. As one of the fastest-growing trees, it can shoot up to 30 feet in its first two decades. With most of its vertical growth in the first century, a redwood can reach 300 feet in two hundred years. The world's tallest trees scrape the clouds at 360 feet. With a diameter of more than 20 feet, a single redwood can contain 360,000 bf—enough to build 22 five-room houses and equivalent to five acres of old-growth Douglas fir. Although the giant sequoia, with a 20-foot diameter and 52,500 cubic feet of volume, is considered the world's largest living tree (and largest nonclonal organism), many coast redwoods—before being cut down and turned into lumber—exceeded 63,000 cubic feet. Several stumps of coast redwoods measured more than 30 feet in diameter; one such stump in Humboldt County was even used as a bandstand. In 1911, Northern Redwood Lumber Company announced it had discovered the "Biggest Redwood in the World" on its timberlands. Humboldt County residents wished to preserve it, but lumbermen could not resist and cut it down and turned the giant into planks. The stump measured 32 feet in diameter.[16]

The early twentieth century witnessed a variety of perspectives toward the ancient forests of the Pacific Northwest. From a lumberman's viewpoint, the quality of lumber in these huge trees was unsurpassed. The trunks rose 150 feet before even the first branch, thus producing incredibly straight and clear-grained lumber. From a forester's perspective, these "decadent" trees should be cut down and replaced by plantations of younger, fast-growing Douglas fir. An aesthetic perspective, however, regarded ancient forests, especially the redwoods, as "cathedrals." Theodore Roosevelt articulated this latter view, stating, "We should not turn into shingles a tree which was old when the first Egyptian conqueror penetrated to the valley of the Euphrates." He summed up his position: "There is nothing more practical in the end than the preservation of beauty."[17]

Even the lumber industry acknowledged the aesthetic value of ancient redwoods. An 1884 booklet promoting California's redwood industry admitted, "Almost the first thought passing in one's mind, as he enters a virgin forest of redwoods, is one of pity that such a wonderful creation of nature should be subject to the greed of man for gold."[18]

Just as wood supplies the framework of a house, living trees provide the foundation of an entire ecosystem perched in the sky with its own soil, water system, and biota. Only in recent years have scientists realized the complexity of life in the forest canopy, which is especially developed in ancient redwood forests. Near the canopy, winds often sheer off the crown. The tree then forms new trunks, as many as sixty, from the break.

Some of these appendages can reach five feet in diameter, large enough to be substantial trees in their own right.[19]

This structural diversity in turn supports a surprising abundance of plant and animal growth. Horizontal branches and tree crotches provide a platform for the buildup of soils and humus up to six feet thick. Retaining water for weeks after a rain, these canopy soils provide sites for numerous plants, such as Douglas fir, Sitka spruce, and hemlock, all growing on the branches of a redwood. In addition, researchers have discovered that redwood canopies contain beetles, crickets, earthworms, millipedes, mollusks, and amphibians, including the clouded salamander, which nests 120 feet high. Altogether, the redwood region currently houses scores of rare and endangered species—134 plants and 42 vertebrates. More than 1,500 species of invertebrates can reside in the canopy of a single stand. So many specialized invertebrates inhabit old-growth forests that biologists discover dozens of new species every time they look and now believe that "a wave of invertebrate extinctions already has accompanied the catastrophic loss of coastal old-growth forests."[20]

In many ways old-growth trees function like coral reefs in that they support a multitude of other organisms, most of which reside in the soil. A single square meter of old-growth soil can house tens of thousands of mites, beetles, centipedes, springtails, and spiders, with an arthropod diversity rivaling that of tropical rainforests. Researchers estimate that eight thousand species of soil arthropods, only half of which taxonomists have described, exist within the sixteen-thousand-acre Andrews Experimental Forest in the Oregon Cascades. But more important for sustaining trees and other vegetation is the extensive network of mycorrhizal fungi that supply essential nutrients to tree roots. Much more than dirt, old-growth soil is really a teeming mass of life functioning in a symbiotic relationship with the trees above. This underground biotic community allows the forest to regenerate following a wildfire or other disturbance. Clear-cutting and heavy slash burning, however, can severely damage and even destroy this life-support system.[21]

The two highest-profile species dependent upon old-growth habitat are the northern spotted owl and the marbled murrelet, both of which the U.S. Fish and Wildlife Service listed as threatened species during the 1990s, setting off a still-unresolved firestorm of controversy. After a century of heavy logging of old-growth forests, both species suffered a precipitous decline in population. Their remaining habitat is now so highly fragmented that their long-term survival is questionable. Less well known, but far more endangered, is the Humboldt marten, which may already be extinct. Extensive surveys in 1990 and 2002 turned up no sign of the weasel-like animal.[22]

The most salient features of the forests of the Pacific Northwest, whether dominated by Sitka spruce, coastal redwood, Douglas fir, or western hemlock, were the sheer size of the trees, their longevity, and their long-term stability. Ecologist Reed Noss noted, "In the absence of human activities . . . the late-seral redwood ecosystem is one of the most stable on the planet, achieving a self-perpetuating steady state with a low incidence of severe fire." Large-scale catastrophic disturbances were extremely rare and recurrent disturbances such as fire, floods, and wind, were largely local. Following the California gold rush, however, logging introduced an entirely new disturbance on an unprecedented scale and rate.[23]

By the late nineteenth century, lumbering had developed into a well-coordinated endeavor that combined skilled human labor and raw animal power. Swinging double-bitted axes and using crosscut saws, a team of two men working in tandem could chop down a five-foot-diameter Douglas fir in less than three hours, although it could take three days to bring down a large redwood. To avoid the swollen, pitch-filled base, which would sink the log if it was dropped into water, fellers inserted a springboard six to fifteen feet up the trunk to provide a chopping platform. Dropping the tree required great skill for if it landed wrong and splintered, the whole day and tree would be wasted. To prevent such breakage, the crew removed the smaller trees surrounding their target and prepared a bed of boughs on which it could land. Such practice, while selective in theory, actually removed a substantial quantity of forest. One eyewitness described the sound of a redwood falling as a dry, tearing sound "as though the clouds were being ripped apart. Then there comes a brief moment when this tearing, ripping, splitting noise has an undertone of a great swishing, like a hurricane being born. And then the long, rumbling crash that booms in your ears and sends a vast noise echoing through the hills."[24]

After the branches were limbed, a team of peelers moved in and, standing atop the huge logs, stripped the thick bark with long crowbars as if they were peeling blubber from a whale carcass. Selecting only the choice sections, sawyers wielding twelve-foot-long "misery whips" then cut the tree into manageable lengths that they could hitch to an ox team. By cutting so high above the swell and leaving the top behind, loggers hauled only about 65 percent of the tree to the mill. In his 1884 *Report on the Forests of North America,* Charles Sargent deplored this practice as "waste" that contributed to forest fires. This accumulation of fuel indeed contributed to the intensity of fires in drier regions such as the Intermountain West. On the rainy West Coast, however, leaving behind this woody debris may have inadvertently maintained nutrients, such as nitrogen, and provided structural diversity to the forest and wildlife habitat for small mammals.[25]

In the redwood region, lumbermen developed the practice of setting fire to the forest after logging. Because of the fire- and rot-resistant quality of redwood, loggers typically cut a year's supply of logs in the spring and summer. Then, in the fall they set fire to the downed logs to burn away the slash and bark, leaving the winter rains to wash off the soot. Although these slash fires made it easier for logging crews to haul out the logs, all too often the fires burned off the protective duff and humus, leaving the "countryside barren and incapable of natural reforestation."[26]

Since each tree required substantial investment of time and effort, loggers removed only the most favored trees. Furthermore, the contract system of paying loggers for the board feet they delivered to the mill reinforced their selecting the choicest logs. As ring rot was often prevalent in Douglas fir over three hundred years old, loggers viewed these trees as too decayed to be worth the trouble and left them standing. Such trees provided seed sources for cutover areas. Therefore, this selective process did not alter the reproduction and function of forests to the extent that later practices would. Similarly, the biggest redwoods were simply too large to handle with animal power and were left intact as well. Ecological studies have concluded that animal logging left the forest in a productive condition. One study reported that 92 percent of areas logged by ox teams between 1860 and 1890 regenerated successfully. In later years, loggers returned to older cutting areas to find ample merchantable timber.[27]

Although environmental historians have characterized animal logging as "ecologically benign," other evidence suggests substantial impacts. While constrained in scope and scale, animal logging greatly altered the forest composition, as loggers selected the more desirable Douglas fir but left hemlock and spruce standing. As the Douglas firs crashed to the ground, however, they often scarred the thin bark of the nearby hemlock and Sitka spruce. Such bark damage fostered the growth of rot fungus, which eventually killed these trees as well. In other areas, the openings created by selective logging promoted the growth of alder, choking out Douglas fir saplings and allowing the hemlock to outcompete the remaining Douglas fir. Animal logging thus shifted an ecosystem dominated by Douglas fir to one dominated by spruce and hemlock. Eventually, lumbermen recognized the economic value of these species and returned to cut them as well.[28]

With animal logging restricted to within a mile or two of a river, riparian areas suffered greatly, and effects reverberated throughout the ecosystem. Logging activity along rivers on the West Coast reduced fish populations, extirpated many species, and pushed California and Oregon's salmon populations toward extinction. Impacts on fish were perhaps even greater on the West Coast than in New Brunswick because of the compressed time span in which logging took place. Furthermore, West Coast rivers were relatively short and steep; therefore, whatever logging occurred upstream quickly affected the entire watershed.

The wet winters and unstable geology of the Coast Ranges caused the region to experience some of the highest recorded rates of natural erosion and sedimentation. Fortunately, fallen trees in the streams slowed the water flow and mitigated landslides and flooding. River drives, however, converted structurally complex rivers into conveyor belts for logs, dramatically increasing sedimentation and erosion downstream and resulting in heavy losses in aquatic habitat. Despite the obvious impact to the region's fisheries, the economic value of river drives was such that in 1889 Oregon allowed counties to declare nonnavigable streams "log highways." Not until the 1950s did environmental considerations lead states to pass legislation that finally ended the practice.[29]

Logging in the riparian zone transformed a stream filled with pools, riffles, moss-covered rocks, and down trees into a shallow, muddy torrent. Logging in the redwood region increased stream sediments by 89 percent. These fine sediments, while increasing short-term nutrients, settled over the gravel streambeds and eliminated salmon spawning sites for decades, possibly centuries. Removing the large overhanging trees along a river further reduced fish and amphibian habitat. The lack of cover increased the water temperature, often enough to kill salmon. By the beginning of the twenty-first century, these impacts resulted in the federal government listing Chinook, chum, coho, sockeye, and steelhead as either endangered or threatened species in Oregon and California.[30]

Amphibians—excellent indicators of ecosystem health—have also declined dramatically throughout the redwood region as a result of logging. Redwood forest streams contain a variety of amphibians: up to eighteen species. Splash dams from historical logging eliminated at least one—the torrent salamander—from the Mattole watershed. Although preindustrial animal logging was confined to relatively narrow corridors along rivers, its effects far exceeded the removal of a few trees.[31]

Cutting trees was a skilled but straightforward process that remained nonmechanized until the 1930s. Limited by the physical ability of animals, getting the logs to the mill presented loggers with the biggest obstacle. Both oxen and horses tired quickly in the summer heat and could not work in the cold or muddy conditions of winter. Transporting logs also caused significant environmental impact. Skidding logs across bare ground tore out vegetation and plowed a furrow into the soil two to four feet deep, leading to erosion and soil compaction, all of which prevented regeneration. In the Northeast and Midwest, timber crews hauled out logs in the winter by horse-drawn sleighs that slid over the ice or snow, causing minimal soil damage. On the West Coast, where winter rain replaced snow, lumbermen had to devise another method.[32]

Skid roads—small logs half buried crosswise, like railroad ties, over which the logs could slide—provided the solution. With chains clanking and men cussing, oxen dragged the chained-together logs over the skid road, usually down a gulch, to the river, where in winter the floods would carry the logs to the mill downstream. Limited by their own energy as well as the terrain, the oxen could haul the logs only a mile or so. Yet an oxen team could haul about ten thousand board feet every trip and make four mile-long trips per day.[33] Inadvertently, skid roads prevented deep soil scarring and ruts. Multiple passes of oxen and heavy logs no doubt resulted in soil compaction. Skid roads, however, distributed the weight and probably resulted in less damage that would have occurred otherwise. Eventually skid roads rotted back into the soil, providing nitrogen and retaining water for new vegetation.

During the 1880s advances in one technology spurred others. When Hammond acquired the Northern Pacific contract, he built mills using the newest technology—steam power and band saws—which increased a mill's production from 40,000 to 250,000 bf. By 1904, 90 percent of West Coast mills operated under steam power. As mills cut lumber faster, demand for raw logs increased, giving rise to logging railroads to move the logs to the mill. Horse and oxen teams could scarcely keep up with the increased pace in getting logs to the railroad landing. Then the steam donkey arrived, revolutionizing logging and turning it into an industrial activity.[34]

In 1881 William Carson's partner, John Dolbeer, modified a ship-loading steam engine, combining an upright boiler with a capstan gypsy head on a horizontal shaft. Using a 140-foot-long manila rope, this device towed logs to the yarding or staging area. Mounted on a wooden platform, the Dolbeer steam donkey could winch itself up a steep hill to a logging area. Another donkey at the yard then pulled logs in from the surrounding area and loaded them onto railcars. In 1892 steel cable replaced manila rope, and the next year the "bull donkey" appeared on the scene. With a larger, more powerful engine and with twin spools that could pull in the haul-back line while towing the log, the bull donkey eliminated the need for animal power altogether. The bull donkey could haul up to twenty logs bound together from two thousand feet away, greatly increasing the scope of a logging operation. As donkeys became more powerful, operators no longer needed to negotiate live trees or stumps and simply uprooted both with the incoming logs.[35]

When Hammond purchased the John Vance Mill and Lumber Company in 1900, a few oxen teams still remained on the coast; horses had become more common, but the trend toward steam donkeys was well underway. Oregon camps contained 35 steam donkeys, California had 61, mostly in the redwood belt, and Washington led the Northwest with 293 donkeys. Because of the size of the trees, the well-capitalized redwood industry had largely shifted to steam logging, employing both donkeys and gypsy locomotives. The gypsy, another of Dolbeer's inventions, wedded a steam donkey and a geared locomotive into a tiny workhorse.[36]

For Hammond and other lumbermen, the mechanization of logging had multiple advantages. Using donkey engines, loggers could haul ten logs down a skid road with the same effort it took to remove one from the bush by horse or oxen. Worker productivity rose fivefold, from 3,000 bf per day with horse logging to 15,000 bf with mechanical skidding. Since steam donkeys did not need to be fed (fuel was lying on the ground all around) or cared for during the off season, they reduced the cost of a logging operation by more than half.[37]

In 1909 a typical Hammond logging operation would punch in a railroad to the mouth of a gulch, where a crew would build a two- to three-hundred-foot-long landing. Skid roads branched out and up the gullies. At the top of each road, smaller donkeys pulled logs into loads of fifteen to thirty logs, which were bound up and dragged to the landing by the bull donkey. Weather ceased to be a factor as well. In 1911 Hammond became the first redwood lumberman to run his logging camps year-round instead of shutting down during the rainy season.[38]

Mechanization, however, required significant capital investment. A donkey engine cost more than $10,000, and a logging railroad ran $30,000 to $60,000 per mile to build. Removal of all merchantable timber thus became an economic necessity. Unlike horses and oxen, donkeys could easily haul out the largest trees, as well as everything else. By 1917 a market for spruce and even hemlock had developed, and loggers left behind only the smallest trees. In a self-reinforcing cycle, increased capital investment required maximum production to meet bond payments, and in turn, more production required more capital investment.[39]

The pace and scale of logging increased dramatically, with profound ecological impacts. With donkeys and branch railroads, the industry was no longer limited to the forests along rivers but could now access remote and mountainous terrain. In 1909 a forest official conceded that the steam donkeys caused "a much greater menace to the forest than team logging."[40] A 1932 Forest Service study confirmed that of the stands that had germinated following steam donkey logging, only 59 percent could be considered "well stocked," as compared to the 92 percent regeneration following animal logging.[41]

Cable or ground-lead yarding by steam donkeys tore up the forest floor, knocked over young trees, and left massive slash piles, increasing the likelihood and intensity of fire. One study found ground skidding denuded 22 percent of the surface and destroyed 72 percent of trees smaller than 3.5 inches in diameter. Nevertheless, "Logging was confined to relatively small areas open to reseeding from adjacent stands."[42] Ground-lead logging was still selective to a degree, leaving behind small-diameter trees, and the slash fires were largely restricted to the actual operation.

Within a few years, however, increased production demand led to the highly destructive practice of high-lead logging. Introduced to the West Coast in 1906, high-lead logging became widespread as it increased output tenfold over ground skidding. Typified by the Lidgerwood skidder, this technique required a spar tree equipped with a complex system of block and tackle and a spiderweb of steel cables that enabled the donkey yarder to hoist logs through the air across a large area. Although some systems carried the logs in a carriage high above the forest floor, most simply dragged the logs along the ground with one end hoisted up. Incoming logs became "enormous battering rams," uprooting stumps, saplings, rocks, and anything else in their path. High-lead logging required total clear-cutting so that logs could be yanked into the air and moved across the landscape unimpeded. Dragging logs to a central point also prevented natural regeneration as it uprooted saplings and scraped the soil clean. These large clear-cuts, with no trees remaining, prevented regrowth and produced the infamous "logging deserts," a seemingly endless landscape of stumps where nothing could grow.[43]

High-lead logging increased the speed of operations. Machines, rather than workers or the terrain, now set the pace. With the ability to haul in a log in less than five minutes, a Lidgerwood skidder could quickly clear an area up to 1,500 feet around the donkey. Such high-lead logging pulled in one hundred logs per day, compared with thirty per day with a ground lead. It also reduced the need for skilled labor that had evolved with ground skidding. A proponent of this new system proclaimed that it eliminated skid roads, swampers, chasers, snipers, barkers, and the skilled hook tender "upon whose caprices hangs the day's output."[44] One observer described a high-lead operation as "a somewhat disconcerting maze of large and small cables running through the air in every direction, and the ground covered with stationary engines, pumps, wood-bucking power saws, steel rails, switches, locomotives, cars, telephones, humidity gauges, movable power plants, and traveling machine shops. It is, in short, a gigantic factory without a roof."[45]

Industrial logging also reversed the forests' relationship with fire. At the same time that lumbermen tried to prevent low-intensity fires in mature forests, they created conditions for widespread conflagrations. High-lead logging left behind massive quantities of slash—tree tops, branches, and stumps—up to twenty-four thousand cubic feet per acre. Recognizing a high correlation between slash and catastrophic wildfire, lumbermen pushed for and achieved legislation in Oregon that required controlled slash burning to protect the forest. They believed that slash burning would prevent reburns. Clear-cutting and slash burning went hand in hand and became standard industry practice.[46]

While slash burning reduced the immediate fire danger, it hindered regeneration. Because these fires burned so hot—up to 1,841 degrees Fahrenheit—they eliminated the protective duff layer and future sources of organic matter. Readily absorbing sunlight, the blackened soil could produce surface temperatures of 140 degrees, killing off 45 percent of Douglas fir seedlings that had managed to find their way into the clear-cut. Not only did these extremely hot slash fires eliminate nearly all the of soil's organic matter; they also shifted the pH from acidic to alkaline and increased nitrogen content. The excess nutrients favored rapid colonization by pioneer species, such as alder, and caused Douglas fir seedlings to develop large crowns and shallow root systems, making them susceptible to drought.[47]

The extensive soil damage was perhaps the longest-lasting legacy of industrial logging. Stripping steep mountainsides of trees contributed to erosion and frequency of landslides in a region with already-unstable slopes, increasing sedimentation of streams and rivers. A recent study in the Cascades found that clear-cutting and road building directly caused more than half of all landslides.[48]

Despite all the ecological studies and evidence that indicated problems with clear-cutting, many twentieth-century foresters endorsed it as a silvicultural practice to foster regeneration of Douglas fir. Clear-cutting on public lands was a dramatic departure from practices in the East, where horse logging prevailed and foresters opposed the use of steam donkeys. But on the West Coast, foresters maintained that steam logging was the "only practical method" because of the rough terrain and large trees. Foresters such as E. T. Allen also developed a scientific rationale for clear-cutting, maintaining that good forestry "coincides with common logging practice." Allen declared that Douglas fir was shade intolerant and therefore the best practice was to clear-cut, burn the slash to expose the mineral soil, and "let the seed stored in the ground sprout." Not surprisingly, the Western Forestry and Conservation Association, a lumber industry trade organization, hired Allen as its executive director.[49]

Despite Allen's contention, by the 1920s it was apparent that cutover lands were not regenerating and that clear-cutting had severe ecological consequences. Later studies demonstrated that the high surface temperatures in clear-cuts hindered regeneration and that Douglas fir was actually moderately shade tolerant, growing best in about 50 percent full sunlight. Foresters began to advocate leaving behind mature seed trees for regeneration, yet wind or slash fires soon toppled these trees. In 1922 forester Thornton Munger offered a more candid assessment in his report on forest conditions on private land in the Douglas fir region, concluding that "whatever reproduction takes place does so, for the most part, in spite of present methods, not as a result of them." Five years later, he examined logging on Forest Service lands and found that 40 percent of logged land was completely barren. He noted that the reproduction that did take hold occurred in areas that were logged prior to the advent of high-lead logging.[50]

Ironically, these findings prompted a move toward total clear-cutting followed by artificial regeneration. Foresters saw old-growth forests as "over-mature" and "decadent" and an obstacle to sustained-yield management. They continued to justify clear-cutting as the first stage in scientific management. By liquidating the mixed hemlock/Douglas fir old-growth stands, foresters could replace them with perfect, even-aged stands of Douglas fir.[51]

Not only did industrial logging conveniently inform the scientific community, but the introduction of steam power and subsequent speedup in production forever changed the nature of work. Previously, the pace of logging was no faster than a man could chop or an ox could walk, and a bull puncher could shut down the whole operation if he walked off the job. By the early twentieth century, however, the factory model, in which machine-dictated production stripped workers of their power, provided yet another motivation for owners to adopt mechanization. Not only did technology increase the pace of logging and milling, but as companies became larger and more consolidated, they needed to increase the rate of throughput to maintain profits and pay off bondholders. All of this increased the pressure on men to work harder and faster, leading to more deaths and injuries. When workers protested the new technology and ramped-up production, owners simply replaced them with unskilled labor. In Humboldt County, this issue came to a head in 1907.[52]

CHAPTER 16

The Great Strike of 1907

By all indications 1907 promised to be a most prosperous year for A. B. Hammond. He and his family had survived the San Francisco earthquake and fire; both his house and business stood intact. In fact, the rebuilding of the city turned into a financial boon with the demand for pilings and lumber. Over the previous ten years, Hammond had acquired tens of thousands of acres of timberlands for a fraction of their worth. He owned one of the largest and most advanced lumber mills on the West Coast and had assembled an extensive transportation network. Furthermore, lumber prices were advancing steadily. To top it off, the recurring bouts of rheumatism that had plagued him seemed to have dissipated. Yet discontent rumbled from deep within his empire.

Hammond's trouble began in 1903, three years after he purchased the Vance Lumber Company. Upon updating the Samoa mill with the latest machinery, he installed a corporate management system. The system suited Hammond, who was operating from the distant financial capital of San Francisco, but it was less amicable to his employees, who were accustomed to the paternalism of other Humboldt County lumbermen. Pioneer lumber barons such as Carson, Vance, and Noah Falk lived in the same town as their workers, walked the same streets, and greeted their employees by name. Men knew who they worked for, and if they had a complaint they knew who to see about it. In contrast, with Hammond's arrival, improved technology and a new bureaucratic structure coincided to alienate workers.

Improvements to the Samoa mill allowed it to run twenty-four hours a day. Workers living in Eureka needed to take the ferry across the bay to the Samoa mill, and unaccustomed to working nights, men resented having to leave their homes early in the evening and not return home until the next morning. Furthermore, their ten-hour shift did not begin until 7:00 P.M., yet the *Antelope* (the Hammond ferry) left Eureka at 5:45. When workers asked to have departure pushed back to 6:30, the company ignored them.

One evening in late October 1903, night-shift workers gathered at the Eureka dock for the *Antelope* to carry them across the bay to Samoa. After they had waited in the cold for an hour, a company representative informed the men that dense fog had grounded the ferry and they should wait until launches returned with the day crew. Several men simply grew tired of waiting and returned home. Others decided it was too dangerous to board the crowded boats in the dark and fog. When they showed up for work the next night, twenty-three men found themselves unemployed.[1]

The fired workers' primary complaints stemmed from the impersonal corporate nature of the Hammond Lumber Company. They had received no response to their request for a later departure time; also, they had $1.50 taken out of their monthly wages for boat fare regardless of how often they used it—missing the boat was a not-infrequent occurrence as it often left early. They also objected to the nonresident owner of the company and the convoluted bureaucracy involved in getting their paychecks when fired or laid off. The incident marked the beginning of an ever-widening fissure between workers and owners in Humboldt County.[2]

In Eureka, and across America, workers struggled to come to terms with the new corporate reality that redefined their relationship with employers. Because of its isolation Eureka remained an "island community" where workers still believed they had some degree of control over their lives and could determine the work pace and job organization. Furthermore, workers in Eureka could count on the support of local businessmen in their struggle with outside capitalists.[3]

Eureka accepted a strong tradition of political dissent that originated in the labor and populist movements of the late nineteenth century. Concerned about the concentration of economic power and indignant over land frauds, during the 1880s residents of Humboldt County formed local chapters of the International Workingman's Association and the Knights of Labor. Labor historian Daniel Cornford argued that the populist tradition politicized Humboldt's working class and provided a sense of legitimacy. In addition, the labor theory of value—which holds that the value of a good is proportional to the labor required to produce it and therefore workers are entitled to the profits received—permeated the city's consciousness. Cornford noted that this paradigm "provided union men with a profound belief in the righteousness of their mission" in opposing capitalism. Although many Eureka residents retained this ideology, the depression of the 1890s had dissolved these early labor organizations.[4]

By 1900 the global building boom had stimulated Humboldt's economy and created a labor market that reinvigorated Eureka's unionization effort. This reflected a national trend; in the United States, membership in trade unions jumped from 440,000 to more than 2,000,000 between 1897 and 1904. In Eureka, the 1905 Labor Day parade drew 2,500 people representing twenty-five unions. By the following year, nearly every occupation in Humboldt County had a union. The Eureka Trades Council, chartered by the American Federation of Labor (AFL), boasted 4,000 members, and most had gained wage increases and formal collective bargaining agreements. Most important, Eureka's craft unions had won the eight-hour day. The Trades Council had its own newspaper, the *Labor News,* one of California's few labor newspapers outside of San Francisco, owned and edited by Joseph Bredsteen, an avowed socialist. In 1906 Eureka unions began building their own hospital, and the newly formed International Brotherhood of Woodsmen and Sawmill Workers (IBWSW) boasted more than 2,000 members. Before long, unionism spread from Eureka into the backwoods timber camps.[5]

The economic boon in the early twentieth century was not the only factor in Humboldt's labor revival. Cornford pointed to the county's social, cultural, and political institutions that gave unions broad appeal, in addition to the deeply felt grievances many workers held against lumber barons such as Hammond. Cornford concluded that Humboldt's unionization was almost entirely a grassroots effort. With virtually no help from national or state organizations, Humboldt's labor movement "became a formidable bargaining force that was able to secure increases in wages and improvements in working conditions."[6] Nonetheless, the lack of outside support meant Humboldt's unions were no match for the organized power of employers.

Humboldt County thus provides an excellent case study of how workers attempted to exert power over their lives with relatively little influence from either the conservative AFL or the emerging, revolutionary Industrial Workers of the World (IWW). While both of these organizations eschewed political involvement in the early twentieth century, unions in Eureka and San Francisco showed how labor could be a significant player in municipal politics. In Humboldt, workers formed a branch of the Union Labor Party (ULP), which also dominated San Francisco's politics from 1901 to 1911. The ULP advocated better schools, compulsory education, civic improvements, parks and playgrounds, public ownership of utilities, abolition of child labor, direct election of the president and senators, immigration restriction, improved workplace conditions, and women's suffrage. By 1908 Eureka's ULP had merged with the local Socialist party and succeeded in electing a Socialist mayor in 1915, pioneer mill owner Elijah Falk. While far from politically homogenous, Eureka maintained a strong sense of community and displayed a remarkable degree of support for unionism and third-party politics, to the extent that it transcended class boundaries. Eureka countered Eugene Debs's contention that "the choice is between the AFL and capitalism on one side and the industrial workers and socialism on the other."[7]

Most of Humboldt's lumbermen, of course, were not like Falk. While they did not support unionism, they seemed resigned to it and did little to oppose it. A. B. Hammond, on the other hand, refused to accept the growing power of labor and viewed the developments in Humboldt County with mounting concern.

Hammond was in San Francisco for scarcely a year before a labor conflict erupted. In January 1901, San Francisco's sailors, longshoremen, teamsters, and six other unions organized into the City Front Federation led by a gaunt Norwegian sailor named Andrew Furuseth. Scrupulously dedicated to addressing the plight of his fellow sailors, Furuseth became secretary of the Coast Seaman's Union in the wake of the disastrous sailor's strike of 1886, after which triumphant shipowners lowered wages and instituted a system of grade books that tracked each sailor's employment history. In 1891 Furuseth folded the Coast Seaman's Union into the Sailor's Union of the Pacific (SUP). In his position as secretary of the SUP, Furuseth remained the head and face of the union for the next forty-four years. For thirty-five of those years, Furuseth battled A. B. Hammond in a protracted struggle.[8]

Living in virtual bondage, with wages as low as twenty dollars a month, sailors were among the world's most exploited workers. Ship captains wielded absolute authority—it was against the law to disobey a ship's officer—and often subjected sailors to physical and psychological abuse. Because captains could imprison them for desertion, sailors could not quit if conditions became intolerable. In 1896 the Supreme Court ruled that the Thirteenth Amendment, which abolished slavery and involuntary servitude, did not apply to sailors.[9]

In Andy Furuseth, Hammond met his equal. None of his other opponents—William Sparks, Marcus Daly, William Carson, or Joseph Bredsteen—proved as formidable, determined, and tenacious as Furuseth. Only the scowling, hawk-nosed Norwegian sailor could match Hammond's stubbornness and intransigence. Nonetheless, the two men eventually came to respect each other. They occasionally engaged in cordial conversation, and Furuseth even personally informed Hammond when the sailors would strike. The two Andrews represented two sides of a deep chasm between labor and capital that divided America for more than fifty years. The core issue was who should control the means of production—the workers who provided the labor, or the industrialists who provided the capital? The debate threatened to tear the nation apart.

Furuseth, however, was not quite Hammond's ideological opposite. Big Bill Haywood and the radical IWW, which advocated revolution and the overthrow of the capitalist system, held that honor. In contrast, Furuseth's primary concern was to improve the lot of seamen, and in this he was absolutely dedicated. As a "bread-and-butter" unionist, Furuseth aligned the SUP with the conservative AFL and its president, Samuel Gompers. Hammond, however, saw little distinction between unions and later lumped all—AFL and IWW alike—as radical "Bolshevikis."[10] Fundamentally, Furuseth and other union leaders subscribed to the labor theory of value and advocated the right of workers to organize and bargain collectively, a right not accorded in the American legal system until the Wagner Act of 1935. Hammond and other capitalists, however, insisted that no one could tell them how to run their businesses. Generally speaking, they did not oppose men seeking higher wages *individually*. It was the idea of organized labor and collective bargaining that they found reprehensible. While Hammond and other industrialists decried the tactics of unions—boycotts, picketing, and strikes—they engaged in boycotts of their own, as well as lockouts and importing strikebreakers. Although owners opposed government intervention in their businesses, they saw no contradiction in their willingness to enlist the power of the state to defeat organized labor and often used federal troops and court injunctions to break strikes.

Above all, businessmen feared a general strike that could paralyze a city and shut down commerce. In the spring of 1901 the City Front Federation strike appeared imminent in San Francisco with the rising tide of union power. As a preemptive measure, fifty of the city's preeminent business leaders formed the Employer's Association to resist the unions. Although membership was kept secret to avoid boycotts, Hammond was likely part of the Employer's Association, as he became president of its successor, the notoriously antiunion Citizen's Alliance.[11]

In July the Employer's Association forced labor's hand with a lockout of the teamsters, informing them they must either quit the union or their jobs. In response, 6,400 teamsters went on strike and were soon joined by some 20,000 workers from other trades. The strike shut down the port and brought the West's largest city to a standstill. The ensuing violence between strikers and scabs on San Francisco's waterfront left four men dead and hundreds wounded. Finally, Governor Henry Gage brokered a secret agreement that resulted in the creation of a virtual closed-shop city and catapulted organized labor into the city's political sphere. The ascendency of union power allowed San Francisco's Building Trades Council to refuse to handle any building material originating in a nonunion shop or one working longer than an eight-hour day. While this gave a guaranteed market to the Bay Area lumber mills that agreed to union demands, it shut all of Hammond's finished wood products out of San Francisco, a situation he resolved to reverse.[12]

The City Front Federation strike also provided Andrew Furuseth with the opportunity to negotiate a contract with West Coast shipowners who guaranteed the SUP sailors a minimum wage of forty dollars a month. For the next five years, sailors and owners maintained a cordial relationship, largely because Furuseth insisted that sailors abide by the contract's prohibition against sympathetic strikes. Then, in the spring of 1906, Furuseth and the shipowners opened negotiations for a five-dollar increase in monthly wages. By this time, however, Hammond had entered the coastal shipping trade and become a prominent member of the Steamship Owners Association. Taking a lesson from the City Front strike, Hammond implored shipowners to band together and hold other owners accountable should they give in to union demands. The shipowners joined the United Shipping and Transportation Association (USTA), which required members to post a $60,000 bond to prevent them from granting wage increases without the consent of the executive board. Such associations had become a commonplace means to thwart unions. Indeed, Hammond employed the same strategy with the Humboldt Lumber Manufactures Association (HLMA) to keep other lumbermen from raising wages.[13]

But in April 1906, the San Francisco the earthquake brought labor and capital together, briefly. In light of the crisis, shipowners suspended their planned increase in freight rates. Likewise, the SUP agreed not to charge overtime for relief work. A month later, however, Furuseth resurrected the demand for increased wages and improved working conditions. He pointed out that wages in San Francisco had risen considerably as result of the earthquake and that to keep seamen on the job they, too, should receive higher wages. Furuseth called for a wage increase for sailors to take effect May 29.[14]

The executive committee of the USTA—composed of Hammond, other shipowners, and Southern Pacific Railroad officials—refused the SUP demands. As committee spokesman, Hammond accused the union of trying to take advantage of the situation and using the earthquake as leverage. Meanwhile, Hammond and other industrialists had realized huge profits from the disaster. In response to the shipowners' refusal, sailors walked out, idling seventy-seven steam schooners in San Francisco's harbor.[15]

Despite the strong-arm tactics of the committee, a number of owners, who could scarcely afford to let their ships sit idle, broke with the association and agreed to the new pay scale. Hammond, however, remained steadfast. The *Labor News* framed the conflict as "one between labor and capital; the question at issue has ceased to be one of wages and other conditions of labor and has become one between trade-unionism and the 'Open Shop.'" Hammond agreed with this assessment as he sought to create open-shop conditions along the entire West Coast.[16]

The sailors' strike dragged on through the summer of 1906 and spread up and down the coast. In Los Angeles, Hammond's ideological twin, Harrison Gray Otis, editor of the *Los Angeles Times,* cast the conflict as a "Bloody Fight for Freedom" against Furuseth, the "pirate" and "despot of the Sailor's Union." At Los Angeles's port of San Pedro, where Hammond had recently begun to build his massive retail and wholesale lumberyard, the longshoremen's union supported the SUP strike and refused to unload nonunion ships. In Eureka, however, longshoremen continued to work with nonunion sailors, creating conflict between the two unions. As the strike dragged on, however, longshoremen on Humboldt Bay finally aligned with the sailors.[17]

In the meantime, Hammond secured a permanent injunction against the SUP from interfering with his ships. Twelve years earlier, a conservative U.S. Supreme Court had upended the Sherman Antitrust Act by ruling that strikes were a disruption to interstate commerce and therefore corporations could use court injunctions to break strikes. From 1880 to 1930, U.S. courts issued more than four thousand injunctions against striking workers. Furuseth and his sailors, however, refused to be intimidated by the legal system and repeatedly violated the injunction, insisting that it was contrary to "constitutional guarantees of individual freedom."[18]

In carrying out the strike, Furuseth hit upon a tactic that proved particularly effective and aggravating to shipowners. While in port, union sailors would hire on to a nonunion ship. After the captain assembled the crew and prepared to hoist anchor, the sailors would grab their duffels and jump ashore. This forced the captain to spend several more days in port finding another crew, at considerable expense to the shipowner. Hammond and other shipowners responded by importing nonunion sailors from the Great Lakes, Mexico, and Australia.[19]

After five months of frustration, all but two of the shipowners who still remained in the association conceded to SUP demands. The Hammond

Lumber Company and the Union Lumber Company, owned by Hammond's good friend Charles Johnson, refused to capitulate. Hammond complained to C. H. McLeod, "The business men seem to have surrendered unconditionally to the labor unions . . . about one half of our organization has flunked and a majority of the balance has not much fight left in them." While Hammond was clearly disappointed in what he saw as lack of courage on the part of other shipowners, he admitted that with the high prices of lumber "the temptation to run their boats . . . and sell their lumber was too great for most operators to resist."[20] Although he lost nearly half a million dollars as a result of the strike, Hammond refused to yield. While the association dropped its claim against SUP for violating the injunction, Hammond persisted and won a decisive court victory against Furuseth. When the *San Francisco Bulletin* asked Furuseth about the possibility of jail for contempt of court, he famously replied in his heavy Norwegian accent, "I would put the inyunction in my pocket and go to yail, and in yail my bunk would be no narrower, my food no worse, nor I more lonely than in the forecastle [*sic*]."[21]

Hammond was winning on the ground as well as in court, and he confronted the now-united front of sailors' and longshoremen's unions. When Hammond's *Ravalli* attempted to dock in Eureka, longshoremen prevented the ship from landing. The captain, however, simply sailed across the bay to Samoa, where Hammond's mill workers unloaded the ship. Other lumbermen did not have this option and had to hire union crews. William Carson reasoned that it was "a bitter pill for me to take but we have too much property unprotected here." Still, he admitted that "the union men are not all outlaws." Hammond, however, would make no such concession and imported fifty "roughnecks" from San Francisco to break the strike. But upon discovering that Hammond intended to use them as strikebreakers, most left Samoa for Eureka. Finally, to break the longshoremen's hold on the waterfront, the HLMA formed the nonunion Humboldt Stevedore Company to unload ships. Tensions mounted between the two groups of workers and ended in tragedy.[22]

On the night of January 9, 1907, a group of union longshoremen confronted fourteen strikebreakers. A gunfight broke out, resulting in the deaths of strikebreaking stevedores William and Albert Jenks and seriously injuring one of the longshoremen.[23] Each side blamed the other.

The *Labor News* charged that Hammond imported "gunfighters" who brandished their weapons "while they were working as sort of an object lesson and eye-sore to any union men that might be around."[24] Hammond, in turn, blamed Joseph Bredsteen of the *Labor News* for instigating the violence. Hammond publicly declared that Bredsteen's "teachings influenced the miserable wretches guilty of the crime to commit the deed. . . . I deem it my duty to expose this anarchist who is masquerading as a Socialist."[25] In the ensuing trial of the longshoremen, the HLC general manager, George Fenwick, testified that "the strikers went around in gangs of from eight to twenty. I have seen thirty strikers together going around intimidating the non-union men."[26] Nonetheless, a Eureka jury acquitted the accused longshoremen. The bloodshed shocked Eureka and helped workers realized that their grievances were not with each other. A few weeks after the incident, the longshoremen agreed to work alongside nonunion Humboldt Stevedore Co. employees. Hammond, however, would for years continue to cite "the murder of the Jenks brothers" as evidence of union violence and anarchy.[27]

For many workers, the longshoremen's union had carried solidarity too far. In a letter to the *Labor News,* eighteen employees of Humboldt Stevedore Co. elaborated on their experiences working on the waterfront, "Under instruction from the union, each man did as little as he could so as to make a little work go a long ways. . . . We as employees are proud to say that we are loading vessels for less money, and in less time than ever was done under union rule." Stating that they preferred to work under open-shop conditions, they noted "the closed shop kills ambition," a common refrain among businessmen.[28] While this letter might have echoed the sentiments of some workers, the wording sounded suspiciously like Hammond's; it also appeared the same week that he made one of his rare visits to Humboldt County.

The meaning of "open" and "closed" shop varied considerably across the economic and political spectrum. For the Elk River Mill manager, Irwin Harpster, running an open shop meant that "We are perfectly free in the employment of men not asking whether they are union or nonunion, nor have the Unions made any demands upon us to employ strict union men."[29] Publicly, Hammond professed a similar position; privately, however, the open shop meant "we will not have a union man in our employ."[30] For labor organizers, the open shop meant a worker could join a union without discrimination.

In a closed shop all of the workers had to join the union. Labor organizers believed that only through such solidarity could workers confront the enormous power of owners. A union was virtually meaningless without the solidarity that could force employer recognition and collective bargaining. For sailors and loggers, the closed shop had the added benefit of union hiring halls, which eliminated the parasitic practices of crimps and employment agencies. To Hammond, the closed shop meant giving labor organizers like Furuseth "the exclusive right to make rules as to what you shall wear, what material you shall use in building your house, who shall work, who shall not work, and to enforce these rules through the boycott and by violence and intimidation."[31] Hammond and other businessmen maintained that the closed shop stifled free trade and economic growth. Certainly, in Hammond's case, San Francisco's closed shop had shut out his lumber as a boycott against his labor policies.

In his first major battle with organized labor, Hammond emerged victorious. More important, he asserted the depth of his anti-labor commitment and discovered a successful strategy in confronting organized labor: using strikebreakers and injunctions while purging union members. Hammond's creation of the Humboldt Stevedore Company effectively broke the longshoremen union's hold on the Eureka waterfront. His steadfast refusal to negotiate with the SUP ensured that the HLC would remain the only nonunion shipping company on the West Coast. Although the coalition of shipowners collapsed during the strike, Hammond remained convinced that if owners and businessmen united and organized, they could defeat unionism. He soon got his chance to prove it.

As a young man, Hammond had witnessed how Pope and Talbot effectively controlled its workforce at Port Gamble, where "isolation precluded outside interference, by government agents, business competitors, or early-day labor organizers."[32] While Hammond replicated this model in Bonner, Montana, Samoa proved the exemplary company town. Although just across the bay from Eureka, the peninsula was accessible only by the *Antelope* or the Oregon and Eureka Railroad, both owned by the HLC. As was the case at Port Gamble, Samoa's isolation helped Hammond maintain control over his workforce.

When Hammond purchased Samoa in 1900, the town site included a cookhouse, several cottages, and a large bunkhouse. He put his brother Henry in charge of building a new company town. In Henry's first action, he broke with the larger community by moving the company's main office from Eureka to Samoa, allowing the HLC to operate independently from the political and social influences of Eureka. In contrast, William Carson valued his community ties and hired local men as much as possible. Hammond had no such loyalty, especially when those workers belonged to a union, and he imported laborers whenever possible.[33]

In a spurt of activity, Henry oversaw the construction of nearly a hundred houses, a new bunkhouse and cookhouse capable of seating the entire workforce, and a general store. To accommodate the clerks and managers arriving from San Francisco and other points of Hammond's rapidly expanding empire, Henry also built a small hotel that included a private dining room and suite for A. B. and a social hall on the second floor. While his main office remained in San Francisco, Hammond made Samoa his operational headquarters and flagship mill.[34]

The construction more than doubled Samoa's housing capacity to 250 and boosted lodging at the cookhouse to 350. The purpose of housing workers in Samoa was not to eliminate their commute from Eureka but to create a captive workforce entirely dependent upon the company—not just for jobs but also housing, food, education, recreation, and a social network. From Hammond's point of view this would reduce worker turnover and the potential for labor unrest and strikes. Although workers repeatedly cited poor working conditions, low pay, and long hours, Hammond was convinced that labor unrest was a result of "outside agitators." At Samoa, the HLC could easily discover and expel such "undesirables." Furthermore, all workers, especially those with families, would be highly reluctant to go on strike as they could lose their homes as well as their jobs. Samoa, however, was no gulag. The size and quality of houses compared with any in Eureka. The streets were wide and clean. By 1910 Samoa boasted parks, a school, a company baseball team, fraternal societies, and a recreational hall that sponsored weekly dances.[35]

Hammond's new corporate model demanded efficiency, and one way to cut costs was to centralize cooking and food production. Not far from the cookhouse, the company established a poultry farm with more than one thousand chickens, turkeys, and ducks. Nearby, a hundred hogs gorged on the cookhouse scraps. The company also installed a ten-thousand-square-foot cold storage facility that brimmed with beef, pork, eggs, chicken, fruit, and vegetables. This supplied not only a retail butcher shop, the cookhouse, and woods camps but also the seagoing vessels that docked at Samoa's port. On the HLC's cutover lands, men raised cattle to supply the store and camps, while the company orchard provided fresh fruit to the logging camps. The HLC also installed a bakery capable of baking four thousand loaves at a time.[36]

While centralization reduced costs, workers were less enthusiastic about such "efficiency." Instead of having camps cook order food, the HLC hired an overseer to lower expenses. The loggers complained that the meat was moldy and unfit for consumption, and in July 1906 the workers at two of Hammond's camps walked out. Even the camp dishwashers left, saying they did not get enough to eat.[37]

Although living conditions in the company town were satisfactory, those in the logging camps were appalling. To meet the increased production demands and keep expenses low, lumbermen built primitive camps that housed hundreds of men. Crammed into vermin- and lice-invested bunkhouses, men bedded on dirty straw mattresses and filthy blankets, often two to a bunk. With no running water or bathing facilities, men returned muddy and wet from the day's work and donned the same damp clothes day in and out. Camps lacked even the most basic sanitation. Fly-infested open-pit latrines squatted next to the unscreened cookhouse. In its report on timber camp conditions, the American Red Cross pointed out that not even "police regulations" could maintain hygiene when thirty-five men were housed under one roof."[38]

Poor working conditions in the woods led to incredibly high turnover. This suited most lumbermen, as they viewed loggers as chattel to be worked to exhaustion and then replaced by a fresh crew. Loggers joked with some truth that camps had three crews at any one time: "one coming, one going, and one working." The apparent collaboration between foremen and employment agencies to split hiring fees infuriated workers. Agencies charged those seeking employment an advance fee and thus were financially motivated to send as many men as possible into the camps and mills. Receiving kickbacks from the agencies, foremen had little desire to retain workers. When workers arrived they often found the higher paying job they had been promised, such as carpenter or sawyer, was taken and the only jobs left were swamping and bucking. Given the miserable work conditions, low pay, and long hours, many simply asked for their time and moved on.[39]

Not surprisingly, considering their immense caloric needs, loggers' primary concern in camp was the quality and quantity of food. In fact, Humboldt's first strike occurred in 1881 when John Vance's woods crew walked out in protest over poor food. Quickly capitulating, Vance responded by firing the cook. Such small-scale, spontaneous walkouts were not unusual, and owners generally conceded to workers' demands over issues like food or an abusive foreman. When dissatisfied with their working conditions, the independent-minded loggers simply "struck with their feet" by packing up and looking for work elsewhere. Such independence and transience made labor organizing in camps especially challenging.[40]

The timber industry was extraordinarily labor intensive. Even as new technologies decreased the number of skilled positions, companies still required substantial manpower to extract trees from the forest, bring them to the mill, and turn them into finished products. Labor was one of the industry's highest costs, but unlike the relatively fixed cost of a mill or a railroad, lumbermen could ratchet wages up or down through layoffs and rehiring, depending on market demands. Although lumbermen needed dedicated and productive workers to maintain consistent production, their own policies promoted a highly transient and seasonal workforce. Hammond, in particular, embraced both archaic labor practices and bureaucratic efficiency.

The shift to the corporate organizational model aided Hammond in stitching together far-flung enterprises, but workers resented his new bureaucratic structure. By 1906 workers could discern a substantive difference between the paternalistic style of William Carson and the corporate approach of the Big Three (HLC, Pacific Lumber, and Northern

Redwood). Under Carson, workers sensed a personal connection with their employer, an almost reciprocal relationship between labor and capital. Workers held a degree of control over working conditions, hours, and pay. Although Carson and other pioneer lumbermen refused to recognize unions, they usually conceded the wage and hour conditions that workers demanded. In the new corporate model, owners lived elsewhere and ran their companies through general managers and corporate hierarchies. Workers were at the mercy of foremen and had little recourse for their grievances. In addition, Carson's employees lived in Eureka, while each of the Big Three built company towns that largely circumscribed workers' lives. Carson sought loyalty, and many of his workers stayed with his firm for decades. Hammond, in contrast, experienced rapid turnover and imported workers every year to replace those who quit or were fired. While the paternalism of many mill owners derived from their own experiences in working in camps and mills, Hammond was less sympathetic. Predictably, he maintained, "I worked at the hardest kind of work in logging camps when I was sixteen years old and did a man's work at very low rates, and it did me good."[41]

Despite Carson's ability to retain workers, the increasing scale of lumber operations compromised paternalism. Companies that once had seventy-five employees exploded to seven hundred. Hammond's own logging superintendent, W. W. Peed, identified the impersonal aspects of such rapid growth as the source of labor troubles. In a speech before the Pacific Logging Congress in 1910 he noted that "the large and complex organizations, scattered over a more or less extended territory" precluded "the close relations" that previously existed between managers and the workers.[42]

Hammond's anti-labor practices stood out even in an industry renowned for its poor labor relations. While union leaders assailed lumber companies in general, they often targeted Hammond in particular, and he became the lightning rod for labor's wrath, a role he heartily welcomed. He acknowledged the war between the classes and intended to win it unconditionally.

Although Hammond fell short in his attempt to organize the redwood industry into one giant combination, he succeeded in coalescing the largest producers around an open-shop campaign. By 1904 the

Big Three accounted for more than 60 percent of Humboldt's lumber production. With Hammond as the ringleader, the Big Three acted in concert to dominate the redwood industry. As the primary stockholder, Hammond wielded total control over his company and could engage in polices of his own creation, while both E. A. Blockinger (manager of Pacific Lumber) and Henry Jackson (president of Northern Redwood) answered to their directors. Jackson, however, appeared so much in line with Hammond's directives that the *Labor News* rhetorically questioned who he actually worked for—Hammond or Northern Redwood. Hammond also cemented a close friendship with Charles Johnson, owner of the Union Lumber Company, the largest in neighboring Mendocino County. Indeed, the two men repeatedly discussed merging their companies.[43]

The *Labor News* hinted that Hammond deliberately goaded workers into striking so he could crush unions in their infancy. However, the desire to achieve efficiency, standardization, and cost-cutting measures were probably what led Hammond, Jackson, and Blockinger to announce a universal wage scale for HLMA members in January 1905. Regardless of their intentions, the plan antagonized workers. Not only did it cut wages 10 to 15 percent; it also contained a fifteen-dollar monthly deduction for cookhouse meals. Traditionally, lumber companies provided room and board without charge in the logging camps. But loggers consumed a prodigious amount of food—some six thousand to eight thousand calories a day, double that of a normal adult. Breakfast alone consisted of bacon, eggs, biscuits, potatoes, coffee, doughnuts, and pancakes. At dinner, each man consumed an average of one pound of meat and another pound of potatoes. To reduce such expenses, the new plan required workers to pay for their meals whether they worked or not. On many days operations shut down because of weather or equipment failure; the men received no pay on those days but still needed to eat. Mill employees in the company towns also opposed the board plan since the company charged them whether they ate at the cookhouse or at home. Workers were livid at having their pay slashed since both the cost of living and the price of lumber had increased over the previous year.[44]

William Carson also deplored Hammond's plan as "just as vicious a scheme as any that labor unions try to force." He recognized that it would quickly lead to a worker revolt and averred that Hammond and Pacific Lumber were doing "more to foster the growth of labor unions here than anything else in the County."[45] The manager of Falk's Elk River Mill, Irwin Harpster, agreed with Carson's desire to maintain present wages and not adopt the HLMA's proposal. Suggesting that keeping wages steady would prevent a strike against their company, Harpster informed his partner in San Francisco that "it was common talk on the streets of Eureka yesterday, 'That Carson's and the Elk River Mill Co. were the only ones who treated their men fairly.'"[46]

Nonetheless, both Carson and Harpster worried that the actions of the Big Three would provoke an industry-wide shutdown and that their mills would be dragged into the conflict. Unlike Hammond, Harpster was reluctant to stir up trouble among workers, noting, "There is considerable unrest among the masses in this country."[47] As an astute businessman, Hammond was aware of the volatile labor situation yet was willing to risk confrontation. Hammond clearly wished to crush organized labor, as the *Labor News* had charged.

By the spring of 1906, it was apparent that Hammond was deliberately provoking a strike to purge the union from Humboldt County's lumber industry. He began flooding Humboldt's labor market to drive down wages and replace union workers. Directed to train the inexperienced men, six of Hammond's older loggers quit in protest. The Big Three advertised for workers in Midwest newspapers, offering a twenty-five-dollar refund of the fifty-dollar fare to Humboldt after six months of work.[48] The HLC produced an elaborate twelve-page pamphlet to entice workers to Humboldt County. Shameless in its boosterism and outright falsehoods, it pictured a "bachelor's cabin" with drapes and a rocking chair out front. The text read, "There's a good job for you in the land of sunshine, fruit and flowers. . . . California is the Garden of the Pacific and Humboldt County is the Paradise of California. Here it never snows or freezes in Winter, or scorches in summer. Fogs are unknown. . . . Work under pleasant conditions among agreeable surroundings for better wages, with bright prospects for the future. . . . If he is single he is furnished with a cabin rent free and his board at the camp restaurant, where the best of food is supplied costs him 50 cents a day. . . . A common laborer can save $45 a month clear."[49]

The reality was quite different. Upon arrival, workers found summer heat, winter fog, and heavy physical labor working around deafening steam engines. Meals in the "camp restaurant" were wolfed down in twelve minutes. Workers also discovered that after deductions for board, transportation, and purchases at the company store, they were lucky if they had ten dollars in their pockets at the end of the month.[50]

The workers in the timber camps began to organize. Despite the widely scattered camps and long work weeks that made it difficult to attend meetings, by April 1906 more than one thousand had joined the nascent International Brotherhood of Woodsmen and Sawmill Workers (IBWSW). In August they received their official charter from the AFL. Instead of preventing organized labor, Hammond had galvanized a movement, just as Carson had predicted.[51]

Organizing continued throughout 1906, and the heightened rhetoric and promises from the union increased worker expectations, yet they saw little change in their conditions. Daniel Cornford pointed out that the IBWSW was caught in precarious position: its leaders realized that a confrontation with the lumber barons would be risky with only half of Humboldt's lumber workers as members, yet "workers were anxious to redress their grievances." Some took matters into their own hands and set fire to the Samoa lumberyard. This only hardened Hammond's resolve. By November he was Humboldt's only employer who had not granted a wage increase in either his planing or shingle mills.[52]

At the start of the 1907 logging season, the IBWSW drew up a list of demands: restoration of free board, a return to 1904 wages, a minimum wage of forty dollars per month for mill workers, time and a half for overtime, and, for woods workers, breakfast no earlier than 6:00 A.M. and dinner no later than 6:00 P.M. The union asked the mill owners to carefully consider their proposition and "inform us whether or not the enclosed schedule is acceptable."[53]

Under Hammond's direction, the HLMA not only flatly rejected all the demands but refused to recognize or meet with any union representatives. As Daniel Cornford suggested, "The humbling of the once powerful Longshoremen's Union emboldened the lumber owners to take a tough stand against the IBWSW."[54] Other lumbermen attempted to defuse the situation. Pacific Lumber granted a $5/month wage increase, but any worker with a pencil could figure out that this still resulted in a net reduction of $5 to $10 a month from the 1904 level. Carson, who had neither reduced wages nor imposed the board charges, responded to the growing discontent by raising wages $2.50 a month.[55]

On April 28, 1907, Humboldt's local unions voted nearly unanimously to strike. Three days later, on International Labor Day, 2,500 men walked out in the largest strike in the history of Humboldt County. Hammond's manager, George Fenwick, declared, "Any of our men who want to quit are certainly privileged to do so." Fenwick expected 10 to 15 workers to leave, stating that "their places can be readily filled." No doubt Fenwick was surprised to find himself short 400 workers the next day. Hundreds of men, with all their belongings packed into their bedrolls, flooded into Eureka from the company towns and woods camps. Flush with their final paychecks, strikers filled the streets and hotels. Yet the exodus of workers proceeded in orderly fashion. The *Humboldt Times* reported that "peace and order have prevailed" and "the saloons have not done an overwhelming business." By the end of the week, 600 men had boarded steamers to seek employment elsewhere.[56]

Hammond had prepared both his business and his fellow lumbermen for the strike. The lumbermen had a strategy as well as an underlying philosophy. Parroting his boss's position, Fenwick stated, "It is simply a question of running our own business and now is just as good a time as any to determine that question."[57] Furthermore, the mill owners had been stockpiling lumber and logs. Although only six men remained in all of Hammond's camps, the Samoa yard had enough logs to continue running for two weeks with a skeleton crew. Other mills simply intended to shut down and wait it out or take the opportunity to make repairs. Nonchalant, Hammond announced plans to build the world's largest sawmill in Astoria, Oregon.[58]

Lumbermen viewed the strike as a battle for the open shop, although the strikers were not asking specifically for union recognition but simply a restoration of the 1904 conditions. The HLMA issued an ultimatum signed by ten of the county's largest mills (both Falk's and Carson's excepted), declaring that "under no circumstances will [we] recognize the unions or treat with them." The power of the association over its members became evident when pioneer lumberman Isaac Minor, with many orders on hand, reached an agreement with the strikers but quickly rescinded it after Henry Jackson pressured him. The Eel River Valley Lumber Mill was also willing to settle the strike but claimed that it was unable to grant the demands without the association's approval.[59]

Jackson attempted to bring Carson into the fold so as to present a united front against the union, but the old lumberman declined. While Carson certainly did not support unions or strikes, he maintained that "some of those large companies (especially the Hammond company) have not treated their employees fairly and for that reason D and C don't care to assume any other troubles at this time as we have many friends among the labor unions when Hammond seems to be the bad man."[60] To Carson's relief, the union singled out Dolbeer and Carson and Falk's Elk River Mill as exempt from the strike. The day before the walkout, union organizer Ernest Pape, "a very warm friend" of Carson's son, Milton, informed the lumberman that his company "had already met the conditions proposed by the Union" and he "considered the differences settled."[61] Nonetheless, Carson fretted that the union would realize it had made a tactical mistake by not calling a strike at all the mills and camps.[62]

By the end of the week, the IWW (or the Wobblies, as they were known) had decided to join the strike as well. Because of the radical IWW's fundamental differences with the conservative AFL unions, the owners had assumed they would refuse to respond to a strike call. All workers in the lumber industry, however, faced the same conditions and pragmatically looked to whichever organization might improve their conditions regardless of its ideological orientation.

As the walkout surged to 3,500 workers and other Eureka unions threw their support behind the strike, Ernest Pape raised the stakes. He called for an eight-hour day as a condition for the men to return to work, arguing "men employed in the woods usually meet with accidents towards the close of the day and . . . shorter hours are really what are needed to prevent such losses of life."[63]

The hardline stance of Hammond, Jackson, and Blockinger contrasted sharply with that of Carson and Falk. With nearly all the Big Three's employees walking out, the strike "made almost a clean sweep" of those companies, yet not a single worker left Dolbeer and Carson.[64] Similarly, at Falk the men declined to strike because the company granted all of the workers' demands. Like Carson, Harpster stated, "The secret of the whole matter is we have tried to be fair with our men." Both companies operated from the core principle that "if the men feel they are treated fairly they . . . will return value received in the way of honest labor [and] . . . fair treatment lessens the chances of unionization."[65] Harpster noted that although the unions had not approached Elk River for recognition or an increase in wages, but if they did we "would be inclined to give it . . . for surely the mills have reaped a gold harvest the past year. And in all fairness, are not the men who helped to roll up the profits entitled to some of the good things as well? . . . [A] small advance over what they have been receiving would not at all cripple the Manufacturers."[66] The pioneer lumbermen apparently regarded workers as collaborators in their businesses, while the corporations saw labor as nothing more than overhead.

The conflict became rhetorical, with each side attempting to cast itself as winning. Some mill owners "declared they would just as soon see the mills closed down for five months" as capitulate to the unions. Unwilling to concede even this much power to labor, Hammond kept the Samoa mill running. Fenwick reported that thirty nonunion men had returned to work and two Hammond camps had reopened. The *Labor News* countered Fenwick's claim, pointing out that the HLC had only twenty-two men left in its camps and that foremen and bosses were running the steam donkeys and even greasing the skid roads. Ten days into the strike the HLC announced it was opening two camps with three hundred men and that all prospective employees should show up at the wharf in Eureka on Sunday afternoon. Although a crowd gathered to see what would happen, no one boarded the *Antelope* for Samoa. The union was holding the line.[67]

The *Labor News* blamed Hammond personally for the strike and claimed he had twin motives for prolonging it: "one is to fight his lifelong enemy, organized labor, and the other is that this will give him a chance to run competition to the wall."[68] While Hammond certainly had both goals in mind, his primary competitors, Dolbeer and Carson and Elk River, continued to operate. Yet Carson continued to fret over the larger implications. He noted, "There is no question in my mind that if the mill owners in this strike hold together they will burst the union."[69] Hammond, too, knew that the longer the strike went on the better his chances of breaking the union.

The strike was larger and lasted longer than the owners anticipated. Lumbermen admitted that their greatest loss would be the experienced workers—head sawyers and camp bosses—but fully expected the men would return. Owners attempted to retain these employees by offering them a significant increase in wages at the expense of less-skilled laborers, thinking this would also undermine worker solidarity. When this strategy failed, owners imported replacement workers. Working-class solidarity, however, proved resilient, for when these men discovered they were to be used as strikebreakers, many returned to San Francisco.[70]

Two weeks into the strike, owners circulated reports that the strike was settled, men were returning to work, and the camps were running full blast. The *Labor News* urged the strikers to disregard the rumors and hold out. But as the Hammond camps, operating with skeleton crews, continued to produce one or two trainloads of logs a day, the strikers grew anxious. More important, owners offered workers everything they demanded if the men forsook their unions and returned to work as individuals. Accepting the offers, men began trickling back to work. The HLC, however, blacklisted many of its former workers for being union members and refused to rehire them.[71]

Then on May 22, the IWW reached an agreement with the Eel River mill for a wage increase and free board, although not union recognition. The Wobblies returned to work, effectively breaking the worker coalition. The IBWSW held on for another ten days, until June 4, when its membership voted to end the strike. Both sides claimed victory; strikers received increased wages on an *individual* basis and free board at many mills. Solidarity had failed, however, and many workers tore up their union cards in frustration. The IBWSW gradually disintegrated and folded for good in 1911.[72]

Labor News editor Joseph Bredsteen attempted to put a positive spin on the strike, positing that workers had discovered their strength and begun to develop a class consciousness. He wrote that the strike "made a profound impression on the men and several thousand are thinking as they never thought before."[73] Consciousness aside, what workers discovered was that their unions were ill matched against the power of the mill owners. Furthermore, the capitulation on the part of the IWW in breaking the strike plagued their reputation and doomed their subsequent attempts to attract Humboldt's workers into the "one big union." Although their organizations were broken, workers were far from complacent. The deplorable working conditions that precipitated the strike remained and even worsened with the HLMA now securely in control of Humboldt's lumber industry. The fight was far from over.

Like many employers, Hammond refused to believe that his employees would willingly go on strike. Instead, he insisted that the 1907 strike was the work of outside agitators and that the strike was settled by a fistfight in which his foreman beat up the union organizer. Hammond, who was infatuated with Napoleon, couched the labor conflict in military terms. Referring to the setback of the union organizers, he wrote, "General Zandt having abandoned his troops, General Pape decided to surrender unconditionally."[74]

If anyone could claim victory, it was A. B. Hammond. He not only purged his company of labor "agitators" and union organizers but, more important, he effectively eliminated organized labor from Humboldt's lumber industry. Hammond had proved to other lumbermen that if they held firm and refused to negotiate with organized labor, they could continue to dictate the terms of employment and working conditions. Not until long after Hammond's death would unions in Humboldt again have the strength to challenge mill owners.

Having firmly established the open shop in Humboldt, Hammond resolved to carry it to the rest of the West Coast. As the 1907 strike suggested, Hammond's extreme antiunion position was based less upon achieving industrial efficiency and more upon a personal obsession in defense of a principle. Although he had won a decisive battle in Humboldt County, nationally the forces of labor and capital were gaining strength and headed for a showdown.

CHAPTER **17**

Socialists and Progressives

> Labor is prior to, and independent of, capital. Capital is only the fruit of labor, and could never have existed if labor had not first existed. Labor is the superior of capital and deserves much the higher consideration.
>
> —Abraham Lincoln

Having crushed the lumber strike, A. B. Hammond seized the offensive. For Hammond, Humboldt County became the front line of a national campaign for the open shop. While the company town of Samoa was firmly under Hammond's control, across the bay, Eureka continued to be a hotbed of radicalism, a place that enthusiastically welcomed IWW leader Big Bill Haywood following his release from an Idaho jail. Hammond could blacklist union leaders, impose "yellow-dog" contracts (which prevented workers from joining a union as a condition of employment), and use company spies to ferret out labor agitators or unfaithful workers, but Eureka eluded his grasp. Although Hammond was the county's largest employer, many men still worked for Carson, Falk, or smaller outfits; others worked as contract loggers, in shingle mills, or as farmers, carpenters, merchants, or lawyers. While economically dependent upon the lumber industry, Eureka retained its political independence as a thriving community with a well-established pro-labor atmosphere, a situation that aggravated Hammond to no end.

During the first two decades of the twentieth century, three great social forces—conservatism, socialism, and progressivism—battled for political and economic primacy in America. Industrialists, like Andrew Carnegie, and giant corporations, such as Standard Oil, clearly held the upper hand as they gobbled up greater and greater slices of the economic pie. Farmers, workers, and artisans in the independent-minded American West, however, continued their populist resistance to corporate domination. Simultaneously, immigration and the demise of family farms led to a greater supply of wage laborers, who were becoming increasingly alienated from their work and their employers and began to look toward socialism, radicalism, and labor unions to improve their conditions. Meanwhile, upper and middle-class reformers, loosely called Progressives, acknowledged the inequities spawned by industrialization and advocated government intervention to address the nation's social, economic, and environmental issues. Historian Gabriel Kolko argued that Progressivism was essentially a conservative response to preserve America's economic structure by granting small reforms to stave off more radical change. This helps explain why some corporations and businessmen supported the Progressive agenda.[1]

Many of America's second-tier capitalists, however, felt trapped between the large corporations and labor unions and strenuously resisted any reforms. Hoping to preserve the conditions of the Gilded Age that had engendered their success, conservative businessmen like Hammond desperately clung to the status quo and regarded the Progressive movement with nearly the same distain and hostility that they held toward Socialists. Hammond even considered Progressive Republicans, such as California Governor Hiram Johnson and Wisconsin Senator Robert La Follette to be Bolshevik sympathizers. Up until the end of World War I, the outcome of this three-way struggle among socialists, progressives, and conservatives for America's future remained an open question. Firmly in control of the nation's economic system, industrialists had the upper hand.[2]

Like many capitalists, Hammond's conservative politics were largely based on an economic self-interest that informed his ideology. Although lacking a formal education above the eighth grade, Hammond saw himself as in the intellectual vanguard of the business community, and he carried out a campaign to stamp out unionism, not only in his own companies but in all industries. Hammond regarded organized labor as a step toward socialism and thus a threat to fundamental American values.

Hammond's battle with labor, however, often ran counter to pragmatic economic interests and assumed the character of a crusade. Humboldt County lay at the forefront.

With a strong craft union tradition, Eureka gravitated toward both trade unionism and socialism as remedies for societal ills. Although nationally at odds, in Humboldt Socialists and trade unions espoused the same goals and often cooperated politically. This cross-fertilization between unions and Socialists, along with the rise of industrial unionism, provided Hammond with plenty of rhetorical ammunition. He wasted little time in putting it to use.

The case of the Union Labor Hospital illustrated the continuing vitality of Eureka's labor movement, even after the failed lumber strike. Local unions and individuals, primarily timber workers, purchased twenty-five-dollar bonds to finance construction of their own hospital. In one of the earliest forms of health insurance, union members paid ten dollars a year for a ticket that entitled them to medical care. In addition to providing a tangible benefit and an incentive to join a union, the hospital had an immeasurable psychological effect. As the *Labor News* noted in 1908, "The Union Labor Hospital awakened a new hope and instilled a new ambition in the hearts and minds of the woodsmen and sawmill workers."[3]

Hospital coverage was particularly important in the timber industry, which saw more injuries and deaths than all other occupations combined. While Humboldt County contained only 2 percent of California's population in 1914, it accounted for 20 percent of all industrial accidents, nearly all of which were in the timber industry. In Oregon, in 1906 there were 151 sawmill accidents, one-fourth of which were fatal, while over half of the twenty-eight logging accidents ended in a fatality. Citing more deaths per month from logging in Washington State than in the Spanish-American War of 1898, the state's Board of Safety declared the timber industry "more deadly than war."[4]

While owners, workers, and government officials all recognized the need for health care for timber industry workers, they championed radically different approaches. Drawing upon their faith in government, Progressives advocated state workers' compensation programs, while unions instituted their own insurance plans and built their own hospitals.

Unwilling to cede control, employers devised their own unique solution: welfare capitalism, whereby owners themselves would provide for the needs of workers. The lumber industry, because of its isolated working conditions and company towns, provided an early form of welfare capitalism. In Port Gamble, for example, the company provided housing, a store, church, and meals. By the turn of the century, welfare capitalism was gaining currency across the country as big businesses recognized that providing benefits could ensure employee loyalty and discourage unionism. In addition to medical care, benefits included housing, recreational facilities, libraries, and profit sharing. But without corresponding wage increases, many workers regarded these benefits as token efforts. Moreover, as the strikes of 1907 demonstrated, loggers viewed room and board as fundamental rights, not added benefits. The real issue boiled down to worker autonomy versus company control.[5]

George Fenwick had advocated a company hospital plan for the HLC for several years. Fenwick and other employers, however, saw the construction of the Union Labor Hospital as a threat to corporate supremacy. In February 1908, Fenwick and the managers of the Northern Redwood and Pacific Lumber companies, Henry Jackson and E. A. Blockinger, met in the back room of the Hammond-controlled First National Bank of Eureka to hatch a plan to destroy the hospital. Intending to undermine Humboldt's labor movement while simultaneously providing employee benefits, the plan imposed a compulsory hospital fee but excluded the Union Labor Hospital from the list of providers. In deducting one dollar a month from each worker's paycheck, the Big Three granted themselves a commission of 12.5 cents from every dollar. For workers, the primary issue was that a company-imposed hospital fee restricted their autonomy, as they previously had purchased plans in the hospital of their choice.[6]

Furthermore, workers regarded the hospital fee as an additional insult on top of the "fleece me" company stores, dangerous working conditions, and unsanitary camps. While a dollar a month might seem an insignificant deduction, it represented another example of corporate domination. From employee paychecks the companies already subtracted merchandise purchased at the company stores, meals at the company cookhouse, rent, utilities, fuel wood for company housing, and transportation on company ferries and trains. Such deductions quickly drained a worker's paycheck. For the month of January 1914, Hammond employee Charles Clay earned $42.80. Subtracting store charges, rent, lights, and hospital fee, he netted $6.60 in take-home pay. If a worker had a family to support, store charges chewed up even more. Another of Hammond's employees, J. M. Wassam, working for $3.40 per day, had store charges of $73.58. After rent, lights, and hospital fees, Wassam collected $1.42 for the month. An extreme case was L. H. Blean, who worked ten hours a day for $1.75 a day. Out of his $49 in monthly pay, $39.95 went to the store, $6.50 to rent, $1.80 for lights and $0.75 to the hospital fee. Blean received an empty paycheck after working 28 ten-hour days.[7]

Nor was it any better in the woods camps. The *Labor News* cited the case of one worker who came from San Francisco expecting to do carpentry but was put to work in the woods. He quit after a week and received his pay of $8, out of which $0.50 per day was deducted for board, $2.75 for two blankets, $1 for a bed tick without straw, and $1 for the hospital ticket. His paycheck left him with two bits. Even the foremen found their take-home pay amounted to only one-third of their $150 monthly salary.[8]

While the company-imposed hospital fee was yet another deduction from worker paychecks, for the Union Labor Hospital it was a matter of life or death. By excluding it from their list of providers, the Big Three, which employed 4,500 out of the county's 6,000 lumber workers, could financially strangle the Union Labor Hospital. Therefore, in response to the new plan, the hospital filed suit against the HLC and six other lumber companies. Employing the same argument that Hammond used against the Sailors' Union of the Pacific, the hospital charged that the lumber companies had conspired to boycott and had coerced and intimidated their patrons, thus injuring their business. After six months of litigation, a jury found that Fenwick and the other defendants "acted maliciously with intent to injure the plaintiff in the hospital business" and were therefore guilty of conspiracy and combination. The judge issued an injunction against the lumber companies from discriminating against the Union Labor Hospital. The Bayside Lumber Company "complied" by posting a notice that read, "We are including the Union Labor hospital in the list from which you many choose and if any employee wishes to change his former selection, and FOR OUR MUTUAL BENEFIT WE TRUST THERE ARE NONE, he will call at the office and make such selection on a blank [form] that has been prepared."[9]

Although the HLC reluctantly complied with the court injunction, Hammond publicly declared, "The Union Labor hospital was promoted by a number of professional agitators, too lazy to labor, who make a living by preying upon the working men of this country." Insisting that it was "a breeding place for anarchy [and] socialism," Hammond stated that "our company will not give it support, and will do all that it legally can to prevent its employees from patronizing it." Indeed, the next summer the HLC announced it would build its own hospital in Samoa. Apparently, this was just a threat, as the company continued to patronize the Sequoia Hospital, a subsidiary of the Humboldt Stevedore Company, which in turn was owned by a consortium of lumber companies, including the HLC.[10]

While economics played a role, for Hammond labor issues became a moral crusade between right and wrong. Hammond put himself on the firing line as he traveled up and down the West Coast, imploring lumbermen to stand with him in his open-shop campaign. The building trade unions had been successful in keeping nonunion-produced lumber out of San Francisco, and Hammond fervently battled for the hearts and minds not of workers but of middle-class professionals and businessmen, who were often sympathetic to the plight of labor.

Visiting Humboldt County for the first time in four years, in June 1908 Hammond engaged in a heated public debate with his ideological challenger, Joseph Bredsteen, Socialist editor of the *Labor News*. Hammond applied his powers of persuasion, which he believed to be superior, to convince Humboldt residents of the righteousness of his position. Employing a combination of economics, patriotism, and morality, he wrote letter after letter, which the local press happily printed in their entirety. While Hammond and Bredsteen focused on a specific place and industry, their views were indicative of a much larger discourse and captured the zeitgeist of the time. Both men realized that this was not necessarily a battle for Humboldt County but one over the future of the nation.

Hammond saw the world as deeply divided. On one side there was prosperity, law and order, and the open shop. On the other lay "the rule of the boycott, with the violence and intimidation with which it is invariably enforces [*sic*]." Bredsteen riposted Hammond's declaration with his own divide. He wrote, "The time is now here when the people of Eureka must choose between a rule of organized labor and a rule of the mill owner's association." He pointed to Samoa as an example of "what company domination means" and accused Hammond's company store of hurting Eureka businesses. He noted, "Practically every penny worth of business that goes into that store would go to Eureka merchants if it were not there." In response, Hammond pointed to the positive aspects of Samoa as a company town, having fewer saloons, no labor unions, and no "Woman's Union Label League" but, rather, being a "community of nearly 500 law-abiding people living under conditions that are apparently satisfactory."[11]

Both men used the deteriorating economic conditions of Eureka to bolster their positions. The Panic of 1907 had hit Humboldt County hard with a drop in lumber shipments. As he had elsewhere, Hammond turned the depression to his advantage, cutting wages and flooding the labor market with imported workers. The *Labor News* blamed the "arrested growth of Eureka," including business failures, population decline, and economic stagnation explicitly on the policies of the Hammond Lumber Company. According to Bredsteen, "When unions are injured the community is injured. . . . When wages are raised [local] business booms."[12] If the plethora of advertisements from local businesses in the *Labor News* were any indication, many agreed with this assessment. Both workers and local businesses regarded the large, corporate lumber companies as alien forces that sucked the economic vitality out of the county.

In a shotgun blast of vitriol, Hammond attacked the principle of unionism. Pinballing from one subject to another, he blamed unions for imposing a closed shop in Eureka that raised lumber prices, accused organized labor of "foully murdering two farmer boys," and insisted that Eureka was beset by Socialists who "intimidated by the threat of boycott" and held the city hostage though political bosses. This, Hammond charged, was the real cause of the decline in population and in real estate values.[13]

Hammond further noted that Eureka's union mills, which ran on eight-hour days, could sell their lumber only in the protected markets of

Eureka and San Francisco. Falling back on the traditional argument used by Pacific coast lumbermen, Hammond insisted that the HLC needed to compete with Midwest mills that ran ten-hour days, and if Samoa went to an eight-hour day, it would put two hundred men out of work.[14]

Bredsteen, meanwhile, traced the labor troubles back to 1905, "when the wages of the woodsmen were reduced . . . [and] Hammond and the rest raised [lumber] prices."[15] Hammond refuted Bredsteen's charges as "an absolute and unmitigated falsehood" and then launched into a fanatical tirade. Hammond denied that he cut wages, insisting that the companies adopted "a daily wage schedule instead of monthly." In justifying this action, Hammond contended, "This arrangement was a satisfactory change for all industrious men. It did not, however, please the loafers who were in the habit of lying around the camp, refusing to work, and eating the employer's food without rendering any services." He then added that even if the companies had cut wages, "they were acting entirely within their rights."[16] Although he presented it as an aside, this was the core question: who would control the conditions of employment and means of production—workers or capitalists?

Hammond declared that he was not opposed to organized labor, just the closed shop, collective bargaining, and boycotts. These, of course, were the very elements that gave unions power. In advocating the open shop, Hammond believed he was supporting the rights of nonunion employees. Like many others infused with the frontier spirit, he was appalled that unions compelled workers to join to secure a job. Insisting that Hammond's open shop weeded out "men and women of independent spirit," Bredsteen declared that in none of Hammond's enterprises was there a single union member.[17] Indeed, eliminating union workers was precisely Hammond's intent.

Like Hammond, Bredsteen understood this debate as part of a larger battle between labor and capital. In one of his editorials, Bredsteen affirmed, "Hammond is merely a type. . . . The real question is not Hammond versus Humboldt but company domination versus the rule of the people." Bredsteen insisted that neither the mill owners nor their capital created any actual wealth for the residents of Humboldt. Instead he attributed Humboldt's wealth to "our bountiful forests of giant redwoods." Bredsteen asserted, "Most of this forest land was fraudulently obtained from our government in the first place."[18] On another occasion, he accused Hammond of extracting and appropriating this natural wealth and further charged that "through his steamboat, bank and the company's stores, he has managed to get back a considerable portion of what he has paid out in wages."[19]

Like many Americans, Bredsteen regarded the growth and consolidation of corporations with trepidation. Bredsteen correctly discerned Hammond's intentions of merging the lumber companies of Humboldt "into one gigantic combination" that would "continue Hammond's tactics until our hills are stripped of their forests, our people are impoverished, local business largely ruined."[20] Echoing Hammond's opponents in Missoula twenty years earlier, Bredsteen accused him of doing "business under so many different names that nobody seems to know where its branches of business begin or end." He warned his readers, "The octopus, having its head across the bay, has been very busy stretching out its multitudinous arms with the greed and perniciousness of a Rockefeller."[21]

Bredsteen called upon organized labor as "the one power that can defeat the aims of A. B. Hammond . . . for this reason he is determined to crush the labor unions." When Hammond threatened to close the Samoa mill as a response to union organizing, Bredsteen suggested that the workers take it over and run it themselves.[22] All across the country ideas such as collective ownership were gaining traction as a solution to the excesses of capitalism. American Socialism had inherited the mantel of the Populist outcry against "monopoly."

Safely ensconced deep within the confines of the Victorian-style company hotel at Samoa, Hammond lobbed verbal missiles across the bay and beyond. Given the accusations of murder and the venomous tenor in these letters, it seems unlikely that Hammond intended to persuade his opponents. Rather, he hoped to use the Eureka press to address the general public, but this was secondary to his enjoyment of a good fight. Like a modern radio talk show host, he relished inciting opposition and thrived on controversy.

After ten days of debate with Bredsteen, Hammond grew increasingly agitated and frustrated. Using the press as a mouthpiece was not working. Despite his illusions, Hammond was not very persuasive in print.

The wild accusations and inflammatory rhetoric in his rambling epistles overshadowed valid points. Furthermore, his insistence on equating unions with violence and lawlessness won few converts in pro-labor Eureka. It was time to try another tack, one he excelled at—behind-the-scenes manipulation. Hammond's loyal cadre of managers, who were far less abrasive, could implement his policies without ideological ranting.

As a union town, Eureka, like San Francisco, maintained a boycott against Hammond lumber. Although the Eureka building market was insignificant to a large company like the HLC, the boycott irritated Hammond. More important, if he could pry open Eureka to nonunion building materials, he should be able to employ the same strategy in San Francisco, the largest closed-shop city in the West. As he controlled the largest bank in Humboldt County, Hammond and his lieutenants coerced local merchants into signing an agreement to support the open shop.[23]

Flexing his economic muscle, Hammond deliberately antagonized organized labor further when he leased the Bendixsen shipyards at Samoa for his shipbuilding enterprise. Pacific coast ship carpenters had already won the eight-hour day, but Hammond announced that he would institute a nine-hour day in his shipyards. The refusal of the carpenters' union to work the extra hour provided Hammond with the excuse to fire union workers. The action prompted an editorial in the *San Francisco Clarion* comparing Hammond to Harrison Gray Otis, the notoriously anti-labor publisher of the *Los Angeles Times*. Noting that "few men have so bitterly opposed the trade-union movement as A. B. Hammond," the paper concluded that "long hours and low wages seem to the Otises and the Hammonds the cure for our industrial troubles."[24]

Hammond indeed joined forces with Otis and Henry Huntington to ensure Los Angeles would remain an open-shop city. Although Hammond could provide dock pilings and raw timber, he was essentially shut out of the San Francisco market for finished lumber because of his anti-union practices. He was determined not to let this happen in L.A., where the Southern California division of the HLC had recently surpassed all others as his primary profit-making enterprise. But in 1910, San Francisco's unions sent labor organizers to Los Angeles, where, like many cities in the West, the Socialist Party was on the upswing; indeed, it appeared poised to capture the mayor's office in the upcoming election. Then, on October 1, 1910, an explosion rocked the building housing Otis's *Los Angeles Times,* killing twenty-one people. A year later, James and John McNamara of the structural ironworkers' union confessed to the crime, sinking any hopes of political victory for organized labor in Los Angeles. From then on, Otis and Hammond used the *Times* bombing as an example of the violence and lawlessness that organized labor could cause.[25]

The year 1910 proved tragic for Hammond as well. Not only had his workers grown alienated, but so had his children. For two decades he had been so intent upon building his empire that he had scarcely recognized their emergence into adulthood. Often as intimidating and demanding with his children as he was with his employees, he could also be quite indulgent. Despite his antipathy toward higher education, Hammond sent his sons, Richard and Leonard, to Harvard and Stanford. Hammond derided other young men who neglected serious employment in favor "of the companionships of young men and young ladies," but he allowed his sons to spend their summers hunting, fishing, and socializing.[26] While Hammond expected his sons would soon join him in the lumber business, he also provided them with plenty of spending money and a comfortable life. In insulating his children from the necessities of life, however, Hammond isolated them from himself.[27]

A. B. and Florence had also grown increasingly estranged, and during his long business absences, she sought refuge in the presence of the children. Together, Florence and the children spent several summers touring Europe. On one such vacation in 1905, Richard contracted tuberculosis in London. Upon returning to the United States, Richard remained frail. Unable to provide a cure, doctors prescribed the dry air of the mountain west. Richard spent the next summer in Montana trying to recover and then moved to Colorado, where Hammond stopped by for a visit on his way to New York in 1910. The following January Richard's condition worsened, and Florence traveled to Colorado to take him to Tucson in hopes that a lower altitude would provide some relief. A week later, Hammond penned an uncharacteristically short letter to his old friend C. H. McLeod: "Richard died at noon today."[28]

Like most men of his background, Hammond had been conditioned by hardships to display little emotion, especially regarding personal matters, and he remarked little on Richard's death. Nevertheless, it affected him profoundly. Now that his offspring were adults, it seemed too late to reconcile the past, and he plunged into his work. With Richard's illness, Hammond had turned to his second son, Leonard, and in 1909 made him secretary and treasurer of the Eddy, Hammond Company, A. B.'s multimillion-dollar holding company. Leonard now bore the full burden of his father's expectations, and while he would prove competent, he also rebelled against Hammond's overbearing authority in his own ways.[29]

Although Richard's illness weighed on his mind, Hammond was further distracted by the old timber-poaching case in Montana. Like Jesus raising Lazarus from the dead, the General Land Office resurrected the investigation of Hammond from a bureaucratic purgatory. Attorney General William Miller had suspended the timber suits in 1889, but in 1907, Louis Sharp, division chief of the GLO, reopened the matter. "After a careful investigation," Sharp concluded that the dismissal had been for political rather than legal reasons.[30]

Moreover, the political landscape had changed significantly since the 1890s. President Theodore Roosevelt and his chief forester, Gifford Pinchot, simultaneously sought to rationalize timber production, conserve forests for future uses, and protect watersheds. Part of their Progressive agenda was to correct past abuses. Roosevelt had already given Hammond fits with his pursuit of the Oregon land frauds, but it was the administration of his successor, William Howard Taft, that caused Hammond's biggest problems. In addition, in 1909 a conservation-oriented Congress appropriated money specifically to examine old timber-poaching cases.

Continuing his investigation, Sharp determined that Hammond had poached twenty million board feet of timber off public lands and recommended filing suit for $211,000. Secretary of Interior Richard Ballinger, however, was more interested in compromise settlements than in criminal prosecution. For example, Kenneth Ross, the manager of the Bonner Mill, now owned by the Anaconda Copper Company, jumped at the offer to settle a $60,000 suit for $45,000. Under pressure, Sharp suggested the government settle Hammond's case for $20,000, in what amounted to a token stumpage fee of $1 per mbf. From Missoula, C. H.

McLeod concurred and advised Hammond to settle as soon as possible "as it will cost a lot of money and trouble to defend this case."[31]

Over the years, however, Hammond had grown increasingly uncompromising. He saw himself as a political power broker who was firmly in control of one of the largest lumber companies on the West Coast. His friends and business associates ranked among the most powerful men in the country. He could count on the support of Senator Thomas Carter, or so he thought, and was not about to let some government bureaucrat push him around. Furthermore, he had successfully evaded prosecution in the Oregon land fraud trials, in which there had been stronger evidence against him.

Indignant that the government could bring up a twenty-year-old court case, Hammond maintained his innocence and blamed the logging on others. He fingered his mentors, Bonner and Eddy, his brother Henry, and his brother-in-law George Fenwick as the culprits and wondered why he should take the heat for all of them. Hammond stated, "In neither of these operations had I any interest, nor did I receive any benefits therefrom."[32] That claim was preposterous, and everyone knew it.

By interviewing witnesses, examining stumps and shipping dates, and surveying timberlands, Sharp concluded that as general manager of the Montana Improvement Company and the Big Blackfoot Milling Company, Hammond "was very active in employing men and contractors and directing their cutting on government land." Sharp reported that Hammond had not only deliberately cut federal timber but also "discharged foremen and others who refused to carry out his orders with reference to government timber."[33]

In the face of such evidence, Hammond's old attorney in Montana, Thomas Marshall, suggested that instead of declaring total innocence, Hammond should rest his defense on the nebulous category of "mineral lands." Marshall argued that since the cutting occurred on public "mineral lands" and was therefore legal under the policy of the time, there was "no liability then, and can be none now." Furthermore, the lumber went to Anaconda for mining purposes within the state, a legally valid use.[34]

Although he rejected this defense, Hammond was not in complete self-denial over his involvement. Writing to C. H. McLeod, he mused over possible options. Declaring that he was not a partner with Bonner and Eddy in the Montana Improvement Company but "simply owned a certain amount of stock in the company," Hammond wondered if "a stockholder in the company is responsible for the acts of other stockholders; in other words would a man who owned one share in the Montana Improvement Company be financially responsible for the acts of the other stockholders."[35] Hammond, of course, owned more than one share of the MIC.

Sharp, too, wished to settle the case without going to court, but Hammond informed him that "under no conditions would we make any settlement with the Government, as we had nothing to settle." The lumberman insisted "that all timber cut on the Blackfoot was cut on Northern Pacific lands or lands on which we had a permit from the government to cut."[36]

Faced with such intransigence, 1910, the government filed suit in June 1910. Hammond confided to McLeod, "I do not think it is going to be a very formidable affair."[37] Nevertheless, he dispatched George McLeod to Missoula to search the papers of the Big Blackfoot Milling Company to find the government permit that had authorized logging along the Blackfoot River. In the meantime, Hammond's attorney and son-in-law, Steward Burnett, pursued a series of legal maneuvers to have the case dismissed. Hammond maintained that the government had to file two separate suits, one each for the Montana Improvement Company and the Big Blackfoot Milling Company, "on account of the two companies having different presidents, directors and stockholders." Yet as the prosecutors were well aware, Hammond controlled both companies, a fact that he steadfastly denied. Burnett built his case on the government's inability to specify which company had cut the timber and where exactly it occurred. Although he continued to maintain his innocence, Hammond also asked C. H. McLeod to keep track of Sharp and his effort to locate witnesses, many of whom still remained in Montana despite the twenty-five-year time lag. One such witness approached McLeod and proposed that if Hammond ponied up $500 he would "make himself unavailable to testify." Believing that the government had a weak case, Hammond declined the offer. But to be on the safe side, he told McLeod that one of the old contract loggers "should be 'braced up' and told not to talk or give any information whatsoever."[38]

Hammond also called upon Senator Thomas Carter, whose political career he had helped launch twenty years earlier. Now, however, Carter was in his second Senate term; firmly ensconced in Washington, D.C., he no longer considered himself "under any obligations whatever to A. B. Hammond or his associates." Furthermore, Carter was not only a "champion of Secretary Ballinger" but also "intimate friends" with Louis Sharp. Nonetheless, Carter assured Hammond that he would discuss the timber suits with his friends in the Interior Department. McLeod, however, believed that "Corkscrew Tom" was purposefully delaying his efforts in order to convince the Missoula Mercantile Company political machine that it needed him in the upcoming election.[39]

By now, Hammond was thoroughly disillusioned with the Republican Party, declaring, "It is about time that the whole nest was cleaned out." In clear denial of his past, he insisted that he never asked for any political favors, but simply aided the party "on general principles . . . and I hardly think it has been appreciated." Despite his claim, for Hammond politics was less about "general principles" and more about purchasing influence, the loss of which led him to conclude, "I do not believe we can be treated worse by the Democratic party than we have been treated by the Republican party."[40]

Despite Hammond's belief of personal persecution, the timber suits were part of a larger effort of the Progressives to mitigate the consequences of the Great Barbeque of the late nineteenth century. Along with the timber theft investigation, Progressives sought to recover lands where the railroads had failed to comply with the provisions of their grants. In 1908 Congress directed the attorney general to recover 2.5 million acres from the Southern Pacific, which had inherited the Oregon and California Railroad land grant. A stipulation of the original grant required the railroad to sell land only to homesteaders, with no more than 160 acres going to any single purchaser. Most of the SP's land sales, however, had gone to large timber companies. Hammond's close ties with the railroad had allowed him to acquire 66,000 acres, making the HLC the second-largest recipient of the SP lands. In addition to seeking forfeiture of the railroad grant, the Interior Department filed lawsuits to recover the lands from Hammond and forty-four other lumbermen, each of whom had bought more than 1,000 acres.[41]

If the forfeiture went through, Hammond would lose not only the purchase price of $7 an acre but also his investment, yet these lands were now worth $200 an acre. More important, the tracts lay in alternate sections purchased to round out other holdings; forfeiting these lands would greatly reduce the worth of all his timberlands. Furthermore, Hammond and other lumbermen had invested heavily in mills and railroads in the area. In comparison, the Montana timber-poaching case was a minor irritation.[42]

Although the Justice Department recognized the lumbermen's plight, the government attorneys lumped lumbermen in with the railroad to avoid compromising the case against the Southern Pacific. To aid the lumbermen, Congress proposed legislation that treated them as "innocent purchasers" and that allowed them, upon surrendering their title, to buy back those same lands for $2.50 an acre, thus satisfying the terms of the grant, albeit not the intent. Fortunately for Hammond, both of Montana's U.S. senators, Henry Myers and Joseph Dixon, sat on the Public Lands Committee, and he dispatched C. H. McLeod to lobby on his behalf. Dixon, a Progressive Republican, threw his support behind the bill and helped usher it through Congress. The Progressives thus succeeded in forcing the Southern Pacific to forfeit the land grant while recasting lumbermen as "innocent purchasers" and providing for a continued timber harvest. Containing vast stands of Douglas fir, these grant lands were now split between large timber companies and the federal government. Neither land owner, however, would prove capable of providing a sustainable yield—the primary goal of Progressive Conservation.[43]

Meanwhile, the Montana timber-poaching case continued to drag on, finally coming to trial in 1913. After two grueling days of testimony on the witness stand, Hammond failed to sway the jury, who had heard depositions of thirty Montanans who admitted to cutting timber under his direction. Nevertheless, Hammond must have made some impression, for while the jury returned a guilty verdict, it reduced his penalty to $51,000, less than a fourth of what the government had sought. This was substantial sum in 1913, to be sure, but only a fraction of the $875,000 profit posted by the Hammond Lumber Company that year.[44]

Hammond immediately filed an appeal. Weary of the case, the prosecutor offered a compromise: if Hammond accepted the verdict, the

Justice Department would give him clear title to the Oregon and California grant lands that he had bought. Confident that he would receive title under the new legislation and still maintain his innocence, Hammond rejected the deal. Although he lost his appeal, Hammond pressured the Justice Department and, with the aid of Senator Myers, finally reached a settlement, agreeing to pay $7,066 for the timber he had cut from the public forests over a nine-year period.[45]

Hammond believed the trial was politically motivated and attributed the renewal of the suits and the verdict to Progressive reformers. He insisted that politicians and the press created prejudices against wealthy and successful businessmen. Popular culture had indeed shifted from the late-nineteenth-century belief in the "self-made man" of the Horatio Alger tradition. Popular opinion of the Gilded Age "captains of industry" was steadily eroding, and the same men found themselves cast as "robber barons" in the Progressive Era. Regardless of the change in social values, the facts of Hammond's case remained clear; his blaming the outcome of the trial on Progressivism was far-fetched.[46]

Hammond believed that he was battling for a principle in the courts, on the waterfronts, and in the woods. Although he enjoyed a good fight, he believed that not only were labor unions and the government allied against him, but so was the general tenor of the times. Wearily, he wrote to McLeod, "There is a growing class hatred, not only among laboring men, but among the small merchants, against those who have been successful." Hammond attributed such animosity to "human nature" whereby men "owe it to themselves to feel this way in order to maintain and preserve their confidence and self-respect." Uncharacteristically, he concluded that instead of enlarging his business, he should sell it to avoid "the responsibility and annoyance that goes nowadays with conducting a large business."[47] Hammond's ennui did not last long, however. The rising tide of radicalism reinvigorated the fight in the increasingly infirm, sixty-five-year-old man who had been contemplating retirement.

CHAPTER 18

Class War and World War

On the eve of World War I, America stood at a crossroads. As alternative economic visions battled for acceptance, the nation's conflicted relationship with laissez-faire capitalism seemed to be at the breaking point. Hammond, for his part, saw Socialism and unions as genuine threats to both his business and his country. Following his timber trespass trial in 1913, he also continued to lament the growing appeal of Progressivism. Six years later, however, organized labor was in retreat, Socialism had been effectively crushed in the United States, the government had imprisoned hundreds of labor radicals, and Progressives had grown more concerned with instituting Prohibition than trammeling capitalism. The war in Europe proved to be the decisive factor in charting America's future.

The eruption of war in August 1914 all but eliminated the export lumber trade, and the industry took a nosedive. At year's end, Hammond had to borrow money to meet his payroll, again relying on the "gilt-edged" credit rating of the Missoula Mercantile to secure a low interest rate. With nearly half the sawmills on the coast shuttered, Hammond believed that production would drop by another 25 percent. As some thirty thousand unemployed men flooded the streets of San Francisco, many coming from the sawmills and lumber camps in Oregon and California, Hammond noted that "the question of taking care of them has become a serious one." Although he had cut production 30 percent and was operating at a loss, he claimed that his concern for his employees, "many of them with families, who would have no way of supporting themselves during the Winter," compelled him not to close his mills altogether.[1]

Hammond, of course, did not keep his mills open out of altruism. Unlike the small owner-operated mills that could simply close when lumber prices dropped, the large mills like Hammond's had fixed costs that needed to be met regardless of production. Shuttering his mills meant docking his steamships, closing his lumberyards, and laying off his sales force. In short, it meant admitting economic failure, and Hammond was not one to concede defeat. Furthermore, he recognized that the mass of unemployed workers presented a potential threat.

Despite his lumber company's losses, Hammond's Columbia River Packers Association "had a very successful season." His shipping line also continued to turn a profit, and after two years of "very hard times in the lumber business" he could report entering 1916 "under much more favorable conditions." As production shifted to meeting military demands, the economy quickly recovered. While employment rose, so did prices, and wages could scarcely keep up with inflation. From 1906 to 1916, wages in San Francisco rose 16 percent, but food costs shot up 39 percent, and workers began to agitate for corresponding wage increases.[2]

Such was the situation up and down the West Coast, and the better-organized unions began to coalesce. In the first unified coastwide strike, 10,000 longshoremen walked off the job on June 1, 1916, after employers rejected demands for higher wages. In Los Angeles, 1,200 longshoremen and mill workers struck, paralyzing the San Pedro harbor, the heart of the Hammond Lumber Company's distribution and retail network. During the strike of 1907, Hammond had replaced union workers with Russian and Italian immigrants, but this time they too walked out. In defiance of the strike, the HLC yard in San Pedro continued to limp along with 40 men remaining on the job. Under protection of deputy sheriffs, the HLC sent in strikebreakers, but after three weeks of clashes, the company suspended operations. Hammond, in New York City on business, hurried to Los Angeles.[3]

Upon arrival, Hammond seized control of the situation. Convinced that the city government and local businessmen were too passive in combating the unions, he joined other lumbermen in getting an injunction against picketing. He then convinced the city council to pass an emergency ordinance creating an armed force of five hundred men to break the strike. By the end of June, San Pedro was essentially under martial law.

Hundreds of uniformed police patrolled the lumber yards and docks, dispersing strikers. The local Merchants and Manufacturers Association trumpeted that Los Angeles would remain an open-shop city.[4]

Two weeks later, however, when the HLC attempted to unload its first ship since the strike began, a riot broke out between longshoremen and strikebreakers. The men hurled stones and swung clubs at each other. As the violence escalated, strikers shot at one of Hammond's steamships and guards returned fire. Inside the locked gates of the HLC compound, Hammond organized his workers into a defensive force as they waited for the police. Hammond's nonunion workers spent the next four days living inside the enclosure and, with forty police officers standing guard, successfully unloaded the lumber arriving from the HLC mill in Astoria. By the end of July, with the help of police, the lumbermen finally succeeded in breaking the strike. The longshoremen discovered that they were no match for the combined power of the businessmen and government.[5]

In Los Angeles lumbermen pressured a compliant city council and relied upon a strident antiunion propaganda machine in Harrison Gray Otis's *Los Angeles Times* for help in crushing the strike. San Francisco, also shut down by the waterfront strike, proved more problematic. Mayor James "Sunny Jim" Rolph balked at the lumbermen's demand for five hundred police to protect the waterfront; instead, he ordered officers to search strikebreakers for concealed weapons. Rolph's refusal to back employers provoked a dramatic, impromptu meeting of the San Francisco Chamber of Commerce.[6]

On the afternoon of July 10, 1916, some two thousand white men in dark suits and starched collars crowded around the marble columns on the ground floor of the ornately decorated Merchants Exchange Building. Declaring themselves "victims of a labor tyranny and domination that threatened the entire community, its welfare, progress and future prosperity," the business leaders quickly adopted three resolutions: integrity of contracts, maintenance of law and order, and unequivocal support for the open shop. To carry out these resolutions, the Chamber of Commerce appointed a Law and Order Committee, and within five minutes businessmen pledged $200,000 to support the effort. Busy confronting the strike in Los Angeles, Hammond missed the meeting, but he soon joined the committee to help with its campaign. It looked as if his dream of a unified business front in San Francisco had finally come to fruition.[7]

Since his arrival in San Francisco, Hammond had implored Bay Area businesses to take a strong stand against organized labor, yet many employers continued to sympathize with workers. When the 1901 City Front Federation strike ended in a stalemate, many businesses engaged in collective bargaining with unions, discovering in the process that union agreements worked to their advantage by setting uniform wage rates and preventing wildcat strikes. Furthermore, local businesses also benefited from the restricted competition brought about by San Francisco's closed shop. Such acquiescence to unions frustrated Hammond and led the National Association of Manufacturers to describe San Francisco as "a city where employers spinelessly accepted dictation from union leaders and lived under a labor government."[8]

Partly in response to such criticism, Bay Area businessmen—including Hammond and his good friend Charles Johnson of the Union Lumber Company—formed the San Francisco chapter of the Citizen's Alliance. By 1903 they had amassed $500,000 to battle organized labor. The Citizen's Alliance began one of the first modern publicity campaigns, hiring professional organizers and placing ads in newspapers to systemically promote the open shop. It also employed a fleet of lawyers to besiege the courts for injunctions. As early as the fall of 1904, the alliance boasted sixteen thousand members, the largest such organization in California. But after a two-year campaign, the alliance failed to defeat organized labor, and by 1907 it had dwindled to fewer than one thousand members. As president of the alliance during its decline, Hammond was unable to rally Bay Area businesses to break the union boycott of nonunion lumber in San Francisco.[9]

The 1916 longshoremen's strike presented anti-labor forces with another opportunity to break the union lock on San Francisco. The strike primarily affected lumbermen, who provided the core opposition to the unions. California's largest corporations—Standard Oil, Wells Fargo, and Southern Pacific—also joined in. Smaller firms, however, remained reluctant to battle organized labor as this might result in more violence and property destruction, rather than an accord with unions. To win the support of small businessmen and the general population, the Law and Order Committee engaged in a massive public relations campaign.[10]

Although technically under the purview of the Chamber of Commerce, the Law and Order Committee overshadowed its parent organization. Taking a dim view of the committee, Mayor Rolph informed the committee's president, Fredrick Koster, "It is my profound conviction that the union of labor makes for the moral uplift of the country as a whole and places the prosperity of all on a fair basis. The system of collective bargaining is the essence of commercial progress." With such opposition from the city government, the committee faced an uphill battle.[11]

But, as if to reaffirm the committee's position, an event occurred that shocked the Bay Area and sent reverberations across the nation. On Saturday, July 22, the Chamber of Commerce sponsored a "Preparedness Day" parade to demonstrate its support for U.S. entry in the European war. In the summer of 1916, organized labor still opposed American participation, and the parade devolved into a symbolic fault line between business and labor. Although five thousand people had protested the event two days previously, ten times that many showed up to express their patriotism in San Francisco's largest parade up to that time. At 2:00 P.M. a bomb exploded in the crowd, sending body parts in all directions. The blast killed ten people and wounded forty.[12]

The Law and Order Committee wasted no time in capitalizing on the tragedy, equating the strike violence on the waterfront with the bombing. By implying that organized labor was responsible for both, the committee channeled the general hysteria into its campaign for the open shop. Five days after the bombing, the Chamber of Commerce sponsored a mass meeting, and six thousand irate citizens turned up. Taking advantage of the outrage, Koster proposed the formation of the Committee of One Hundred to aid the Law and Order Committee. This intentional allusion to frontier vigilante justice underscored the city's apparent impotence in addressing the situation. Besides condemning the bombing, Rolph pointedly told Koster, "The attitude and the activities of you and your particular group have done much, in my opinion to engender industrial unrest and class hatred."[13]

To raise its legitimacy, public profile, and financial resources, the Chamber of Commerce nominated San Francisco's leading businessmen and public figures to serve on the Committee of One Hundred. Selected from the top echelon of the city's elite, many of these men were also members of San Francisco's Bohemian Club and were neighbors as well as business associates. Reading like a social register, the committee included such notables as J. D. Spreckles, D. Ghirardelli, E. W. Hopkins, and William Crocker, along with lumbermen Robert Dollar, James Tyson, George Pope, and, of course, A. B. Hammond, who took a leading role.[14] While these lumbermen savagely undercut each other in business competition, they unified against a common enemy: organized labor. Whatever their differences, faced with the prospect of class warfare, these men formed an impenetrable front. For the working class, on the other hand, ethnic, religious, and cultural bonds were often a greater glue than class consciousness. The difference proved crucial.

With strikes breaking out in the transportation, restaurant, and steel industries, the Law and Order Committee jumped on each one as an opportunity to challenge organized labor. Rather than resolve disputes over wages or hours, the committee's goal was to break the unions and establish an open shop in each industry. To this end, it actively sought court injunctions against strikers and promoted boycotts of closed-shop businesses. Firms that signed agreements with unions suddenly found their supplies dried up and banks calling in their loans. Employing Hammond's model from Humboldt County, the committee also formed the American Stevedore Company to supplant union longshoremen with nonunion workers. In defiance of Rolph's authority, the committee also supplied its own armed guards to protect strikebreakers.[15]

The Law and Order Committee's aggressive agenda soon prompted a backlash. On August 16 a group of Methodist ministers condemned the committee, and the following week, 279 businessmen signed a formal statement opposing the open-shop drive. The committee responded with a massive membership campaign. By September 1916, the Chamber of Commerce membership jumped 300 percent, from 2,500 to 7,300, making it the largest in the United States. By the end of the year the chamber had stockpiled a million-dollar war chest. Top contributors included Standard Oil, the Southern Pacific Railroad, C & H Sugar, Wells Fargo, and the Pacific Improvement Company, of which Hammond represented 50 percent of the ownership.[16]

In November the Chamber of Commerce placed a measure on the ballot to ban picketing within city limits. The chamber employed a modern get-out-the-vote campaign and, using the new technology of the telephone, hired four hundred operators to call every voter in San Francisco. Equating unions with violence and crime, the Law and Order Committee also placed full-page newspaper ads claiming, "There is no such thing as peaceful picketing any more than there is lawful lynching or peaceful mobbing." The ads insisted that picketing was "an un-American method by which small business men are ruined, the city kept in turmoil, and the city streets used for private strife."[17]

Although it was the Law and Order Committee that actually kept the city in turmoil, San Franciscans approved the initiative by five thousand votes. The victory proved ephemeral, however. The more the committee increased its political activities, the more members resigned from its parent organization. Like earlier vigilante groups, the Law and Order Committee often operated outside the legal apparatus, but it employed subtler and more insidious methods than lynching. When police arrested Tom Mooney and Warren Billings for the Preparedness Day bombing, the committee pressured the district attorney to prosecute the two innocent men in what proved to be one of the nation's worst travesties of justice. By 1917 the Law and Order Committee had become an embarrassment to the Chamber of Commerce. The committee's battle with organized labor demonstrated the tight linkage between ideology and economics; as such it corresponded to the larger contest unfolding on the national stage.[18]

America's entry into the war in Europe in April 1917 created immense demand for lumber and catapulted the industry out of the doldrums. The U.S. Army needed lumber to build cantonments, other military facilities, airplanes, and especially ships. As part of the mobilization effort, President Woodrow Wilson created the government-owned U.S. Emergency Fleet Corporation to build one thousand wooden steamships to offset Allied shipping shortages. Although the Fleet Corporation's directors "detested the idea of wooden ships," they recognized the wartime scarcity of steel and the abundance of timber, especially in the Pacific Northwest. The announcement of lucrative government contracts drew a plethora of applicants, some 80 percent of whom had no shipbuilding experience. Each of the 3,500-ton ships required 1.7 million bf of lumber, including forty-foot timbers and ninety-two-foot masts. The redwoods and Douglas firs of the Pacific coast were ideal for these requirements. Hammond was thus well positioned to profit from the war. Not only did he own lands capable of producing such timbers, but he also possessed mills, transportation networks, and shipbuilding facilities, and thus secured several government contracts.[19]

Unfortunately, the HLC's lease on the Bendixsen shipyard near Samoa was about to expire. Even worse, Rolph, the mayor of San Francisco, bought the shipyard and "declared with great fanfare that he would only employ union labor." Nothing could have irritated Hammond more. Believing that Rolph purchased Bendixsen to increase his political support in Northern California for his upcoming gubernatorial bid, Hammond accused the mayor of political pandering to organized labor.[20] Hammond publicly referred to Rolph as "a weak, self-seeking, hypocritical demagogue, who, masquerading under the guise of altruism and respectability, is endeavoring to keep in the lime light."[21]

Rolph, who had his own shipping company, was probably more attracted by the lucrative government contracts than by making political hay out of Bendixsen. Hammond, however, refused to attribute economic motivation to Rolph, insisting that politicians like Rolph acted only out of political expediency. He complained to C. H. McLeod, "A lot of these fellows think more of their political jobs than they do of their country, and that, in my judgment, is what is the matter with democracy."[22] Equating economic development with patriotism, Hammond believed that only the captains of industry acted in the true interests of the nation.

Despite Hammond's accusation that he was catering to unions for votes, Rolph had assumed a progressive position regarding labor long before entering politics. The antipathy between the two men extended back to 1901, when Rolph broke from the Shipowners' Association to negotiate with the Sailors' Union of the Pacific and thus undermined any hope of a united employer front. Just as Hammond was convinced of the rightness of his antiunion position, Rolph sincerely believed that union recognition and collective bargaining were beneficial to both worker and employer. Such a stance helped Rolph win election as mayor of San Francisco in 1911. While Hammond could chalk this up to city politics, he regarded Rolph's entry into Humboldt County as a personal intrusion into his fiefdom. With the open shop at the core of their dispute, the two men were on a collision course.

Although Rolph took over the Bendixsen yard, Hammond controlled most of the timberlands in northern Humboldt Country. Assuming that, for Hammond, business trumped politics, Rolph proposed a cooperative agreement with Hammond whereby the HLC would supply the lumber and Rolph would assume the contract to finish the vessels under construction at Bendixsen. Hammond flatly refused, primarily because Rolph insisted on using union labor. Instead, Hammond intended to use his timber for his own endeavors and built a new shipyard next to his Samoa mill "on modern lines," meaning nonunion. Bidding on three more government contracts, Hammond believed he could net more profit in shipbuilding than in running his lumber mills.[23]

Meanwhile, Hammond had two half-built ships to finish at Bendixsen. Three months earlier, in January 1917, Hammond's ship carpenters had walked out, demanding an eight-hour day in accordance with the Pacific coast standard. The HLC, of course, refused the demand, but the wartime labor shortage forced it to pay overtime to its highly skilled workers. To compensate, General Manager George Fenwick filled the ranks of unskilled laborers with men from the logging camps. With Rolph's purchase of Bendixsen and his commitment to the eight-hour day, the two crews worked side by side, although Hammond's arrived an hour earlier and stayed an hour later. Not surprisingly, many of Hammond's workers migrated to the Rolph crew. The lumberman angrily accused the mayor of stealing his workers, reminding him that these ships were for the war effort and that Rolph had a patriotic duty not to interfere.[24]

Rolph countered with his own patriotism, telling Hammond that he wished only to build as many ships as possible since "with every hull that slips into the water, the work of the Enemy U-boat is nullified by just one more degree." The mayor also appealed to Hammond's economic self-interest in explaining his position on organized labor. He wrote, "The man who sells his labor is selling not a commodity but his life . . . in my own business I have found that organization of men in Unions . . . has helped me quite as much as it has helped the men."[25]

Hammond refused to accept this "Marxist twaddle." After all, nine years earlier, the Big Three lumber companies had driven unions out of the Humboldt lumber industry, and since then Hammond's one thousand workers in Humboldt had "been enjoying the benefits of the open shop"—that is, until Rolph showed up and began to reestablish unionism and its "regime of bloodshed."[26]

In responding to Hammond, Rolph reiterated his support for the rights of workers to organize and bargain collectively. His rather mild statement—"I am also sorry that my views on labor matters do not coincide with yours"—sent the old lumberman into a fit of rage. Conveniently ignoring facts, Hammond lectured the mayor, claiming that in 1907 union longshoremen had murdered the Jenks brothers because they were nonunion employees. According to Hammond such "brutal violence" compelled Humboldt mill owners "to rid themselves of this tyranny, and their efforts in this behalf were strongly endorsed by the decent, law-abiding people." Referring to the Preparedness Day bombing and implying Rolph's complicity, Hammond rhetorically asked the mayor, "Do you suppose for a moment that our company . . . would favor the re-establishment of that reign of bloodshed . . . and impose . . . the terrorism and brutal violence which have characterized the city of San Francisco?"[27] Rolph, a thick-skinned politician, simply ignored such inflammatory rhetoric.

Hammond's statements mirrored those of other conservative businessmen. As the open-shop movement had become a nationwide phenomenon, employers across the country parroted each other in their attempts to equate violence and anarchy with organized labor. Lumbermen, in particular, drew parallels between patriotism and the open shop. Unloading on Rolph for his pro-labor stance, Hammond told the mayor, "Your insipid truisms amount to this: You believe that no American citizen has the right to work in your shipyard unless he pays $30 for the privilege of joining a labor union. *I* believe that any decent, law-abiding citizen of this country has the right to labor in our shipyard regardless of whether he belongs to a labor union or not, and that the Union of these United States of North America is a good enough union for an American citizen."[28]

Echoing his boss, George Fenwick asserted that Rolph's union shop would bring the nation's most radical union—the IWW—into Humboldt. In contrast, Rolph and the other Progressives understood that

Hammond's labor practices fertilized radicalism and that the surest way of countering the IWW was to improve working conditions and recognize the conservative AFL. Yet lumbermen appeared oblivious to this rather obvious conclusion. While Hammond dug in his heels, the IWW began to organize.

For Hammond, 1917 was like a game of Whac-A-Mole as he raced up and down the Pacific coast confronting strikes in what became the largest labor disturbance in the history of the lumber industry. In other industries, too, strikes broke out across the West as wartime inflation and labor shortages provided workers with both motive and opportunity. Nationwide, 4,500 strikes involving one million workers occurred in 1917. In confronting organized labor, Hammond and other lumbermen drew the U.S. Army into the fray, and the timber camps of the Pacific Northwest became the home front in World War I. The IWW appeared to be at the center of the maelstrom.[29]

Across the nation, industrialization alienated workers from their labor and from their employers and produced a large underclass without skills or education. Migrants from European and American farms flooded into cities only to discover most jobs already taken by skilled workers. Before long, America's unskilled workers became a "womanless, homeless, vote less" labor force drifting between jobs—baling hay, picking apples, logging, or loading ships. As itinerant workers, they could not vote and thus regarded the ballot box with suspicion. For these men, capitalism did not offer the wide-open opportunity it had for Hammond's generation but, rather, a meaningless and miserable existence with no hope of improvement. In addition, they were often the victims of state violence, whether it came from local police or federal troops crushing strikes. Eventually, these conditions provoked a radical response that culminated in the formation of a new industrial union.[30]

Unlike the AFL craft unions, the IWW was open to all workers regardless of skill, job, sex, or race. Formed in 1905, its mission was to overthrow the capitalist system and replace it with "industrial democracy" wherein those who produced the labor controlled the means of production. The Wobblies attracted many followers who had experienced the dark side of capitalism, but they also irritated and frightened more-established citizens. While Wobbly rhetoric was revolutionary, few actually engaged in violence or sabotage. For the most part, they stood on street corners, passed out pamphlets, and spoke out on the evils of capitalism.

The logging camps of the Pacific Northwest were a perfect breeding ground for the IWW. The migratory nature of the lumber industry, lumberjack culture, hazardous work, and miserable living conditions contributed to turn the region into the nation's hotbed of worker radicalism. The very nature of the industry generated a workforce without homes or families. As the timber camps moved across the landscape, the loggers moved with them. Not only was logging seasonal, but wages rose and fell according to lumber prices. With labor as the primary nonfixed cost of lumber production, companies frequently laid off workers as demand waned.[31]

Isolation from society created a lumberjack culture and a level of class-conscious solidarity that transcended ethnic divisions to a degree unusual in other industries. As they worked tougher, ate together, slept together, drank together, and whored together, men in the camps internalized the IWW's basic premise: "An injury to one is an injury to all." After a long workday, loggers had no homes to go to; instead, they sat around the bunkhouse stove and swapped stories. Discussions of the "outrages of capitalism" pervaded the dank and overcrowded bunkhouse. Dog-eared books and pamphlets passed from hand to hand, with the selections becoming increasingly radical after 1900. Works by Upton Sinclair, Voltaire, Rousseau, and Marx, along with labor journals and IWW pamphlets, ranked among the most popular reading materials. Immigrants, especially those from Italy and Eastern Europe, also brought revolutionary ideas that independent-minded and utopian westerners readily embraced.[32]

While infused with romantic and radical ideas, the vision and theoretical underpinnings of the IWW remained inchoate, a factor that eventually contributed to its demise. Nonetheless, the organization, with its interchangeable union membership and rejection of electoral politics, resonated with transient workers, who lacked voting rights. With welcoming union halls and underclass solidarity, the IWW provided a sense of belonging and witnessed explosive growth in the woods camps, where the conservative AFL had difficulty organizing.[33]

Ultimately, loggers flooded into the IWW because of the atrocious living and working conditions. Concern over such conditions led the Industrial Relations Commission to investigate the logging camps of the Pacific Northwest in 1917. The commission discovered that bedbugs invested half the camps, one-third had unusable toilet facilities, and only half contained showers. The commission also found that "forty loggers occupied a bunk house that should not have accommodated more than a dozen—the men sleeping two in a bunk, with two more in a bunk on top; a stove either end, sending the steam rising from lines of wet clothes strung the length of the room; beds made in many cases by dumping hay into a wooden bunk; food that was unsavory; the crudest kind of provision for cleanliness and sanitation."[34]

Lumbermen were particularly appalled at the IWW's revolutionary rhetoric. But blinded by stubbornness and independence, they failed to recognize that their own labor practices and violent reactions had given birth to the IWW and continued to sustain it. Increased mechanization meant that timber workers became increasingly disposable. With turnover approaching 1,000 percent annually, workers charged that management intentionally created unstable conditions. Employers countered that the loggers were filthy degenerates who would quickly contaminate a clean camp. Refusing to address worker complaints, employers simply fired anyone suspected of attempting to organize a union. The open hostility of lumber companies toward labor drove more men to the IWW. Lumber company files burst with reports from company spies who searched cabins, opened worker mail, and often acted as agents provocateurs to draw out antiemployer statements. Spies provided lists of IWW suspects whom bosses promptly fired. Companies shared these reports with each other, blackballing men throughout the region, which only increased the appeal of radicalism.[35]

As the AFL struggled to gain a solid foothold in the Pacific Northwest, the Wobblies' antiwar stance resonated with workers, and the IWW doubled its membership to 100,000 in anticipation of strikes against the essential war industries of lumber and copper. Sounding more like a trade union than revolutionaries, the IWW demanded an eight-hour day, a minimum wage, better working conditions, sanitary camps, abolishment of hospital fees, and no discrimination against its members. But unlike traditional unions, and in keeping with its independent spirit, the IWW did not demand a closed shop. The lumbermen, however, refused to negotiate, fearing that any capitulation to worker demands would propel them down the road to union recognition.[36]

The labor situation in the lumber industry reached the boiling point in the early summer of 1917. As walkouts and wildcat strikes began breaking out in woods camps and mills, the IWW moved to control the situation and called for a general strike on June 20, bringing logging to a virtual standstill east of the Cascades. The strike quickly spread westward and paralyzed the lumber industry from Montana to the Pacific coast. Humboldt County, where Hammond had effectively suppressed the timber worker unions, remained a notable exception.[37]

At first, most of Hammond's mills and camps seemed largely unaffected by the strike, and he bragged that "there are mighty few, if any, I. W. W.'s working for the Hammond Lumber Company." Hammond believed that lumbermen brought the strike on themselves by carelessly employing union workers and Wobblies. He stated, "Any logger or millman who employs that class of labor to carry on his operations deserves about all that he gets."[38] Even at the height of the strike, with more than half the camps and mills on the coast shuttered, Hammond could crow, "All of our operations, including Logging Camps, Saw Mills, Planing Mills, Lumber Yards, Shipyards and Steamships, are running at full blast." Hammond attributed this to his strict orders to dismiss "labor agitators of any description" and his company's explicit open-shop policy, which he believed attracted loyal, hardworking men. Like many industrialists, Hammond believed he was protecting his employees from being coerced into joining a union. To counter the appeal of unions among his employees, Hammond raised wages. With cost-plus government contracts, he simply passed the increase on as part of the price of manufacturing lumber.[39]

Despite Hammond's precautions—including hiring illegal workers from India—in September 1917 he found his mill in Astoria subject to a mass walkout. Just three years earlier, Astoria had hosted a gala reception honoring A. B. Hammond on the twentieth anniversary of his undertaking to build the railroad linking the city to the rest of the country. By 1917, however, Astoria's high proportion of Scandinavians, who embraced the IWW, had recast the city's electorate so that even the mayor openly professed Wobbly sympathies. Predictably, Hammond's affection toward Astoria vanished. With the strike threatening his mill, Hammond declared he was "not going to stand for much monkey business from the I. W. W's at Astoria" and appealed to Oregon governor James Withycombe for protection. The governor responded by sending in the National Guard to clear picketing strikers from the streets. With nearly half its workforce out on strike, the HLC continued to operate with the help of troops guarding the plant.[40] While Hammond admitted he could use one hundred more men, he noted that "we have to be very careful whom we employ."[41]

Gradually Hammond and the other West Coast lumbermen began to marshal their strength. No doubt inspired by the Law and Order Committee's success in presenting a united employer front, lumbermen convened in Seattle on July 9, 1917, to form the Lumbermen's Protective Association. Pledging to defeat the unions and the eight-hour day, the association threatened to boycott firms that refused to join and agreed to impose stiff penalties on any members who acceded to union demands. Despite their traditional independence, some sixty lumber firms signed up. Some lumbermen proclaimed that only the threat of the IWW compelled them to join.[42]

Hammond and other lumbermen regarded themselves as patriots helping to defeat the Germans by supplying much-needed lumber, but they were stymied by the IWW. Hammond informed the War Department, "Labor unions, IWW and other enemies [of] our country [have] created conditions [in] which troops . . . are protecting our Astoria mill which is being operated under great difficulties in endeavoring to supply ship timbers [to] fourteen government hulls."[43] Branding the IWW un-American for striking during the war, employers painted the Wobblies as unpatriotic at best and as Bolshevik agitators, anarchist revolutionaries, or Germans spies at worst. Not only did lumbermen ignore the ideological differences between Bolsheviks and Germans; they also saw little reason to distinguish between conservative trade unions and the IWW. Hammond's statement—"The I. W. W.'s and the Labor Unions all look alike to me. I see little difference between them"—reflected a prevailing attitude among Northwest lumbermen.[44]

While Hammond professed that his patriotism and ideology transcended economic concerns, he was not above renegotiating government contracts to increase his profit margin. In June 1918, he sent his son-in-law Steward Burnett to Washington, D.C., to secure additional shipbuilding contracts. Burnett succeeded in obtaining up to $50,000 more per hull.[45] Ultimately, as it had for many American businesses, U.S. involvement in the European conflict solved many of Hammond's lingering difficulties, notably his labor issues and access to remote timberland. The solution to both problems arrived in the form of a singular tree species—the Sitka spruce.

When the United States entered World War I, the War Department regarded the new technology of aerial warfare as crucial to winning the war. Construction materials for aircraft were exceedingly scarce, however. Only lumber from the Sitka spruce had the requisite qualities of strength and lightness, but the trees were restricted to a narrow coastal strip from Alaska to Northern California. Unlike other conifers, Sitka spruce has limited tolerance for drought and can survive only along the moist coast and in river valleys of the Pacific Northwest. Highly susceptible to the white pine weevil, the tree is further limited to the fog belt, where cool summer temperatures provide insufficient warm days for the beetle to complete its life cycle. Sitka spruce evolved a high degree of salt tolerance, allowing it to thrive along the rocky coastline, where sea spray prevents other trees from growing. On the Oregon coast this resulted in scattered, localized groves of Sitka spruce that most lumbermen thus far had largely ignored. Not only was spruce difficult to retrieve, but until this point it had had little commercial value and lumbermen had made very little effort to access their stands. Victory against Germany, however, appeared to rest on the spruce production of the Pacific Northwest. The U.S. Army identified Hammond as the "largest individual owner of spruce in the world" and requested his presence at a meeting in Washington D. C.[46]

Hammond informed the War Department that although he would be "most happy [to] cooperate," labor troubles prevented him from leaving Oregon. He intended to build a railroad up the Necanicum drainage to access spruce, but labor disturbances had halted construction and logging operations. In his telegram Hammond argued that he could not produce spruce unless the government would ensure protection and "give its unequivocal support to those who are endeavoring to maintain law and order [against] the mob I. W. W.'s, radical labor unions, pro-Germans, and the rotten press."[47] Hammond, like other lumbermen, exaggerated the labor strife in hopes that the army would provide a free security force to suppress labor unrest. The IWW, however, had pledged nonviolence, and the strike was remarkably civil. Furthermore, by the time Hammond telegrammed the War Department, the strike had been over for a month. Nevertheless, the War Department took the labor threat seriously.

By the fall of 1917, strikers had drifted back to work but, deploying a new strategy, engaged in a deliberate slowdown and often walked off the job after eight hours. This "strike on the job" allowed the Wobblies to draw wages while avoiding arrests, court costs, and confrontations with strikebreakers. More important, the workers could control the rate of lumber production, now a crucial war industry. Spruce production dropped 25 percent, prompting concern in the War Department. In addition to supplying security, lumbermen hoped the army would purge the woods of Wobblies as well as provide strikebreakers in the form of uniformed soldiers. Ultimately, as historian Robert Ficken pointed out, Sitka spruce became more important "as a weapon to be wielded against labor than as a weapon of war."[48]

Investigating the holdup of spruce production, the War Department discovered several bottlenecks. First, spruce was difficult to obtain, and lumbermen were reluctant to invest in building the transportation and production infrastructure for what might be a short-lived market. Second, lumbermen—well familiar with the dangers of overproduction—were quite satisfied with the high prices that resulted from restricted supply. But nearly everyone, from Progressive reformers to army investigators to the lumbermen, agreed that the labor issue posed the largest barrier to increased production. They did not agree, however, on a solution.[49]

Lumbermen like Hammond saw the war as an unprecedented opportunity to crush the Wobblies once and for all. In no particular order, the lumbermen advocated conscription, deportation, vigilante action, exposure, and a coordinated effort to drive the subversives out of the camps and out of the country. To this end, they relied upon their access to political power, especially at the local and state levels. Lumbermen called upon governors for National Guard troops to protect their property and disperse strikers and drew upon the extralegal, quasi-vigilante

American Protective League (APL). Sanctioned by the Wilson administration's Justice Department, the APL consisted of businessmen, often lumbermen, devoted to ferreting out German spies. But with spies in short supply, the APL turned its attention to the Wobblies and aided the Justice Department in dozens of raids against IWW halls. As most loggers belonged to the IWW, such raids inhibited rather than increased spruce production.[50]

In contrast to lumbermen, Progressive reformers like Carlton Parker, a professor at the University of Washington, saw the war as an opportunity to initiate social reforms. Parker also wished to eliminate the IWW, but he regarded radicalism as a symptom rather than a cause of substantial social ills. After investigation, Parker concluded that the IWW had legitimate grievances and that intolerable working conditions constituted the real threat to production. Parker pointed out that drafting subversives made little sense, as this would simply reduce the labor supply; it would be far better to make the Wobblies into patriotic citizens by improving their lives. Parker called for a reversal of the government policy of outright suppression and instead advocated the creation of a federal agency that would provide workers an eight-hour day, decent food and housing, protection from company spies and from price gouging by company stores, and, most important, a sense of dignity and self-worth. Such actions, Parker maintained, would increase spruce production and eliminate the threat of a worker's rebellion.[51]

Secretary of War Newton Baker agreed with Parker's conclusion that addressing working conditions was the key to increasing spruce production. In the fall of 1917, Parker accompanied Colonel Brice Disque, who had assumed control of the army's Spruce Production Division (SPD), on a tour of the lumber camps and mills of the Pacific Northwest. Disque was shocked at the conditions, noting, "We treated captured Moros better in the Philippines during a war." He reported most of the camps would fail any army inspection. One millworker "told him that the notches cut on the housing around a saw represented arms and hands that the huge blade had cut from workers." Disque concluded, "My wonder was not that production was low but that there was any production at all."[52] While Disque conceded that working conditions needed vast improvement, he disagreed with Parker's view that material conditions lay at the root of labor unrest. Instead, like Hammond, Disque saw IWW ideology as the problem. But unlike the lumberman, he believed Wobblies could be reformed.

Despite their differences, Disque and Parker hatched a remarkable plan that combined U.S. Army discipline with Progressive reforms. Under the auspices of the Spruce Production Division, they forged a unique organization: the Loyal Legion of Loggers and Lumbermen. The legion, or 4L, was the first and only time the U.S. Army formed a labor union. Under the direction of the army, the 4L supplanted traditional unions and provided patriotic lumbermen with a palatable substitute. Furthermore, since it was composed of both workers and their employers, lumbermen could still maintain some control. With the 4L, Parker had the opportunity to enact a Progressive agenda, while Progressives in other parts of the country could only hope that the war would push the government into implementing social reforms.[53]

The army, however, faced a manpower dilemma. From France, General John Pershing demanded a million trained men for the American Expeditionary Force (AEF). At the same time, he believed Disque would need 125,000 troops to check insurgency in the Pacific Northwest. Meanwhile, Washington governor Ernest Lister announced he would keep the National Guard at home to prevent "internal troubles." But if the 4L could alleviate the labor strife, it would free up much-needed troops for the AEF.[54]

Armed with rifles and axes, twenty-five thousand soldiers moved into the woods camps to provide both labor and security. Already relying upon the Oregon National Guard to quell strikers at his Astoria mill, Hammond wholeheartedly endorsed the notion of bringing soldiers and army discipline into his camps. He and other lumbermen saw the soldiers as strikebreakers who would drive the Wobblies into the Pacific Ocean if necessary.[55]

Disque, however, used his military and economic leverage to persuade lumbermen to make reforms. To qualify for the soldier-loggers, camps had to meet minimal army standards, which included adequate sanitation, housing, and acceptable food quality. Suddenly, for the first time for many, loggers slept on laundered sheets and clean blankets, had access to daily showers, and even received dental care.[56]

Unlike Parker, Disque believed that material conditions alone would not dispel radicalism. Instead, he regarded patriotism or, as he called it, "The Religion," as the remedy. The army, through the 4L, would permeate the workers' consciousness and "bring each rootless logger into a direct connection with patriotism." Army intelligence reports maintained that in late 1917 the labor situation in the Northwest was "a smoldering volcano" checked only by "The Religion" of the 4L. The legion decked bunkhouses with bunting, flags, and patriotic slogans. As Disque predicted, workers found dignity by participating in military parades around the woods camps. The presence of armed troops no doubt helped convince the more reluctant. In addition, Disque could always count on local draft boards to conscript revolutionary Wobblies who failed to convert to patriotic citizens.[57]

Lumbermen, sensing a comrade in Colonel Disque, began to circle closer. Russell Hawkins, a prominent antiunion Oregon lumberman, became Disque's assistant, while F. W. Leadbetter, a Portland lumber broker, served as Disque's liaison with the War Department. Hawkins and Leadbetter also served as conduits between Hammond and the War Department, advocating for the lumberman and passing along confidential information. Furthermore, Disque moved his headquarters to the Yeon Building in Portland, where the HLC had its Oregon office, giving George McLeod, Hammond's manager, easy access to the colonel. Surrounded by lumbermen, it was hardly surprising that Disque absorbed many of their ideas and values. In turn, they saw him as an ally in their war against organized labor.[58] At the same time, Disque resented Parker's civilian oversight, and the two men became increasingly estranged. It certainly looked as if Parker's attempt to use the 4L to enact reforms would fall flat.

As 1917 drew to a close, spruce production began to rise and labor strife seemed to dissipate. Nevertheless, the nagging issue of the eight-hour day remained. The Wilson administration was firmly committed to the reform, while lumbermen adamantly opposed it. Disque supported an eight-hour day in principle but hesitated to impose it dictatorially. Rather, he implied that lumber companies operating on an eight-hour basis would receive the most 4L assistance and priority shipping on railroads. Although lumbermen had united on the issue just a few months previously, the eight-hour day now divided them.

While many lumbermen saw the eight-hour day as inevitable, they wanted it to unfold on their terms. Believing that labor unions would claim credit for the reform, Hammond vigorously opposed adopting the eight-hour day. Furthermore, he believed that instituting the reform in the Pacific Northwest would give an unfair advantage to lumber companies in the South, which still operated on a ten-hour day. But when the Western Pine Association announced it would adopt the eight-hour day, other lumbermen began to cave and Hammond could no longer hold them in line. After much debate, sixty-four voted in favor of a resolution approving the eight-hour day for lumber manufactured during the war. Nine, including Hammond, opposed it, thereby blocking any agreement.[59]

While lumbermen failed to reach consensus, their lopsided vote indicated that resistance to reform was eroding. They agreed to hold another meeting to make a final, binding decision. After a contentious, all-day summit, at midnight on February 25, 1918, the last holdout finally acceded, and by a unanimous decision lumbermen accepted the eight-hour day in every mill and camp in Oregon and Washington. Hammond was absent, and was instead represented by the more moderate George McLeod.[60]

In his history of the 4L, Harold Hyman noted that the lumbermen would never have accepted the eight-hour day if it had come from the Wilson administration or from the unions. The army, however, was an acceptable mediator, and Hyman credited Disque's diplomacy, gentle pressure, and patriotic appeals with winning the day. Whatever the method, the shorter workday, along with Disque's efforts to improve working conditions, successfully defused the labor unrest. But as Hammond had predicted, the unions—specifically the AFL—claimed credit for the reforms and began to organize in the camps. Disque, however, resisted the AFL intrusion into "his" union—the 4L—and for twenty years after the end of World War I, the legion served as a company union dominated by employers rather than by workers.[61]

The Spruce Production Division not only curtailed union organizing during the war but also provided lumbermen with lucrative government contracts for supplying spruce to a guaranteed market. Although the army saw labor issues as the primary barrier to increased production, it also recognized the difficulty of accessing spruce and the reluctance of lumbermen to invest in infrastructure. To overcome this obstacle, lumbermen suggested that the government underwrite the cost of building railroad spurs. In September 1918, George McLeod negotiated a contract with the SPD to build a railroad up the Necanicum River and deliver airplane-quality spruce. Hammond, however, decided "that the work could not be done at the price the Government wished to pay, and withdrew the contract." He renegotiated and won more favorable terms: Hammond would receive more money per log, and the government would pay 65 percent of the $1 million it would cost to build the railroad. Hammond justified the increase "on account of the high wage rate, inefficient labor, and the eight-hour day."[62] In spite of these claimed difficulties, Hammond ran his Astoria mill day and night, cranking out 150,000 bf of government spruce every twenty hours. Furthermore, with the shortage of labor, the HLC used SPD soldiers to build the railroad and work in the mills, discovering "they make excellent labor."[63]

November 1918 brought a sudden end to the war and subsequently to spruce production. Even though the government canceled its contracts, Hammond had a railroad up the Necanicum built primarily at government expense. Now he could finally access remote stands of valuable, old-growth Douglas fir as well as spruce and hemlock. He also had two expanded and upgraded lumber mills in Astoria and, most of all, had a compliant labor force—for the time being.

West Coast lumbermen learned two important lessons from the war. They discovered that if they banded together in a united front, they could gain concessions. Even more important, they realized that government "could be the servant rather than the antagonist of business."[64] Such servitude had limits, however, for Progressive reformers within the government learned that they could defuse a potential revolution by mediating between workers and employers. While the army and lumbermen united in the Pacific Northwest, in another labor dispute far to the south, Hammond and the lumbermen clashed head-on with President Wilson's Labor Department.

Shortly after the lumber workers in the Northwest drifted back to work, a strike erupted in October 1917 when more than four hundred of Hammond's Los Angeles mill workers walked out. Hammond blamed the strike on mediators from the Department of Labor, in particular,

Harris Weinstock, a wealthy San Francisco merchant sympathetic to organized labor. Counterpoised against Hammond's involvement with the Citizen's Alliance and its campaign to eliminate unions, in 1903 Weinstock had founded the San Francisco Commonwealth Club to find solutions to labor strife through cooperation between business and labor leaders. In 1914 President Wilson appointed Weinstock as a member of the Commission on Industrial Relations, which was charged with investigating the causes of labor conflict in the United States. After 154 days of hearings, the commission concluded that unions were the solution to industrial unrest and that anti-labor employers were "a bar to social tranquility and a detriment to the economic progress of our country."[65] Clearly, Weinstock, the embodiment of a Progressive businessman, and Hammond were headed for a showdown.

To alleviate the nation's labor troubles, in 1917 Wilson authorized the President's Mediation Commission, with Weinstock as California's representative, to go beyond investigation and actually mediate labor conflicts. Discovering that lumber mill workers in Los Angeles received, on average, one-third less pay and worked longer hours than did other mill workers in California, the commission concluded that "the men had just grievances." Weinstock lectured Hammond that the only way for him to prevent strikes and maintain an open shop was "by treating his employees as fairly, as liberally and as generously as they are treated by Union employers elsewhere." Hammond, of course, was not about to be told how to run his business by government officials whom he considered to be "labor agitators and self-seeking politicians."[66]

The ideological difference between Hammond and Weinstock degenerated into a tug-of-war over the workers themselves. When Weinstock informed mill workers of their right to organize and of the mill owners' refusal to engage in mediation, the men went on strike. However, thirty-three of Hammond's employees informed the mill superintendent that they "were perfectly satisfied with conditions in general. We were sorry that we had to walk out." Four days later, presumably under Weinstock's influence, they retracted their statement.[67]

Privately, Hammond admitted that the eight-hour day was inevitable. Noting that the cost of living had increased, he acknowledged that his workers were entitled to higher wages and instructed Harry McLeod, his manager in Los Angeles, to advance "the wages for your best men" but avoid "having anything to do with the representatives of the Unions." To Weinstock, Hammond insisted that the mill owners were preparing to raise wages before the government "butted in."[68]

Weinstock defended his involvement in the Los Angeles labor dispute in terms of the "world's greatest war." He informed Hammond that the government could fix wages and prices, commandeer factories, and "butt into your private affairs." Evoking the ever-present theme of patriotism, Weinstock told the mill owners, "Stop our wheels of production and at once we are at the mercy of our enemies. . . . As mediators we feel we are performing the highest sort of patriotic service." Although Hammond employed such rhetoric himself, he remained unswayed by Weinstock's appeal. Charging the commissioner with using "intemperate" language to magnify the conflict, Hammond insisted that this was "a minor squabble over a local planning-mill company . . . that was not engaged in furnishing Government supplies of any kind," Hammond angrily added that while he was busy building ships and producing aircraft spruce for the government, mediators were interfering with his business. He concluded by accusing the Department of Labor of impeding the war effort.[69] Although the commission succeeded in some sectors, it ran into a brick wall in the lumber industry, as Hammond and other lumbermen flatly refused to meet with mediators or labor organizers.

Despite the efforts of Weinstock and the Mediation Commission to encourage union organizing, men gradually drifted back to work, effectively ending the strike. Harry McLeod informed his boss, "A number of these were union men, and from talks I have had with many of them I learned that they are disgusted with unionism." The strike gained nothing; although some skilled workers eventually received pay raises, the mill stayed on a nine-hour basis. Hammond could thus point to the workers themselves as validating his antiunion position. But just to be sure, McLeod insisted that the men destroy their union cards to be rehired.[70]

The Wilson administration often operated at cross-purposes, demonstrating the ambivalence of many Progressives. While the newly formed Labor Department actively supported union activity, the Justice Department under Thomas Gregory raided, tried, and imprisoned hundreds of

IWW members and condoned the extralegal anti-labor activities of the American Protective League. Meanwhile, War Secretary Baker forged ahead with a government-sponsored union, the 4L. Recognizing the fault lines between government agencies, Hammond used them to his advantage. By cooperating with the War Department, he could effectively offset and undermine the efforts of the Labor Department.

During the northern timber workers' strike in August 1917, Hammond had trouble fulfilling his government shipbuilding contracts because of his antiunion stance. Upon completing the ship hulls in Samoa, Hammond contracted with the Union Iron Works in San Francisco to install the machinery. The union workers, however, refused to work on HLC ships, as the company was on the "unfair" list. Hammond informed the Emergency Fleet Corporation that this would force him to cancel his contract to furnish government ship timbers and seriously interfere with supplying spruce lumber for government airplanes. In playing one government agency against the other, Hammond again accused the Labor Department of interfering with war effort. He then enlisted fellow lumbermen in the War Department for assistance, and in November the labor secretary informed union leaders that discriminating against nonunion materials, such as Hammond's ships and lumber, "in the time of war would be intolerable." The HLC, however, would be expected to pay union wages.[71]

Hammond's conflict with the Union Iron Works made it into the national press when *Collier's Weekly* printed an article on Pacific coast shipbuilding. The author, William Wolff, characterized Hammond as "a dour, two-fisted fighting man" who was "more interested in beating the unions than in beating the Germans."[72] Naturally, the lumberman objected to what he saw as an attack on his patriotism. He insisted "more than one-half of the men in our shipyard are members of the labor unions, but are good, reliable, patriotic citizens, and not agitators, and we have never had any trouble with them."[73] Hammond, however, would have been hard pressed to find even one union employee if anyone had challenged that claim.

Although he was easily offended himself, Hammond enjoyed being a gadfly, and provocative correspondence with newspapers, magazines, politicians, and government officials became something of a hobby for him. To C. H. McLeod he admitted, "I have had considerable fun with Collier's." All across the country, he admittedly enjoyed harassing and antagonizing labor supporters and derived "satirical merriment" from it. Hammond took pride in his intellectual dogfights and distributed copies of his letters to his managers and friends when he appeared to be winning. Often his letters were filled with anger and vindictiveness, especially toward Progressive reformers and fellow businessmen, such as Rolph and Weinstock. When the strikes were over and the dust settled, Hammond continued to needle his opponents. In Hammond's mind the Progressives, with their notions of reform, pro-union sympathies, and affinity toward government regulation, were taking America down the wrong path. Although Rolph and Weinstock were ultimately correct in viewing organized labor as the antidote to industrial strife, Hammond's sheer tenacity kept unions out of his enterprises for another seventeen years.[74]

Unions, however, emerged from the Great War in their strongest position yet. Under the Wilson administration, which recognized the right to collective bargaining, organized labor made significant advances. The Clayton Antitrust Act of 1914 ostensibly barred the use of injunctions against labor unions. The Seamen's Act of 1915, championed by Andrew Furuseth and Robert La Follette, greatly improved working conditions for sailors, while the Adamson Act of 1916 extended government authority into labor relations. Emboldened by such progress and beset by inflation, American workers walked out on an unprecedented scale in 1919, with some four million workers participating in 3,500 strikes. In Seattle, a five-day general strike in February paralyzed the city.[75]

Humboldt County was not immune to the wave of labor unrest and prolonged strikes that paralyzed the steel, stockyard, and railroad industries. Nevertheless, Hammond quickly crushed the strike that broke out at his Samoa mill. In a steady stream of letters to the Eureka mayor and the local press, Hammond insisted on the linkage between unionism, and "IWWism," and equated IWW membership with a "readiness to commit murder . . . as a means to an end." Substituting proverbs for logic, Hammond further insisted, "The leopard cannot change his spots." Therefore, according to him, it was only a matter of time before

Humboldt labor organizers would incite mass violence, social upheaval, and revolution.[76]

Around the country, many industrialists equated union activity with Communists and anarchists, no matter how spurious or absurd the connection. Nonetheless, massive labor unrest and a rash of bombings in the wake of the Bolshevik Revolutionof 1917 instilled fear into many that America was on the verge of a great upheaval. Employers, politicians, and the media fanned public passions by associating organized labor with Communism, thus instigating the Red Scare of 1919. Painting organized labor as "red" allowed Hammond to impose an official company policy of refusing to employ any union labor. Similarly, Hammond and other employers began to refer to the open shop as the "American Plan." By capturing the rhetorical high ground, employers cast any opposition to the open shop, such as collective bargaining, as "un-American" and therefore "Bolshevik."[77]

For years Hammond had trumpeted the dangers of the "red menace." Along with other businessmen, he actively fed the mob mentality of the Red Scare as überpatriots broke up Socialist meetings. Such hysteria reached the highest levels of government as Wilson's new attorney general, A. Mitchell Palmer, carried out his infamous raids against the IWW, radical organizations, suspected Communists, or anyone else who questioned the juggernaut of industrial capitalism. Making Woodrow Wilson's New Freedom campaign rhetoric seem positively Orwellian, the Palmer raids jailed hundreds of dissidents, while the New York State Assembly expelled duly elected Socialists.[78]

Finally, the pendulum of national sentiment began to swing back to Hammond's favor. By the end of 1919, the largest strikes had collapsed, and the backlash against organized labor sent unions into retreat. From a peak of five million in early 1920, union membership dropped by a third over the next three years. Although the Gilded Age was long past, the era of the businessman was about to begin.[79]

CHAPTER **19**

The Age of Consolidation and Cooperation

Although it took a long time to build to a frenzy, the Red Scare ebbed rather quickly. Nonetheless, it left an indelible mark on America, and one of its collateral casualties was Progressivism. Following the war, Progressives brightened at the prospect of continued government intervention in economic and social affairs. Woodrow Wilson had temporarily nationalized the railroads as a wartime measure, and many looked forward to a continuation of this policy. The Red Scare, however, marginalized any such propositions. Nevertheless, political reform efforts persisted throughout the 1920s, culminating in Wisconsin senator Robert La Follette's Progressive Party bid for president in 1924. The first line of the party's platform read, "The greatest issue before the American people today is the control of government and industry by private monopoly."[1] By "monopoly" Progressives meant large corporations that dominated a particular industry.

By 1920 a general conservatism pervaded the country. Americans, by and large, had accepted the permanence of such corporations. More and more people worked for wages and as middle managers, while farm and small-business ownership declined. Although the transformation from an agricultural/producer society into a consumer/corporate one had begun decades earlier, the 1920s witnessed the full flowering of consumer culture, a faith in unlimited material progress, and, perhaps most important, the fusion of big business and government. No longer antagonists over natural resources, these two became mutually reinforcing and interdependent. Summing up the decade, Hammond declared, "This is the age of consolidation and cooperation."[2]

The Great War made lumbermen aware that cooperating with the federal government brought constancy and profits to the timber industry. Industry trade groups, such as the West Coast Lumbermen's Association, attempted to stabilize the market through price fixing and lobbied state and federal governments for fire protection, lower taxes, and import duties. Long hampered in their cooperative efforts by the Sherman Antitrust Act, industry groups persuaded Congress in 1918 to pass the Webb-Pomerene Act, which allowed companies to fix prices on exports.[3]

Regardless of lumbermen's efforts to put cooperation ahead of competition, the industry in the 1920s remained chaotic, thus serving as a potent indicator of the state of the American economy. Ever since its origins on the Atlantic coast, the lumber industry had suffered from chronic volatility, illustrating the inherent instability of unfettered capitalism. Following World War I, the industry performed its usual gyrations—a short building boom and rising prices triggered overinvestment and overproduction. Expensive new technologies, such as diesel donkeys and Lidgerwood skidders, engendered faster and more-efficient production but created even greater debt. By now a familiar cycle, overproduction led to a glut and falling prices, which prompted producers to crank out even more lumber on an ever-decreasing profit margin to service their debts.

Combining his astute modern business sense with nineteenth-century social Darwinism, Hammond noted, "In the lumber business there are too many sawmills and the business will not be profitable until the industry drifts into the hands of large companies through the inexorable course of the survival of the fittest."[4] Hammond, like many businessmen, economists, and historians, regarded consolidation and combination as inevitable. Neither evolution nor economics, however, proceeds along a preordained and predictable path; both depend upon contingencies and individual action.

Conveniently, Hammond's ideology of economic Darwinism put him at the top of the food chain. Nevertheless, his determination to become an industrial capitalist was informed by the world in which he lived. In another age Hammond might have been a general, a politician, or even a football coach—anything that would allow him to track or enumerate his victories. But in America's industrial age the businessman embodied the paragon of accomplishment. With the waning of Progressivism in the 1920s, the media even dropped its muckraking portrayals of "robber barons" and instead heralded "captains of industry" as the models for society. Profits proved a quick and measurable indicator of success. Yet, as he entered his seventies, Hammond no longer sought personal wealth. Instead, sheer ambition drove him to make the HLC the largest redwood lumber company in the world, a mission to which he devoted the rest of his life.[5]

During the 1920s, the U.S. economy was on a decade-long roller-coaster ride. One industry and region could be up while others were down. The Great War had pushed up prices and employment. Following the armistice, returning soldiers swelled the ranks of the unemployed, yet prices continued to climb. After a brief depression in 1921–22, the nation witnessed rapid growth in the electronics and auto industries concentrated in the Great Lakes region, invigorating midwestern cities. But farmers in the South and on the Great Plains suffered from low commodity prices and drought. As the primary provider of building materials, the lumber industry bounced and tumbled with every bump on the ride. Nonetheless, much of the country enjoyed a few years of apparent prosperity before the long slide into the Great Depression.[6]

Following the war, industries on the West Coast reveled in a giddy surge of business, which appeared as if it would continue despite strikes and unemployment. Buoyed by a rising tide of construction in Southern California, Hammond saw his Los Angeles operations post one record-breaking month after another. Looking back at 1919, Hammond crowed, "Upon the whole our business was exceedingly profitable last year." For 1920 he expected sales "between $7.5 million and $8 million" in Los Angeles. Hammond had reason for optimism. In 1915 lumber prices averaged $21.63 per mbf; they more than doubled over the next four years. In March 1920, lumber prices shot to $72.41. Practically dazed by the money pouring in, Hammond informed C. H. McLeod, "All of our concerns are making money much faster than they ever have before." In Montana, meanwhile, McLeod complained that high prices dramatically curtailed sales and imposed hardships on the working man.[7]

In September 1920 lumber sales in Los Angeles broke all previous records, but by January 1921 business had dropped off sharply and 60 percent of the Pacific Northwest's sawmills had shut down. Predicting that 1921 "will no doubt be a very lean year in the lumber business," Hammond rather gloomily concluded, "This year, if we can operate without losing money, I shall feel that we have done very well." Not only was the lumber market slumping, but high expenses, especially labor and freight rates, were eroding Hammond's profit margin.[8]

McLeod, a bit of a pessimist, kept Hammond apprised of the situation in Montana, which had never recovered from the drop in farm prices following World War I. The war in Europe had sent wheat prices skyrocketing. Induced by quick profits and government-guaranteed rates, thousands of Americans drifted onto the Great Plains to take up homesteads. After the war, however, the wheat market collapsed. Busted homesteaders defaulted on their loans, causing rural banks to fail. Across the country, more than one million farmers lost their lands to foreclosures in 1920–21.[9]

The sagging agricultural sector combined with postwar inflation to tug the national economy into a depression in 1921. The combination hit Montana particularly hard, sending the state's economy into a twenty-year downward spiral. Long before the rest of the nation, Montana began to suffer from the combined effects of an overextended economy, low commodity prices, harsh winters, and droughts. Banks were hard up, merchants lost money, and property values dropped. Pessimistically, McLeod noted that five years of drought "has placed the state in a position where it is going to take a long while to recover and a great many banks and business men are going to be wiped out." In addition to crop failures, the national depression closed the copper mines and sawmills. McLeod summed up the situation: "Montana is broke."[10]

McLeod believed that at best the Merc would break even in 1921. When Hammond suggested cutting costs, McLeod resisted and informed him that with the rising cost of living, "men cannot work and support their families on the salaries they formerly received." Nonetheless, by the year's end, the MMC had turned enough profit to post its usual dividend of 10 percent. Ever conservative, Hammond, however, advised against any disbursement that year.[11]

Nationally, other businesses were in far worse shape than the Merc. Sears, Roebuck and Co. was $16 million in the red, and the meatpacking giant Armour and Company lost a whopping $32 million. In contrast, the Hammond Lumber Company largely on the strength of retail sales in Southern California, continued to post profits in the millions.[12]

Although the agricultural sector languished throughout the 1920s, the rest of the economy rebounded, at least for the time being. Los Angeles surpassed San Francisco as the largest city on the West Coast and ranked as the second-fastest-growing city in the nation. In 1922 alone, eighteen thousand houses sprang up in Los Angeles and sold as fast as they were built; demand for lumber and construction materials boomed. Hammond, by luck or foresight, had banked on Southern California's growth, and by 1923 the HLC claimed the world's largest retail lumberyard. Located at 20th and Alameda Streets, it supplied homebuilders and contractors with a variety of planks, boards, and sheathing of Douglas fir, spruce, hemlock, and redwood. As the yard was also the largest roofing supplier west of Chicago, a builder could also load up on trusses and shingles. A nearby factory supplied both standard and custom flooring, trim, doors, and windows. Neatly displayed were paint and all the hardware—nails, screws, latches, and locks—everything necessary for construction except plumbing and lighting.[13]

Likewise at San Pedro, Hammond expanded his operations into the West Coast's largest wholesale lumberyard. The *Los Angeles Times* described it from a bird's-eye view as "one vast wilderness of building material. . . . The harbor appears to literally gorge with lumber . . . being rapidly absorbed by local builders." In Los Angeles alone, the two yards and office employed nearly two thousand men and women. In addition to the houses, apartments, and office buildings, Hammond Lumber supplied Hollywood movie sets, while aqueducts and irrigation projects in the arid Southwest required redwood for pipes and storage tanks. Furthermore, the oil boom in Southern California led to a profusion of wooden oil derricks, each needing 30,000 to 40,000 bf; copper mines in Arizona chewed up another 150 million feet a year.[14]

Here then was the most crucial factor in Hammond's success: vertical integration. Certainly other lumber companies owned timberlands and mills, but few appreciated the importance of retail markets. Hammond had learned the value of an aggressive sales division in the 1880s when Marcus Daly pulled the Anaconda contract. Forty years later, Hammond had established sales offices across the United States, while the home office in San Francisco handled the export business. By 1920 more than 80 percent of HLC sales from the Los Angeles yard were retail. Throughout 1923 the Los Angeles yards continued to post record-breaking profits, boosting the HLC's net for the year to nearly $5 million and prompting Hammond to buy two more steamships. One of these Hammond renamed the SS *Missoula* and transformed it into "the world's largest lumber carrier" with a capacity of 4.5 million bf. From Astoria, the Hammond fleet brought both raw and finished Douglas fir and then stopped off at Samoa to load redwood. One after another, the ships unloaded at San Pedro, where the lumber was sorted and then reshipped by rail or ship to customers from Australia to New York. Such integration allowed the HLC to weather the upcoming economic storms. As overproduction drove prices below the cost of production, Hammond often found it more profitable to buy raw lumber from other mills and resell it through his retail network than to cut and mill his own timber.[15]

The HLC stood out as Southern California's largest lumber supplier. Whether through direct ownership or as a dealer, the company controlled sixty-five lumberyards in Los Angeles and another three hundred in the surrounding region. Controlling one-third of the business, Hammond could dictate prices. When other manufacturers refused to maintain high prices, the HLC simply cut rates until its competitors declared bankruptcy, after which the HLC raised prices back to its desired level. This practice, in Hammond's mind, amounted to the "survival of the fittest" and underscored the need for industry consolidation.[16]

Hammond set his sails with the prevailing winds, as he always had, and applied knowledge gained from past successes to changing economic conditions. Although his need for control remained unabated, Hammond embraced new technologies and began to reorganize his business along more modern lines. Hammond's close managerial oversight, ruthless ambition, and tight financial control enabled the HLC to post profits and ensured its continued success, even in the worst of times, while other lumber companies struggled to stay afloat. Essentially, Hammond's personality traits translated into three essential business elements: organizational efficiency, vertical integration, and lack of debt.

Hammond's upbringing in New Brunswick and experience in the Panic of 1893 taught him the dangers of becoming financially overextended.

Thus, when many corporations with dispersed ownership had to accede to the demands of stockholders to pay dividends, Hammond could reduce or even decline to pay dividends if he felt it prudent to do so. Hammond's fiscal oversight even extended to checking personally the company's telephone and postal charges. Such tight control allowed Hammond to maintain a positive cash balance and slash expenses when needed.[17]

The old Missoula Mercantile played no small part in Hammond's continuing success over the years. Maintaining the impeccable credit rating of the MMC allowed Hammond to access short-term loans to meet payroll and other needs at times when banks were reluctant to extend credit to the floundering lumber industry. Hammond also used the Merc to avoid wholesalers in getting hardware directly from manufacturers and for obtaining clothing and groceries for his company stores. Like everything in his empire, Hammond operated these stores not as an employee service but as a profitable enterprise, averaging $250,000 in yearly sales.[18]

Hammond's need for control, combined with his competitive drive, hard-line managerial style, and insistence upon loyalty led him to view business as a war, with himself as a general deploying troops and making tactical decisions. As early as 1890 the Missoula press referred to Hammond as the "Napoleon of finance," a characterization he, no doubt, fancied. As in his campaign against organized labor, Hammond employed military analogies in addressing management issues. As Hammond aged, he told McLeod, "I will arrange to throw more responsibilities on our field marshal [Hammond's son Leonard] and major generals."[19] Hammond, however, always remained the generalissimo, and his tall, gaunt frame, piercing eyes, and sharp goatee presented an intimidating visage to his managers when he came to check on their operations, which was frequently.

His managers had good reason to worry. Hearing that "old A. B." was on his way to one of his district offices, managers rushed to clean off their desks. If Hammond saw a man's desk covered with papers, he would "sweep them off onto the floor with his cane." They also hurriedly "locked their pretty girls in the office vaults," for since his early days in Missoula, Hammond had had a reputation as a lady's man, and entering his seventieth year, he scarcely slowed his skirt chasing.[20]

Despite his diligent oversight, Hammond credited much of his success to his competent executives, the major generals, so to speak. The three McLeods, C. H., George, and Harry, all proved superior managers of their respective divisions—Missoula, Oregon, and Southern California. In Samoa, however, George Fenwick had "allowed his organization to run down," requiring intervention by the generalissimo. Although Fenwick and Hammond were born the same year, A. B. said that since his brother-in-law was too old to change, "he will be retired with all due honors." Leonard moved into the vacancy, but proved less compliant than the other generals and occasionally clashed with his father on management issues.[21]

Although Leonard shared many of his father's ideals—hard work, thrift, and patriotism—he also embodied the new age. As historian Lynn Dumenil argued, the nineteenth-century emphasis on character gave way to a twentieth-century preoccupation with personality. The rise of a consumer culture with a desire for indulgence and leisure eroded the Victorian ethic of sacrifice and restraint. The emergence of welfare capitalism and spectator sports—two of the more apparent manifestations of this shift—converged at the HLC company town of Samoa.[22]

Leonard Hammond absorbed many of the ideas of welfare capitalism, including filling workers' recreational needs. During his tenure at Samoa, the HLC provided housing and medical care, built a large gymnasium for workers and a women's clubhouse, and organized a baseball team. The Humboldt County league consisted of only four teams—Eureka, Arcata, Scotia, and Samoa—but competition was fierce. Engrossed in his team, Leonard provided incentives to attract top players, including easy work schedules, higher salaries, and new housing. When A. B. arrived on one of his periodic tours, he was aghast at such frivolity. Declaring that Samoa had plenty of housing, he halted construction. After he left, however, Leonard quietly resumed the building. Leonard also made improvements to the Humboldt operation, adding new logging locomotives and overseeing the transition from a steam-powered mill to an all-electric one in 1921. Having proved his worth at Samoa, Leonard moved back to San Francisco to assume the vice presidency of the HLC.[23]

Convinced his business was in competent hands, A. B. Hammond began to contemplate retirement. He still suffered from bouts of illness, each of which left him thinner than the last. Nonetheless, the Astoria newspaper had suggested that he could pass for "a man under fifty." The paper attributed his "eternal youth" to "the fact that he makes play out of work and this is the core of his great success. Follow him for a day and you will come to the conclusion that he is doing big business not for the money that is in it, for he has enough of that . . . but for the keen enjoyment that he takes in seeing the wheels go round."[24]

Similarly, C. H. McLeod fully grasped his old friend's primary motivation. He told Hammond, "I am not as ambitious to control large property interests as you are . . . we all cannot love the game as you do, nor do we all possess the ability, industry and aggressiveness that you do."[25] For half a century Hammond had devoted his life to acquisition, but unlike another business partner, Henry Huntington, who used his share of the HLC profits to finance a world-class art collection and library, Hammond ploughed profits back into the company, buying HLC stock as it became available and purchasing more timberlands and mills. Indeed, when Huntington invited Hammond to tour his art collection, the lumberman displayed little interest.[26]

Nine years Hammond's junior, C. H. McLeod also considered retirement, and the two old friends exchanged their thoughts on the subject. Indulging in a bit of self-reflection, Hammond acknowledged that the accumulation of wealth was no longer of interest. He told McLeod, "For some time I have been divesting myself of my worldly goods." In an extraordinary admission, he stated, "I don't care to make any more money." McLeod, however, did not believe the old lumberman could ever retire. He insightfully pointed out, "Successful operations make men very ambitious, and I can understand . . . why it is that you are constantly looking for more worlds to conquer." As McLeod implied, Hammond could no more retire than he could stop breathing.[27]

Other industrialists, such as Andrew Carnegie and John D. Rockefeller, turned their energies toward charity once they had realized their fortunes. But the notion of giving away money ran counter to Hammond's very soul. Above all, he believed in hard work and making one's own way in the world, stating, "When a man earns something himself, he appreciates it more than when he gets it for nothing."[28] In 1922, when the HLC posted profits of nearly $5 million, the company donated a total of only $37,000, and Hammond regarded the 0.7 percent sum as excessive. Beset by a seemingly endless line of petitioners, Hammond penned a humorous, albeit pointed, letter to Father Joseph Stack's request for a donation:

> The Episcopalians, Methodists, Presbyterians, Catholics, Salvation Army, YMCA, YWCA, Associated Charities, Sisters of Charity, Old Ladies' Homes, Institution of the Blind, Disabled War Veterans, Children's' Hospital, Association for Booming Northern California, the "be a Booster' Campaign, and about two dozen civic associations . . . and other charitable enterprise, etc. etc. raid us at our office, waylay us on our way to lunch and follow us home, and if by any chance we should succeed in eluding any of the pitfalls which they set for us, we are not to escape, because in some instances, as a last resort, fair blondes, dark-eyed houris, or Titian-haired beauties, arrayed in short skirts, high heels and low blouses, greet us with a smile and accomplish our undoing. We are no longer doing business, for under existing conditions the word is a misnomer, and as a consequence I propose to pull down the blinds, close up shop and seek refuge in the delectable isle where they count not the days.[29]

Regardless, Hammond enclosed $200 for the Jesuits, the one charity he could never deny. Whether this was due to his friendship with Father Anthony Ravalli in his early days in Montana or stemmed from his childhood in Madawaska is uncertain. Regardless, Hammond was proud of this epistle, keeping a copy on file and sending it in reply to any future requests for donations he received. As he aged, however, he softened. While Hammond usually turned down requests for outright donations, he occasionally agreed to loan money if it was for a legitimate purpose or to repay a karmic debt. For example, he agreed to help Maude Wilson, the daughter of one of the many Acadians from Madawaska whom he had brought to Montana to work on the Northern Pacific contracts.[30]

As Hammond began to contemplate his own mortality, he moved to consolidate ownership of the HLC. Marcus Daly and C. P. Huntington had died in 1900, followed by grocery wholesaler Francis Leggett in 1909. Their heirs displayed little interest in the operations of the HLC other than receiving dividends and were represented on the board of directors by their proxy attorneys. In 1915 General Thomas Hubbard, former vice president of the Southern Pacific Railroad, died, leaving only John Claflin as one of the original six investors in the HLC besides Hammond himself. Claflin's poor business practices resulted in his bankruptcy in 1926, after which his HLC stock formed the bulk of his family's assets, and he began taking an increasing interest in the HLC's affairs. Hammond, of course, resented such oversight, and to ensure the company remained "under the management of [the Hammond and McLeod families] . . . who have helped me to build it up," he purchased HLC stock from the heirs of the original investors whenever available. Illustrating the pitfalls of a family-owned corporation, internal dynamics within the Hammond family, however, nearly tore the company apart in the ensuing years.[31]

In addition to securing his stock, Hammond continued to buy timberlands and smaller companies to fortify his finances. Since voluntary associations had failed to control the redwood industry, Hammond was convinced that the best way to maintain a viable organization was through sheer size. If one company could dominate the redwood industry, it could set prices, restrict production, and thereby increase the value of its property.[32]

As he aged, Hammond grew ambivalent about his own role in forging such a combination. At times, Hammond felt he "was too old to undertake any new enterprises, and . . . preferred to sell and let somebody else do the consolidating." More than once, he offered to sell off divisions of HLC, timberlands, sawmills, and equipment. In 1914 his price for the Humboldt operations was $11 million. Such a price far exceeded the improvements he had made and represented an elevenfold increase over what he had paid for the Vance Lumber Company fourteen years earlier. Not surprisingly, no one jumped at the offer. Eventually, Hammond realized that liquidating the business would be imprudent and therefore began to reorganize the company so that it would be as profitable "in the future as it has been in the past, regardless of what may happen to me."[33]

To increase his lumber supply and consolidate timberlands, Hammond continuously bought and sold land from 1920 to 1922. By the end of 1921, he had boosted his redwood holdings to 87,000 acres, providing his Samoa mill with an eighteen-year supply at its then-present rate of cutting. With an eye toward consolidating his timberlands in Oregon as well, Hammond sold the remote Necanicum operations to Crown-Willamette Lumber for $1.5 million and used the funds to buy more timberlands near Astoria, where he had purchased the Tongue Point Mill in 1908. Making extensive improvements to the mill, Hammond increased its capacity to four hundred thousand feet per day and built one hundred houses for employees. But then, on September 11, 1922, disaster struck.[34]

By any measure, fire was a lumberman's worst enemy, far more dangerous than the IWW. Surrounded by piles of sawdust and lumber and with sparks flying from machinery and waste burners, sawmills were highly susceptible to fires. While many could be confined or put out quickly, others could wipe an entire operation off the map. At Hammond's Tongue Point Mill, as the day shift was getting ready to leave, a fire ignited in a pile of sawdust. Moments later the entire mill was engulfed in flames. The fire quickly spread to the power plant, which exploded into "a seething, whirling cauldron of orange and yellow flames shooting for hundreds of feet into the air together with black rolling clouds of smoke." The power plant's dozen boilers generated electricity for Astoria and surrounding towns, and the sudden destruction plunged the city into darkness. The fire department and mill workers battled the blaze, finally resorting to dynamiting the lumberyard to protect the company store, offices and docks. The trade journal *Timberman* estimated the loss at $1 million. To McLeod, Hammond dryly noted, "It was well insured."[35]

While Hammond could recoup the loss, Astoria was less fortunate. The fire not only destroyed the city's power supply but also threw six hundred men out of work and eliminated the city's largest payroll. Understandably distraught, Astoria officials telegrammed Hammond wanting to know if he planned to rebuild. After he had put them off for two months, another massive fire hit Astoria, this time wiping out thirty downtown blocks and causing $15 million in damages. Hammond quickly informed the city that he would indeed rebuild the mill and sent a $5,000 relief check. A year later the mill remained in ruins.[36]

Although Leonard Hammond advocated rebuilding the Astoria mill, his father was lukewarm. More than anything else the eventual decision not to rebuild rested on economics. While Hammond had a firm grip on the redwood industry and dominated the Southern California retail trade, he faced serious competition in the Douglas fir industry of Oregon and Washington. Everyone, from the industry giant Weyerhaeuser to small "peckerwood mills," was cranking out lumber, flooding the market, and driving down prices, often below the cost of production. As a result, the Astoria mill was never rebuilt.

During the mid- to late 1920s, the nation enjoyed unprecedented prosperity, yet the lumber industry began to decline once again, as Hammond had predicted. Having experienced the depressions of 1877, 1893, and 1907, Hammond noticed the signs of impending trouble, yet he also saw opportunity in the depression that hit the lumber industry well in advance of the rest of the economy. With falling prices, Hammond looked toward increasing efficiency and restricting production. In 1925 he told C. H. McLeod, "The lumber business . . . is in a rather serious condition and I do not look forward to any revival of trade within the next two or three years at least—perhaps longer; consequently we are reducing our expenses all along the line."[37]

Conditions were especially bad in Montana, which witnessed, in addition to drought, more than 150 bank failures. By 1925 20 percent of Montana's farmers had left the state, and by the close of the decade some sixty thousand people had fled, making Montana the only state to lose population during the 1920s. McLeod noted ominously, "There seems to be a lack of confidence among the people living in the state as to its future." Predictably, Hammond blamed the state's economic woes on the Montana Progressives, who had placed Thomas Walsh and Burton K. Wheeler in the U.S. Senate and Joseph Dixon in the governor's office, men whom Hammond regarded "as being in sympathy with Soviet Russia."[38]

McLeod, on the other hand, fingered discriminatory railroad rates, which made it more expensive to ship consumer goods and agricultural exports to or from Montana than it was from the West Coast. Enlisting Hammond to his cause, McLeod prodded the Northern Pacific into more equitable common point freight rates, whereby the transcontinental rate would be the same regardless of destination. Surprisingly, it was Hammond who wined, dined, and cajoled the railroad officials while McLeod berated company officials and politicians. The HLC, after all, was a major shipper, and Hammond harbored no compunction about diverting half a million dollars of business from one railroad to another. Hammond also repeatedly pointed out to the Northern Pacific and to Montana's senators that it was much cheaper for him to ship lumber through the Panama Canal to the East Coast than to use the railroads. Eventually Hammond's tenacity prevailed and he succeeded in obtaining a universal freight rate for his lumber, earning him the moniker "Common Point" Hammond.[39]

Regardless of his victory, the possibility that Montana's economic conditions were a bellwether for the nation troubled Hammond to no end, and he began to plan for the worst. Preoccupied with his business, Hammond was unprepared when the worst came from a different direction. Florence, his wife of forty-six years, died following a lengthy illness. After spending a few days at home to recover, Hammond threw himself into his business more than ever, zipping off to Samoa and Los Angeles. Feeling lonely, he encouraged visits from friends and relatives. Asking McLeod to bring a photo of Florence taken thirty-five years before, he wrote, "I am all alone, with a big house and plenty of servants, and it will be a great pleasure to have you visit me."[40]

By now the Hammond children were fully grown and scattered throughout California, and without Florence to hold the family together, their bonds to each other gradually unraveled. Of the five remaining children, only Leonard was active in the management of the HLC. Following Florence's death, A. B. took an increasing interest in the lives of his Montana relatives, often providing unsolicited advice, along with his annual $100 Christmas presents. He was especially concerned with the fortunes, or lack thereof, of the family of his older brother, George, who had cajoled Hammond into leaving New Brunswick and seeking their fortunes in the West. George had died in 1904 and left behind a widow and four grown children, whose sole inheritance was a ranch on the Blackfoot River, northeast of Missoula. Perhaps during their travails the brothers had made a pact to look after each other and their families, or possibly Hammond felt he owed George for his start in life. Whatever the reason, he provided money for the education of George's eleven grandchildren. Hammond's support of his relatives, however, was not without tribulations. His nephews, George and Leonard, caused him endless headaches.[41]

While full of entrepreneurial ideas, neither nephew seemed to have the dedication, drive, or the work ethic their uncle valued so dearly. Hammond was especially irritated at his relatives for seeking privileges through the virtue of their name. Both George and his mother frequently purchased goods from the MMC on credit, accrued large bills, and made no effort to pay them. To finance his schemes, George mortgaged the ranch, assuming his connection to the Merc would float his debts. Finally, fed up with George's "improvident and irresponsible" ways, Hammond suggested that C. H. McLeod foreclose on the ranch. McLeod, however, feared this might cause a local social backlash. But in 1922, George was arrested for selling mortgaged property and lost the ranch to foreclosure. Although Hammond loathed the thought, he eventually paid the debts of his relatives.[42]

Giving up on George, Hammond turned his attention to his nephew Leonard and his family, although he had little confidence in him, either, and chastised him for his spendthrift ways. Hammond's financial support came with certain conditions, such as dictating where the family should live and work. He told Leonard's wife, Ann, that he would send them $100 a month for their children's education but they should move from central Montana to Bonner, where Leonard could get a job and the kids could go to school. When Ann sent her children to St. Anthony's so they could learn "honesty, truthfulness and politeness and many little things that they do not learn in public schools," Hammond insisted that they should attend public school instead, where the "County furnished the books." No doubt, Leonard regarded Hammond's interference as an insult to his manhood and ability to provide. But as the economy worsened and his family faced a series of medical bills, he was left with little choice but to depend on Hammond's beneficence. Quite naturally, Leonard and George resented their rich uncle's condescending and paternalistic attitude. Their wives, however, readily kowtowed and responded to Hammond's charity with grateful appreciation.[43]

Hammond's attitude toward education was often ambiguous. College, he believed, produced young men intent on enjoying themselves rather than working. When the daughter of one of his old Missoula friends approached him to set up a $150,000 endowment for a college in Oregon in his son Richard's name, Hammond pointedly informed her that neither

Washington nor Lincoln had had a college education. He added, "I was quite intimately acquainted with Collis P. Huntington, E. H. Harriman and James E. Hill [*sic*], generally admitted to be the three greatest captains of industry in this country. None of them were college graduates; two of them were not even graduates of a high school." Furthermore, with regard to the working world, he told her, "We have in the service of our company ten men drawing salaries ranging from $10,000 to $20,000 a year, only one of whom is a college graduate. . . . We have, however, quite a number of college graduates drawing salaries all the way from $1,500 to $3,000."[44]

Despite his apparent antipathy toward higher education, Hammond sent his grandnephew, George L. Hammond, $100 a month as a college allowance. He pointed out to his nephew that with interest this amounted to $5,415. Criticizing George for his poor spelling and handwriting, Hammond wrote, "The money being put up to keep you at the University should produce return on the investment; the question arises, how are such returns to be made?" Concerned that most of the boys were going to college "for the purpose of playing football," Hammond asked the president of the university to keep him informed of George's grades.[45] Given his love of literature and history, it was not formal education itself that Hammond opposed but, rather, the sense of entitlement that it seemed to engender. Both Hammond and McLeod grumbled that schools and universities turned out men who no longer valued hard work. Hammond, however, discovered that many of his older managers who had been with him for years had lost their edge and drive, and he began replacing them with ambitious young men, regardless of their education.

No matter how much attention and energy Hammond gave to his extended family, he ultimately measured his self-worth through his own business achievements. The more he aged, the more intense became his desire to make that one big deal. As the lumber industry continued on its self-destructive path of overproduction, Hammond regarded industry consolidation under single ownership as the only solution. Beginning in 1926 the HLC, Northern Redwood, Union Lumber, Pacific Lumber, and a newcomer—the Little River Redwood Company (LRRC)—began a series of lengthy negotiations toward a merger of the largest redwood companies. But the issue of control proved problematic for all parties.[46]

Hammond was unwilling to form a partnership and, at first, insisted the others give him a cash buyout of $47.5 million. Failing that, he proposed forming a new firm in which he would have a 70 percent controlling interest. The differing organizational structures of the companies complicated matters. Although he had distributed most of his stock to his children, Hammond was still the president and patriarch and retained full discretionary authority from his board of directors. The HLC was essentially a one-man corporation. In contrast, while both Pacific Lumber and the LRRC began as family firms, by 1928 they had become managed estates composed of multiple stockholders, resembling modern corporations. As such, they had to appoint representatives and gain approval from their boards for each negotiating point. This became endlessly frustrating for everyone, as Hammond, acting on his own initiative and used to doing business one-on-one, kept changing the terms, proposing an option one day and then taking it back or refining it the next. With his father being manipulative and obstinate, Leonard, who was more accommodating and "very anxious" to finalize the merger, assumed the role of lead negotiator and traveled to Detroit, home of the Pacific Lumber stockholders.[47]

The primary point of contention was how to determine the value of each company. In most industries this procedure was fairly straightforward. While it was relatively easy to set prices on sawmills, railroads, ships, and buildings, determining the value of standing timber was much more complex and problematic, as no one really knew how much timber they actually owned. Unquestionably, the LRRC had the best timberlands, which Hammond desired as these were contiguous with his own holdings. Geographically, the HLC was well poised to extend its railroads and operations up the Little River. Both the HLC and Pacific Lumber agreed on an assessment of 100,000 bf per acre, valued at $4 per mbf (except on the Prairie Creek and Del Norte tracts). The LRRC, in contrast, assessed its timber at 151,000 bf per acre at $4.70 per mbf, a reasonable claim, given the high quality of its tracts. Hammond, however, refused to allow joint timber cruises to reconcile the disputed values, reportedly saying he "did not want anyone to tell him how much timber he had." Another sticking point: the LRRC stockholders insisted on receiving at least 40 percent in cash or bonds as part of the merger. Hammond, however, flatly refused any deal that would increase the debt of the consolidated company.[48]

Preservationists further complicated the negotiations, as the Save the Redwoods League (founded in 1918) sought to protect groves along the highway as a state park. The newly completed Redwood Highway ran up the Eel River through the middle of Pacific Lumber's timberlands and exposed the public to the majestic beauty of the redwoods. LRRC officials worried that public opinion would prevent Pacific from logging along the Eel River and therefore drastically reduce the value of the timber.[49]

Ultimately, both Pacific and the LRRC feared they would "be completely dominated by the Hammond organization." Indeed, part of Pacific's motivation for including the LRRC in the merger was to avoid being submerged by the HLC. Hammond had bought up Pacific stock in anticipation or, perhaps, to force the merger. For its part, the LRRC worried that the merger would result in the loss of the "best property in the redwood belt . . . and the identity of our business." Frustrated at the impasse, the LRRC officials agreed that it was "pointless to continue discussions at this time."[50]

Throughout the 1920s overproduction continued to plague the lumber industry, and many companies accrued debt, borrowing against their assets—primarily, standing timber. Not surprisingly a large number of firms folded, and by 1927 the industry was in critical condition. Nonetheless, the HLC continued to operate in the black. Its strong financial condition and its distribution and marketing network, which stretched from Los Angeles to New York, put it in an enviable position. But the following year, Hammond estimated that 95 percent of the lumber companies on the West Coast operated at a loss. Nearly every sawmill cut production or suspended operations. Many simply went out of business; even the HLC lost money on its sawmills. On the strength of his retail operations and shipping line, however, Hammond was still able to post profits, exceeding $1.6 million in both 1927 and 1928. At the end of 1929, Hammond could report, "We are in excellent financial condition," and he had even been able to loan $100,000 to the Union Lumber Company.[51]

Convinced that the depression would last at least three more years, Hammond, once again, sensed opportunity in crisis and started buying.

Although he avoided taking over any companies with large indebtedness, the Whitney Lumber Company at Tillamook on the Oregon coast appeared attractive. It did not have any outstanding debt, and its holdings were contiguous and interspersed with HLC's own timberlands.[52]

Valued at nearly $2.25 million, the Whitney Company owned logging railroads, camp buildings, and one of the largest sawmills on the West Coast, with a 250,000 bf capacity. Whitney had also built a dramatic incline railroad with 60–80 percent grades to access steep slopes of the Coast Range, much of which had already been clear-cut by 1927. Hammond, however, had scarcely touched the thirty thousand acres up the Nehalem River he had acquired from the Southern Pacific Railroad a quarter century earlier.[53]

Rather than buying Whitney outright, Hammond formed a subsidiary—The Hammond-Tillamook Lumber Company, capitalized at $5 million, with the HLC holding 65 percent of the stock. This new company combined Hammond's timberlands with Whitney's Garibaldi mill, railroads, and logging equipment. By all standards this appeared to be a natural and potentially profitable merger. Economics, technology, and the physical environment, however, coalesced to make it one of the biggest mistakes of Hammond's career.[54]

The new company set Hammond back $1 million, with the HLC and—much to McLeod's chagrin—the Missoula Mercantile Company carrying the debt. While he still had no outside debt, Hammond had put himself in the same position as other companies: he needed to cut timber fast and furiously to service the loan. There was a good reason, however, that the Nehalem lands had been spared thus far.

For years the topography of the Oregon Coast Range thwarted logging endeavors. While it contained some of the largest trees on the planet, the stands of Douglas fir, western hemlock, and Sitka spruce grew on exceedingly steep slopes in a crumpled mass of mountains bisected by twisting rivers and narrow canyons. Such conditions made access inordinately expensive. Previously, Hammond had relied on the federal government, driven by its need for airplane spruce, to foot the bill for building a railroad to access his holdings there. Now he was on his own. So, after polishing off Whitney's timber in the Kilchis River drainage, the Hammond-Tillamook Company began building a main line up the

Nehalem River. Using portable lighting, crews worked day and night, removing and filling thousands of cubic yards of earth to form a roadbed. With costs running $25,000 per mile at a time when lumber prices were plummeting, Hammond's new company faced a $193,000 loss its first year. To make matters worse, the stock market collapse of 1929 put the lumber industry in a more desperate condition than ever, and it seemed the only solution was to keep cutting. The Tillamook forest was ripe for disaster, and it arrived in 1933.[55]

In the meantime, as the economy worsened lumbermen realized Hammond was right: financial survival depended on consolidation. Reopening negotiations in September 1929, the HLC, LRRC, and Pacific finally agreed to conduct an independent assessment of their timberlands. But then Pacific Lumber dropped out and the negotiations stalled again. Undaunted, Hammond continued to buy and sell timberlands, consolidating and increasing his holdings. He bought half interest in the Bayside Mill in Eureka while keeping his eye on the Little River lands. Despite the deepening depression, in 1930 Hammond told McLeod, "The time will come when stumpage will bring a good price, and then will be the time to sell and clean up all around."[56]

Instead, economic conditions deteriorated, and Hammond began to cut costs and demanded that each division reduce expenses. The redwood division eliminated train crews and cut back on timekeepers, scalers, and foremen. Mill City cut thirty-six jobs. Los Angeles trimmed expenses by over $100,000, and Hammond even halved his own salary. In addition to lowering expenses and eliminating jobs, Hammond sought increased efficiency from his workforce by "dropping out some of the old men that have been spoiled by the prosperous times which existed during the war and putting younger men that are not so set in their old-fashioned ways in their places."[57] These younger men, of course, also received a lower salary than those they replaced.

By the end of 1930 Hammond had slashed his annual overhead by $1.5 million. Such drastic salary and wage cuts led dozens of workers to quit in protest. Nonetheless, Hammond insisted on keeping all his plants open, albeit running at 40 percent capacity, despite an expected $500,000 loss for 1930. He recognized that if he closed his mills, taxes, insurance, and depreciation would continue. In addition, he noted "there would be the loss of prestige," customers, and business infrastructure. Remarkably, the Merc under McLeod's tutelage continued to make money, pay dividends, and keep its finances healthy throughout the Depression. The LRRC, however, having lost money for several years, had piled up $4 million in debt and faced possible bankruptcy. Thus, in late 1930 it reopened negotiations, and this time Hammond was in a position to dictate the terms.[58]

Hammond agreed to the nominal appearance of a merger and saddled the new company with the unwieldy name The Hammond and Little River Redwood Lumber Company, but it would actually be a subsidiary of the HLC, which took two-thirds of the stock and assumed the debt. While the LRRC property came with a railroad and sawmill, it was the fifty-year supply of old-growth redwood that Hammond desired. He chortled that with the blocked-up land ownership he could effectively isolate other timber owners who would have to "pay the taxes on their property" until the HLC needed their timber. Assuming so much debt was an uncharacteristically risky move for Hammond. Nonetheless, he was confident that the HLC could "wipe out" the $4 million debt within five years. C. H. McLeod, however, was not so sure.[59]

On Christmas Eve 1930, Hammond and the directors of the LRRC signed an agreement creating America's largest redwood lumber company. The merger combined 3,500 employees with an annual payroll of more than $6.5 million and $42 million in assets, including three of Humboldt's largest mills, two railroads, and, most important, 10 billion bf of standing timber on one hundred thousand acres. Hammond then wrote to his principal stockholders, Claflin, and the Huntington, Leggett, and Harriman estates retroactively seeking their approval.[60]

While Hammond found himself with the largest unbroken tract of redwoods left on the coast, he also had to find buyers for $4.25 million in bonds. The HLC took on a fourth of the debt; Hammond then turned to the Missoula Mercantile. Despite his reluctance, McLeod agreed to approve the purchase of another $650,000 in bonds, which still left more than $2 million to be sold to banks.[61] Once again the MMC had provided a significant source of capital to fund western resource exploitation.

Although Hammond had built his success on avoiding debt, his desire to make the one big deal before he passed from the scene clouded his judgment. The HLC now had two subsidiaries—the Hammond-Tillamook and the Hammond and Little River Redwood Lumber Companies—both laden with debt. Under normal conditions, this would not raise concern, but in 1932 the world economy hit rock bottom. The year marked the biggest loss ever for the HLC with no improvement in sight. Furthermore, Hammond was eighty-three years old and many began to wonder if the time had come for the old man to step down. Chief among these was his son-in-law Steward Burnett.

Burnett had long served as Hammond's attorney and company vice president. Intelligent and ambitious—but also prickly and self-aggrandizing—Burnett ran afoul of his father-in-law, who demanded not only loyalty but also a certain degree of obsequiousness. Disgruntled with Hammond's control, Burnett resigned in 1923, only to return the following year. He quit again in 1925 when Hammond "found fault with some of his actions," and reduced his salary. Enthusiastic about the company's expansion, Burnett rejoined the company as a paid attorney to help negotiate the merger with the LRRC, but he insisted on more authority and a salary equal to Leonard's. Hammond gruffly informed him he "would not allow any man to dictate . . . the terms of his employment." Unhappy, Burnett stayed on but began to plot a takeover.[62]

Control of the HLC rested with its holding company, the A. B. Hammond Co., whose stock was owned entirely by the Hammond and McLeod families. Burnett thus regarded dissolution of the holding company as essential to wresting control of the HLC from Hammond. Unable to convince Hammond of the desirability of terminating the holding company, Burnett attempted to force the issue through his veto power as trustee. Despite his age, Hammond was not about to be pushed around, and Burnett backed down, resigning both his job and directorship. But the fight was far from over.[63]

In February 1932 Burnett secretly approached the other stockholders in an attempt to loosen the old man's iron grip on the company. Burnett surreptitiously traveled to Georgia to meet with John Claflin at his winter home, where he attempted to convince the last remaining cofounder that the HLC was being mismanaged. Livid upon learning of the trip, Hammond dashed off a preemptive letter to Claflin warning of Burnett's purpose. The apparent treachery on the part of his son-in-law increased

Hammond's paranoia. But what really infuriated the old lumberman was, as he told Claflin, Burnett's attempt to take control of assets that were "accumulated by me, alone."[64]

Upon returning to California, Burnett sought out another chief stockholder—the Huntington estate. However, its executor, Henry Robinson, had just suffered a serious injury in an auto accident and was unable to meet with him. Undaunted, Burnett tried again a few days later and presented his case while Robinson lay in his hospital bed. Hammond, in turn, dispatched Harry McLeod, manager of the Southern California division, to Robinson's bedside as well. Robinson assured Hammond directly that while he was "in no way disturbed" about the condition of the company, he did want to see the annual statements for the last three years and discuss the issue once he was out of the hospital. Claflin, too, praised Hammond's ability, but he worried that the HLC might keep expanding, putting stock value at risk. Agreeing with Burnett on the need to dissolve the A. B. Hammond Co., Claflin noted that "the Eastern stockholders are practically powerless to prevent new ventures which might prove successful or might prove disastrous."[65]

Any mention of stockholders looking at the records or a change in organizational structure elicited a visceral response from Hammond. He had invested so much of his life and personality into his company that he regarded any suggestions of change as a personal affront and was aghast that anyone should question his judgment. Evoking the ghost of his hero, C. P. Huntington, Hammond reminded Claflin that the HLC had been a success because "the Directors at the time the company was organized took Mr. Collis P. Huntington's advice . . . giving me authority to buy and sell and run the business as I thought best." Noting that the Long-Bell Lumber Company was $42 million in debt and about to default on its bonds, Hammond pointed out that all the other lumber companies except his were mortgaged to the hilt and losing money. While an overstatement, the claim served Hammond's purposes. He further attributed his accomplishments to his intimate knowledge of local conditions and how he had taught young men to build railroads, operate steamships, and manufacture and market lumber. Yes, replied Claflin, that is all good and well, but what will happen when you are gone?[66]

With his own finances in disarray, Claflin was upset that Hammond insisted upon paying interest on bonds while refusing to issue dividends. Claflin suggested they shut down unprofitable plants and liquidate the company's holdings. Seeking to reassure Claflin while also chastising eastern stockholders, "who have no first-hand knowledge of the lumber industry," Hammond listed all of the consequences of mill closures. If stockholders really wanted him to, he said was willing to sell off their entire holdings but only "after the depression has passed and business returns to normal."[67]

The showdown with the stockholders to determine the future of the HLC was set for the annual meeting on May 8, 1933. In the meantime, Hammond maneuvered to prevent "the control from going into the hands of incompetent people." Three of his children—Leonard, Florence Whitesides, and Edwina King—stood solidly behind their father. The McLeod family, too, was unquestionably loyal. Burnett steered Hammond's other two children, Grace (his wife) and Daisy in opposition. The "outside" stockholders, such as Claflin, held 45 percent of the HLC stock and could tip the balance one way or the other. Ultimately, however, as they looked back on thirty years of uninterrupted dividends, they saw little reason for radical change in the organization. Clearly outnumbered, Burnett backed off. Besides, Hammond was eighty-four and his health was in precipitous decline.[68]

A. B. Hammond died at home on January 15, 1934, and the age of the lumber baron came to an end. During his youth working in a soggy timber camp, he had dreamed of amassing great wealth. For half a century, he had looked toward the great industrialists—Huntington, Harriman, and Hill—for inspiration in building a vast West Coast empire of lumber. He achieved it all. Hammond left behind an estate worth $60 million (about $2 billion in today's dollars); he was California's undisputed premier lumberman, one of the West's top industrialists, and three years before his death, had assembled the world's largest redwood lumber company.

But at what cost? Had his desire and ambition to make that one last deal been a mistake? Hammond had built his reputation and empire upon one basic premise: retain control by staying out of debt. Yet he left behind two subsidiaries, both in debt and both losing money. In addition, years of cost cutting had left the HLC with crumbling infrastructure, outdated mills, decrepit railroads, and rusting equipment.[69]

None of the old-time loggers could remember a summer as hot and dry as that of 1933. The ordinarily cool, wet forests of the Oregon Coast Range seemed like a desert as the spongy forest duff turned to powder. The clear-cuts where the men worked were even worse. Without the protective canopy of Douglas fir, spruce, and hemlock, the ground temperature topped one hundred degrees. The understory of salmonberry, ferns, and salal wilted in the heat.[70]

Logging railroads laced a landscape filled with donkey engines and skidders that were converting old-growth forests into a wasteland of stumps. The high-lead logging operations left virtually no living trees. Looking like a scene from the apocalypse, the land was littered with "waste"—logs too small for the mill, tops of trees, and branches. Gullies of black earth streaked down the mountains where skidding operations had uprooted all of the vegetation.[71]

As exceptionally hot and dry weather amplified the possibility of forest fires, Oregon governor Julius Meier closed the state forests to logging. Lumbermen also worried about the possibility of forest fires on their lands. They posted lookouts and agreed to cease logging if the fire danger increased, but lumbermen hated to interrupt production. They needed to pay off their bonds, and even if they were losing money by logging, they could avoid bankruptcy as long as they could pay the interest. As lumber prices continued to slide downward, lumbermen tried to make up for the loss by increasing production. Technological advances such as steam shovels, diesel trucks, more powerful donkey engines, and locomotives allowed logging crews to cut faster and produce more. On nearly every mountainside and drainage of the Oregon Coast Range, logging crews furiously cut trees and hauled timber. Lumbermen encouraged their crews with yarding bonuses, whereby the logging boss would set a high footage mark and, if the crew managed to exceed it, everyone received one more dollar more that day. The woods droned with the chugging and clattering of donkeys and locomotives. The Tillamook, indeed, seemed more like a "factory without a roof" than a forest.[72]

On the particularly hot afternoon of August 14, the fire watchers called a halt to logging. Anxious to get another load in, one crew kept working. As they dragged in a log, sparks from the friction ignited a nearby slash pile. Fire spread to downed timber and then into the forest crown. Despite the heroic efforts of two thousand firefighters, ten days later a gusting east wind exploded the fire into a conflagration that darkened the sky and rained ash across the region and even on decks of ships five hundred miles away. The mushroom cloud of smoke towered thirty-four thousand feet high. Winds generated by the fire itself reached hurricane force, yanking 250-foot trees out of the ground and tossing them aside like toothpicks. In just three weeks, the fire incinerated 380 square miles of the Tillamook Forest before rain finally put it out.[73]

The fire destroyed 13 billion bf worth $275 million and more than $1 million worth of logging equipment. Ecologically, the fire ripped the heart out of one of the most productive forests on earth, an area where Douglas fir, hemlock, cedar, and spruce often exceeded 100,000 bf per acre. Out of the ten major timber owners in the Tillamook, the Hammond Lumber Company suffered the greatest loss. Of its 58,000 acres, all but 2,150 went up in flames, one-third of the company's Oregon timberlands. The loss compelled the HLC to close its Garibaldi mill indefinitely.[74]

Although uncommon in the Oregon Coast Range, with its high rainfall, such catastrophic wildfires were not unknown. While destroying the old-growth stands of Sitka spruce and western hemlock, these fires set the stage for the extensive Douglas fir forest that followed in its wake. Despite the blackened landscape, foresters thus fully expected the Tillamook to recover. Two years after the fire, forester Leo Isaac conducted a study of the natural regeneration and islands of green Douglas fir in the blackened landscape. Such patches, Isaac believed, could provide future seed crops before these isolated trees eventually succumbed to exposure, insects, or disease. On several slopes, the fire had burned hot enough to leave only mineral soil; nonetheless, Isaac concluded that "the seedbed conditions were fairly good."[75]

Unlike earlier fires, however, the Tillamook burned three more times in bizarre six-year intervals. Much of these subsequent burns struck within the perimeter of the 1933 fire, an unnatural anomaly as reburns in the Oregon Coast Range had not occurred before European settlement. In 1939, fire torched 190,000 acres, followed by 180,000 in 1945 and, in 1951, 32,000 acres. These subsequent fires blazed even hotter than the first as a result of the downed trees and snags. The reburns thus wiped out any chance of natural regeneration. The Tillamook fire interval had suddenly shifted from five hundred years interval to six years, with profound ecological and economic repercussions.[76]

All told, the Tillamook fires—consequences of industrial logging practices—destroyed 355,000 acres of timber worth $442 million. After salvaging what it could, the Hammond Lumber Company and other operators abandoned their cutover and burned timberlands. Concluding that "said lands are worthless," the HLC released its deeded lands to the county in lieu of taxes in 1941. The county then turned over the former HLC timberlands to the state to form the bulk of the Tillamook State Forest. Thus, a sizeable chunk of the Oregon Coast Range once again became part of the public domain, albeit in a much degraded condition.[77]

Epitaph

Following his death, A. B. Hammond did not, despite the legend, rise from his coffin and fire his pallbearers. Nonetheless, everyone inside his organization felt his presence, or lack thereof. Hammond had built his company on the force of his personality, and his death decapitated the HLC. The division among the stockholders the previous year, combined with Hammond's own need for absolute control, prevented a smooth transition, underscoring one of the fundamental problems of the legacy of the proprietary business model. Unlike people, however, corporations can live forever, but would the HLC? Leonard, as vice president, would have been the natural heir, but his alignment with his father during the recent conflict alienated the Burnett faction. While the owners dithered over the future of the company, it continued to hemorrhage.[1]

Uncertain of the company's financial condition, the stockholders ordered a report. Noting that the HLC suffered losses of $2 million in 1932 and $1 million in 1933, the report attributed the deficit to "selling lumber below the cost of production." Cognizant of the structural conundrum that plagued the timber industry, the report stated, "Sales prices were demoralized by our competitors who forced lumber on the market at ruinous prices in order to pay off their obligations." To their satisfaction, however, the stockholders concluded that the HLC "physically and financially, was in far better position and condition than any other lumber company on the Coast."[2]

Although the stockholders expressed confidence in the abilities of Leonard Hammond and George McLeod as executives, they also wanted a full assessment of the company's operations and hired Robert Lea as an outside consultant and troubleshooter. After a seven-month investigation, Lea issued a scathing condemnation of everything from logging practices to bookkeeping procedures. Lea was especially critical of the merger between the HLC redwood division and the Little River Lumber Company. He noted that while each company lost about $1 million from 1925 to 1930, the merger greatly increased the new company's debt to $4.2 million and mortgaged all of its timberlands and mills. Instead of combining operations and planning for long-term development, the company continued the same ruinous logging practices: clear-cutting everything as fast as possible. The report concluded that the "merger was an ill-advised and unwise business decision" as HLC "did not need the additional timber" and its facilities were not equipped to handle increased production.[3]

At Samoa, Lea found that the practice of sawing for quantity rather than quality had led to a surplus of lower grades and an accumulation of unsalable lumber, resulting in "inordinate expenses in yarding and handling." Overstocked yards and warehouses had become labyrinths with employees wearing "themselves out working around and climbing over" stacks of "worthless lumber." Lea concluded that the HLC was producing a poor-grade product at high cost. Furthermore, Hammond's parsimony created a situation where "practically every important piece of equipment [was] tattered and torn."[4]

Lea reserved his greatest criticism for Hammond's organizational structure. Hammond's autocratic approach had resulted in a corporate culture where foremen and supervisors "could hold their jobs only by catering to the whims of the boss . . . actual operations were often at variance with the wishes of the San Francisco office; and this fostered the growth of an internal ring of secrecy."[5] Apparently, the practice of concealing inefficiencies and problems from "old A. B." had become pervasive and institutionalized.

To cut waste and inefficiency in the woods, Lea proposed the HCL switch from clear-cutting to selective logging. This required replacing donkey engines and the destructive high-lead logging with tractors. Rather than logging off entire watersheds, tractors, like the old bull teams, could pull out selected trees. Unfortunately, these were the biggest and best redwoods—trees most crucial to maintaining ecosystem integrity. Lea also recommended closing or selling the Oregon sawmills as well as the San Pedro distribution facility. Observing that the sash and door factory in Los Angeles had lost more than $500,000 over the past four years, he suggested closing it too.[6]

Faced with continuing losses, lack of direction, and infighting, the stockholders agreed to appoint Lea as the company president in June 1935 and gave him complete authority to implement his recommendations. Lea quickly closed down the Oregon mills and put them up for sale. Lea's actions ran counter to Hammond's strategy of buying during a depression and selling when prices were high. Questioning the approach of the new president, Leonard intervened before Lea sold the Los Angeles facilities. Nine months after Lea assumed the helm, Leonard convinced the majority of the board of the recklessness of such liquidation policies. Insisting that Lea cost the company "at least a million dollars," Leonard later told C. H. McLeod, "Father . . . would not have sacrificed our assets as Colonel Lea did. He would have held on, head up, and disposed of assets when the market was more favorable." The HLC, apparently, was not quite ready to fully embrace a CEO-style organization, and in March 1936, Lea resigned. Replacing him, Leonard reverted to his father's business approach.[7]

Leonard inherited a company saddled with debt and threw himself into the challenge. As the company's lead salesman, he initiated an aggressive retailing effort. The HLC not only operated marketing offices throughout the United States but also joined with other producers in a "vigorous foreign sales campaign" in thirty-two countries. At the end of his first full year as president, Leonard paid stockholders nearly half a million dollars in dividends, maintained a cash surplus, and claimed that the company was "in a better financial condition than it has been for a long time." In 1939, Leonard acquired the financially strapped Hobbs-Wall Lumber Company, the largest timber owner in Del Norte County, for a mere $150,000. The next year, Leonard halved the HLC's debt and posted a net profit of $586,000 on sales of $1 million a month.[8]

The recovering national economy contributed to Leonard's success. President Franklin Roosevelt's acceptance of Keynesian economics, with its emphasis upon deficit spending, and the issuing of lucrative government contracts to feed the growing war machine in Europe stimulated the economy. Above all, the U.S. entry into the Second World War snapped the lumber industry out of the doldrums. The HLC received massive orders from the government for lumber to build cantonments, nurseries for rubber plantations, and "the first big Japanese concentration camp."[9]

Although Leonard enjoyed a high degree of worker loyalty and popularity, he, like his father, took a dim view of organized labor. While much less vociferous and visible, he continued the HLC tradition of antiunion militancy. Leonard, however, had to take a more clandestine approach, as the passage of the National Labor Relations Act in 1935 recognized the right of workers to engage in union organizing and collective bargaining. With a pen stroke the federal government undid Hammond's thirty-year campaign to keep unions out of the lumber industry, and throughout the Pacific Northwest timber workers began to organize once again. In May 1935, the Lumber and Sawmill Workers Union voted to strike for higher wages and fewer hours.[10]

Unlike previous strikes in Humboldt, this one was instigated by the Carpenters Union in Oregon and Washington and lacked local support. Nevertheless, the Humboldt County local joined the region-wide strike. But the political landscape of Eureka had changed. Eureka mayor Frank Sweasey created a "Committee of One-Thousand," essentially a vigilante force, whose members underwent formal attack training financed by the lumber companies. In addition to hired thugs, the HLC employed company spies and amassed an arsenal of tear gas and clubs for use against striking workers. As in A. B.'s time, the antiunion rhetoric was couched in ideological terms. Communists replaced the IWW as the bogeymen.[11]

A malicious press and lack of worker solidarity led to increasing tension. As the strike began to collapse, the union decided to concentrate its effort on a single mill, and on June 21 two hundred strikers set up a picket line at the Holmes-Eureka mill. Anticipating violence and determined to crush the strike, the police arrived and fired tear gas into the crowd. Instead of dispelling them, this provoked the strikers, who retaliated with rocks and sticks. The police, in turn, opened fire. When the smoke cleared, ten workers had been shot, three fatally, and five officers had suffered beatings.[12]

Perhaps Leonard's role in the violence caused him to reevaluate HLC's strident anti-labor practices, for when the Los Angeles operation voted in favor of union representation a few years later, he became resigned to getting "along with the unions in one way or another."[13] But in Humboldt the HLC resolutely held onto A. B.'s policy of refusing to recognize unions, prompting a short-lived strike in 1942. World War II put the issue on hold, only to have it erupt in force again once the war ended.

Just as A. B. had confronted labor in the wake of the Great War, Leonard faced massive worker unrest following the Second World War. A wave of strikes attempting to address long-simmering grievances swept the country. Ten thousand workers shut down the Ford Motor Company in a ninety-nine-day strike that resulted in formal recognition of the United Auto Workers. At General Motors 200,000 workers went on strike. Then, in January 1946, 750,000 steelworkers walked out. But the longest strike occurred when 4,000 lumber workers in Northern California went on strike against the nine major redwood lumber producers for twenty-seven months.[14] What redwood workers lacked in numbers they made up for in tenacity.

Although the lumber companies agreed to a significant pay raise, they held firm to the late A. B. Hammond's insistence on an open shop. This time, however, unions had legal standing and were better organized. After a nine-month shutdown, the HLC had recruited enough strikebreakers from returning war veterans to reopen the old Bayside Mill in Eureka. Union carpenters retaliated with a nationwide boycott against redwood products. While the big firms held out, the strike broke their stranglehold on the redwood industry. With the postwar housing boom, new, smaller lumber companies using union labor sprang up to fill the gap in production. The number of sawmills in the redwood region exploded from nine in 1944 to two hundred in 1946. At the outset of the biggest construction boom in U.S. history, the HLC was sitting on the sidelines.[15]

No doubt old A. B. would have buckled down and held out against the unions. He might have imported workers from Mexico or India, pounded on the governor's desk, traveled to Washington, D.C., and berated senators, or hired his own militia. Leonard might have followed a similar, albeit less antagonistic, route. Three weeks before the strike broke out, however, Leonard Hammond died of leukemia. He was sixty-one.[16]

Leonard Hammond had more than adequately filled his father's shoes. When he took over the company in 1936, the HLC had missed three years of dividends and was deep in debt. By the time he died, he had retired the company's debt, paid the missed dividends, and cushioned the HLC with a cash reserve of more than $5 million.[17] Had he lived on, the subsequent history of the Hammond Lumber Company could well have turned out quite differently.

Once again the HLC was rudderless in the midst of a major crisis. The acrimony among the stockholders, however, had diminished. Rather than repeat the experiment with an outsider, the board of directors selected seventy-five-year-old George B. McLeod as the fourth president of the HLC. McLeod, the first employee of the Hammond Lumber Company, had served as Hammond's right-hand man for three decades. With McLeod as the new president, it felt as if A. B. Hammond were still in charge, despite being dead for twelve years.

While McLeod shared Hammond's ideology, he was supremely rational and lacked A. B.'s vindictiveness and spite. After thirteen months of severely hampered production because of the strike, McLeod agreed not only to a minimum wage of $1.20 an hour but also, and more important, a union shop. Other mills, however, held out for another fourteen months until the unions finally gave up following the passage of the 1947 Taft-Hartley Act, which restricted union activities. Ironically, the only fully unionized major redwood producer was now the Hammond Lumber Company.[18]

McLeod also deviated from his former boss in his concern for conservation. For McLeod, conservation meant protecting the forest resources from fire and ensuring a continual supply of timber. Following the disastrous 1902 Yacolt Burn in Washington, McLeod attempted to organize an Oregon state board of forestry to patrol the forests for fires. In 1909, McLeod joined George Long of Weyerhaeuser and other leading lumbermen to form the Western Forestry and Conservation Association (WFCA), an industry trade group, to lobby state and federal government for increased fire protection and lower taxes. The WFCA became powerful enough that it could influence, and often dictate, state and national forest policies in the early twentieth century.[19]

In 1950 the HLC entered a new era of "scientific forestry" in which timber management replaced total extraction. As part of a larger public relations campaign to dispel their cut-and-run reputation, the company started calling their timberlands "tree farms." With an eye toward the future, George McLeod declared that the Hammond Lumber Company's new policy was "to manage its lands on as close to a sustained yield production basis as possible, so as to perpetuate our operations in

Humboldt County."[20] Lumbermen like McLeod wished to convey that they had stopped mining the forest and had shifted toward the notion of growing trees as a renewable resource. What exactly they meant by "sustained yield" was vague, however. Certainly, lumbermen did not intend to wait two thousand years for the old-growth redwoods to replenish. Nor should they, according to foresters like E. T. Allen, executive director of the WFCA, who maintained that old-growth forests were "decadent" and should be clear-cut and replaced with tree plantations. With McLeod in charge, "sustained yield" came to mean a planned approach to timber harvest in which the company could log the same site two or three times over a period of ten to twenty years before converting it into a plantation of Douglas fir. Under these guidelines, the HLC continued to cut its old-growth stands, and by 1952 it had logged 47,000 acres of what was previously the world's most intact redwood forest.[21]

A. B. Hammond's death marked the end of the era of the lumber baron. It appeared that just as he had predicted, the age of consolidation and cooperation had arrived. Hammond's successors were well educated and well connected; their success depended on their ability to forge relationships, cajole and convince, manage and administer. And yet, unlike many other industries, the lumber sector seemed to attract forceful and even megalomaniac personalities. For the remainder of the twentieth century pseudo lumber barons continued to appear in the redwood country, but all lacked the staying power of Hammond. The modern corporate structure made the difference.

Understanding the importance of size in the lumber industry—the larger the company, the better its position to set prices and restrict production—Hammond had spent thirty years buying smaller companies and seeking mergers and consolidation. Ironically, his successors in the HLC, by bringing the company out of debt and in acquiring more timberlands and mills, had created a very desirable meal for an even bigger fish.

While the HLC was moribund during the strike of 1946, Owen Cheatham was assembling a system of lumber mills and sales offices to produce and market plywood, which brought three times the price of lumber. Based out of Georgia, Cheatham expanded throughout the southern United States. In 1949 Cheatham's Georgia-Pacific Plywood and Lumber Company (GP) was on par with the HLC in terms of sales and profits. Unlike the HLC, however, the newcomer lacked a substantial timber base.[22]

Seeking lumber companies with old-growth Douglas fir, Georgia-Pacific, backed by massive stock offerings, went on a buying spree in the 1950s. With annual sales approaching $92 million, by 1955 GP had acquired four billion feet of timberlands and still sought more. The following year, GP made two big purchases, one in Oregon and one in California. For $70 million it picked up the Coos Bay Lumber Company, started by Charles A. Smith (A. B.'s old rival). Then, in what was billed as the "largest lumber deal in California history," GP entered the redwoods when the remaining HLC stockholders voted to sell the entire company for $75 million. Sounding like Hammond himself, Cheatham, when asked about his company's rapid expansion, replied, "You've got to shoot when the duck is in your gunsight."[23]

Cheatham's gun, however, was loaded with borrowed birdshot. To assuage investors and ensure short-term profits, Cheatham announced he would immediately liquidate the HLC's thirty-three retail yards, along with the Los Angeles distribution centers and timberlands. The proceeds helped finance GP's 1959 purchase of the Booth-Kelly Lumber Company in Oregon for $93 million. Primarily interested in manufacturing plywood, GP opened a plant at Samoa next to the old HLC sawmill. Six years later, GP added a $30 million pulp mill, the first on the California coast and the first to use redwood chips. In 1966 the company torched the original Samoa mill as part of its modernization program.[24] A new era based upon acquisition and liquidation had arrived.

Likewise, a new era in conservation arose in the 1960s, and public support for a Redwood National Park gained momentum. As early as the 1860s Americans had expressed concern over the felling of the majestic redwoods. In 1880 Carl Schurz, the farsighted interior secretary, had recommended that 48,000 acres of redwood forest be set aside. But by the turn of the century, lumbermen or speculators had snapped up nearly every acre of redwoods. Nonetheless, in 1913 Progressive reformers such as Theodore Roosevelt and California congressman William Kent, along with local women's clubs and chambers of commerce, began advocating for a redwood national park, seeking both preservation and tourism.[25]

While Progressives were hardly hostile to capitalism, they wished to provide a space, both geographical and political, for noncommercial considerations. Part of this entailed preservation of the nation's scenic wonders. As the tallest trees on earth, the coastal redwoods ranked beside the Grand Canyon and Yosemite as part of the nation's heritage. But unlike these majestic landscapes, the redwoods had high commercial value. And so the timeless redwood forest became a highly contested zone between opposing cultural values.

Within the same individuals, commercial exploitation could coexist with the desire for preservation and aesthetic sensibilities. Historian Susan Schrepfer detailed how preservation, science, and commerce all converged into the Progressive mindset. Composed of businessmen, academics, and professionals, many of whom formed San Francisco's elite, the Save the Redwoods League was a case in point. During the 1920s oil tycoon and industrialist Joseph D. Grant served as the league's chairman. Not only did Grant live next door to Hammond, but many of the league's officers were A. B.'s fellow members of San Francisco's Bohemian Club. Despite the group's anti-logging rhetoric, Hammond saw little threat from the league, as its members shared his socioeconomic position and outlook. In 1922 Leonard even became a life member of the league, and the HLC donated several isolated groves of redwoods along the Eel River for inclusion in a new state park.[26]

By keeping the federal government out of the picture and working cooperatively with lumber companies and philanthropists to purchase redwood groves, the league succeeded in stitching together a patchwork of state forest reserves along the Redwood Highway. But the winter of 1954–55 revealed the inadequacy of this approach. Massive floods caused by heavy rains and accelerated logging that had denuded the slopes above Humboldt Redwoods State Park wiped out hundreds of the protected trees.[27]

As the twentieth century progressed, trees continued to fall at an increasing rate, making them more precious to both lumbermen and preservationists. At the close of the 1950s, only 10 percent of the original redwood forest remained, while erosion and highway construction threatened the state parks. By the mid-1960s support for a Redwood National Park had grown loud enough that Congress began to entertain proposals ranging from fifty thousand to ninety thousand acres, much of which included the old HLC timberlands. In 1966 Georgia-Pacific and the other lumber companies agreed to a one-year moratorium on logging within the proposed park while Congress deliberated. But the following year, despite pleas from thirty-four congressmen, GP resumed logging along Redwood Creek, where, three years earlier, a *National Geographic* team had discovered the world's tallest tree. The lumber company's actions prompted the *New York Times* to declare, "Decision by bulldozer is not democracy."[28]

Finally, in 1968 President Lyndon Johnson signed a watered-down act designating 58,000 acres as Redwood National Park. The legislation protected only 11,000 acres of old growth, for by the time the bill passed the rest had been logged. In exchange for giving up 28,000 acres for the park, lumber companies received 14,500 acres of timberlands and $92 million, making this the single most expensive park purchase to that date.[29] Less than one hundred years earlier lumber companies had fraudulently acquired these same timberlands for $2.50 an acre from the federal government; they were now selling them back in a much-diminished condition for $3,300 per acre. But the 1968 bill proved even more costly down the road.

While they stridently opposed the park, lumber companies heavily influenced the act as it wound through Congress. The former HLC timberland along Redwood Creek was the greatest stumbling block and resulted in an oddly shaped, half-mile-wide corridor snaking up the drainage. As soon as the bill was signed, GP resumed cutting. When the Interior Department asked GP to moderate logging near the park boundary, the company informed the government that its experts recommended clear-cutting old growth to alleviate the threat of erosion."[30]

The continued logging prompted conservationists to advocate for park additions. In 1978, with only 9,000 acres of old-growth forest remaining in Redwood Creek, President Jimmy Carter signed a second Redwood National Park Act. This bill added 48,000 acres to the park and cost the federal government $350 million, making the original act of a decade earlier seem like a bargain.[31]

Meanwhile, GP expanded enormously. In the late 1960s company executive Harry Merlo began another round of acquisitions in the redwood country. By 1971 the firm had grown so large that it attracted the attention of the Federal Trade Commission, which sought to limit the company's dominance of the lumber industry. The following year, GP agreed to sell off its Louisiana-Pacific (LP) subsidiary, which included the West Coast and Intermountain Divisions, and Merlo became the new corporation's president.[32]

During the 1980s Merlo appeared to be a reincarnation of A. B. Hammond. An aggressive contrarian, Merlo "shunned large land purchases while rivals bought heavily" but then doubled the company's timberlands, acquiring 750,000 acres, when prices dropped. When most lumber companies negotiated union agreements, LP under Merlo "weathered a bitter strike" that broke the union and lowered wages. Before long, LP surpassed both its parent company and Boise Cascade to challenge Weyerhaeuser as the largest lumber company in the United States.[33]

Then, in 1995 both Merlo and LP came crashing down. The company faced federal lawsuits for willful violations of the Clean Air and Clean Water Acts. The State of Colorado, where LP operated a large mill, brought a fifty-six-count indictment for fraud and environmental violations, and the company confronted a rash of civil suits for selling faulty and toxic products. Moreover, Merlo faced personal charges of sexual harassment. Worst of all, at least from the stockholders' perspective, Merlo had inflated the company's earnings report by $30 million. This last item was the final straw, and the LP board of directors unceremoniously dumped their CEO.[34]

Besides the inflated earnings reports, the essential difference between Merlo and Hammond lay in ownership. Not only had Hammond been the president of the HLC; he had also controlled, even if he did not technically own, the majority stock. By Merlo's time, corporations had grown so large that rarely could a single individual exert the kind of control Hammond enjoyed. Despite his power, and abuse thereof, Merlo, unlike Hammond, was ultimately an employee and beholden to the corporate structure.

In the final analysis, the HLC was not ecologically sustainable. But then, Hammond had never intended it to be. Hammond's enterprise was to conduct a liquidation sale on the ancient forests, converting trees into cash. In slightly more than one hundred years, the HLC and the other redwood companies had all but eliminated the redwood forest from Monterey Bay to the Oregon border. By the 1980s only 2 percent of the original redwood belt remained, most in remnant patches on state and national parks. One large tract of old-growth redwood, however, still lingered in private ownership.[35]

During the 1980s lumber companies—with their high proportion of undervalued standing assets, such as trees—became prime targets for corporate raiders who would quickly liquidate the company's assets and move on. In 1985 the venerable Pacific Lumber Company, with several thousand acres in old-growth redwood, attracted the attention of Houston businessman Charles Hurwitz. Financed with junk bonds floated by Michael Milken and Ivan Boesky (both later convicted of securities fraud), Hurwitz engineered a hostile takeover of Pacific Lumber. To finance the junk bonds before they collapsed, Hurwitz raided the company's pension fund and sold off extra assets. He then doubled the timber harvest and shifted from selective logging to clear-cutting.[36]

Capturing international attention with Julia Butterfly Hill's 738-day tree sit from 1997 to 1999, Pacific Lumber's accelerated harvest became the front lines in the battle to save what was left of the ancient redwoods not already under formal protection. With mass demonstrations and arrests, including the largest anti-logging protest ever, the issue divided local communities, pitting environmentalists against workers and even resulting in the tragic death of one young man. Underscoring the fundamental problem of regarding trees as corporate assets, this "Last Stand" was the culmination of a 150-year struggle over the nation's forests, a struggle that framed A. B. Hammond's entire life.[37]

The battle over the forests of the Pacific Northwest came to a head over two relatively inconspicuous birds. In 1990 the listing of the spotted owl as a threatened species, followed by the marbled murrelet two years later, brought the region's timber production to a virtual standstill. Both birds required coastal old-growth forest for their survival, and their downward-spiraling populations indicated the seriousness of the ecosystem's decline. A century of logging had fragmented both the Douglas fir and redwood forests into tiny blocks. Pacific Lumber's Headwaters Forest contained the largest remaining tract of old-growth redwoods in private ownership. It also happened to house a third to half of California's remaining murrelets.[38]

Unlike in Hammond's era, in the late twentieth century, state and federal governments had instituted laws and regulations that restricted the timber industry. Governments often declined to enforce such restrictions, however, leaving it to citizens to prod them into action. During the 1990s the Environmental Protection Information Center (EPIC), a local environmental group in Mendocino County, filed lawsuits under the Endangered Species Act and California's Forest Practices Act to stop logging in Headwaters. To resolve the impasse, the federal government agreed to purchase 5,600 acres, including 2,700 acres of old growth in the Headwaters Grove, from Hurwitz for $300 million.[39]

Nonetheless, Hurwitz continued to cut whatever remained, regardless of regulations, leading California to issue Pacific Lumber hundreds of citations. In addition to facing multiple civil charges, the company lost seven lawsuits for disregarding environmental laws. Saddled with $700 million in debt, in 2007 Hurwitz pulled the plug on Pacific Lumber and declared the company bankrupt. Before walking away, however, Hurwitz had extracted more than $3.6 billion out of Humboldt County in redwood lumber.[40]

Were Hammond, Merlo, Hurwitz, and the like "robber barons" who greedily exploited the workers and the environment simply to line their pockets, or were they "captains of industry" who hammered American's industrial might out of raw materials? To what degree were they products of the culture of capitalism? Such questions had dogged Hammond since the days of the "Missoula Mercantile Monopoly," when the press and residents protested his dominance of the regional economy. Hammond, however, maintained that his actions resulted in the development of the country. Like many businessmen, Hammond firmly believed in Adam Smith's contention that the pursuit of private wealth resulted in public gain. No doubt, trees provided lumber for a growing America, but just as the forests of New England were cut over and those of the Midwest and South mowed down, the ancient forests of the Pacific Northwest are now all but gone—the ecosystems stretched to the breaking point, the vast carbon sink of old-growth forest transferred into the atmosphere, and timber-dependent communities broken and impoverished.

When Hammond was born, in 1848, industrial capitalism was in its infancy. His father and grandfather grew up in a world in which mercantilism provided the economic foundation. Trees were no less a commodity, and the United States and Great Britain nearly went to war over access to the forests of the north country. Profit margins, however, only marginally affected residents of Madawaska, who were more concerned with religion, community, security, and self-sufficiency. Relationships—both economic and social—formed the basis of the mercantilist system; fostering and maintaining these personal relationships was paramount to financial security. In a region where survival depended upon a large larder, small-scale agriculture and community ties were vital to maintaining food security and self-sufficiency.[41]

Hammond embodied the Protestant values of his ancestors—hard work, thrift, and acquisition—all of which corresponded nicely with the dictates of capitalism. Indeed, he built his entire empire around such values, avoiding indebtedness at all costs—until the end. Furthermore, Hammond entered the business world at a time when strength of character and reputation, rather than potential profits, determined credit and prestige. The Missoula Mercantile Company was a case in point. From its inception, Hammond diligently insisted that it meet all its obligations in a timely manner and pay annual dividends. Such scrupulous attention to the balance sheet provided both Hammond and the Merc with an unassailable line of credit, upon which he drew to purchase two Oregon railroads and attract investors such as C. P. Huntington, Thomas Hubbard, and E. H. Harriman.

Time and again across the West, the pattern repeated. With money from New York, Boston, and London, the great conduits of capital—J. P. Morgan, the Huntingtons, Harriman, and the like—transported it west. Ultimately, however, it was regional entrepreneurs like A. B. Hammond, Marcus Daly, and Samuel Hauser who were largely responsible for the transformation of the western landscape. These men built the railroads, developed the mines and ranches, built the stores, banks, and hotels, logged the forests, and dammed the rivers, changing the West dramatically and irreversibly in a mere fifty years.

While much of Hammond's capital came from eastern investors, a significant portion originated within the MMC and his other enterprises. In Hammond's case, at least, a regional mercantile firm in a midsized town in Montana engendered one of the West's most potent symbols of capitalism run amok: the near-elimination of the coastal redwood forest. Thus, rather than simply being a colony of the East, the West exploited itself.

Hammond's life illustrates how the transformation of the American West corresponded with the transition from proprietary capitalism to the modern corporation and concurrent technological changes. In 1867 Hammond was chopping cottonwood trees with an axe to sell as fuel for riverboats. In 1918 he was using steam-powered donkey engines to harvest ten-foot-diameter spruce trees and loading them onto railroads to be milled into lumber for airplanes. Sixteen years later, Hammond was using tractors to tow two-thousand-year-old redwoods to all-electric mills powered by diesel generators. The lumber was sent south on diesel-powered ships for oil derricks, which extracted more petroleum to be refined into more diesel. The industrialization of the West was complete.

In addition to technology, the structure of capitalism changed dramatically during Hammond's lifetime. The pioneer entrepreneur with little but sheer determination, imagination, and dumb luck could no longer simply appropriate the vast natural wealth of the American West. By snatching up all the cheap lands, men like Hammond ensured that no one would follow in their footsteps. Furthermore, in creating the national forests the federal government asserted its authority and ownership over what resources remained. By the time Hammond died, the corporation, rather than the individual, dominated American business.

Although Hammond was instrumental in building railroads, establishing businesses, building cities and towns—including Missoula and much of Los Angeles—and greatly altering the ecology of the coastal forests of the Pacific Northwest, he was not alone in his activities. While he was the most visible of the lumber barons and the most vehemently antiunion, he was, in the final analysis, a phenotype of American capitalism.

In multiple ways, A. B. Hammond pioneered the way for those who followed, including Cheatham, Merlo, and Hurwitz. Previously, proprietary regimes, such as William Carson's, regarded workers as individuals. Hammond employed a different approach, one in which labor was reduced to an overhead cost of production that could be ratcheted up or down as the situation demanded. Hammond also demonstrated the value of the vertically integrated lumber company: the HLC controlled everything from raw materials to retail sales. While some early lumbermen could express remorse at cutting down ancient redwoods, Hammond embraced a system that valued trees only in terms of their board feet.

But for all his ruthlessness, A. B. was quickly surpassed by the new corporate structure that compelled profits over all else. Under Georgia-Pacific and Hurwitz, the bottom line was not just the primary consideration, as it had been for Hammond: it was the only consideration. Individual capital accumulation, however, depends upon a vast social, political, and economic network to distribute the costs. In Hammond's case these costs were borne by his workers and the forests of the Pacific Northwest—the human and biotic communities that constituted the ultimate source of his wealth. Ultimately, Hammond and the other lumber barons were outward manifestations of an economic system that has grown so pervasive that it not only dominates world commerce and human activity but also has dramatically altered both the nation's forests and the earth's climate.

The underlying structure—the genome, if you will—of capitalism rewarded and reinforced Hammond's business practices, thus molding his actions. At the same time, Hammond imprinted his individual personality on his businesses. While he often based his decisions upon economics, he also imbibed ideological, religious, and cultural values, such as the Protestant work ethic and the views of Adam Smith. Occasionally, Hammond simply acted out of tenacity or spite. In the final analysis, it was individuals—Hammond, along with thousands of others—who made the expansion of market capitalism possible through their actions. Capital is inert—a lump of coal. It can do nothing on its own; it requires individuals to ignite it and blow it into flames.

Notes

ABBREVIATIONS

D&C	Dolbeer and Carson company files, Bancroft Library, Berkeley, Calif.
ERLC	Elk River Mill and Lumber Company records, Bancroft Library, Berkeley, Calif.
FHS	Forest History Society, Durham, N.C.
HEH	Henry Huntington Collection, Huntington Library, San Marino, Calif.
HLC	Hammond Lumber Company files, Oregon Historical Society, Portland, Ore.
HSU	Humboldt State University, Arcata, Calif.
LRRC	Little River Redwood Company files, Timber Heritage Association, Eureka, Calif.
MHS	Montana Historical Society, Helena, Mont.
OHS	Oregon Historical Society, Portland, Ore.
PANB	Provincial Archives of New Brunswick, Fredericton, New Brunswick
UM	K. Ross Toole Archives, University of Montana, Missoula, Mont.

Unless otherwise noted, all correspondence to/from A. B. Hammond is from the C. H. McLeod collection, MSS 001, University of Montana.

INTRODUCTION

Epigraph. Elster, ed., *Karl Marx: A Reader,* 26.

1. McKinney, "A. B. Hammond, West Coast Lumberman," 196.
2. Clark, *Mill Town,* 63.
3. See Robbins, *Colony and Empire,* and Toole, *Montana.*
4. For more on this process, see Cox, *Lumberman's Frontier.*
5. Quote from Klein, *Change Makers,* xiii.
6. In 2012 a new owner reopened the mill on a temporary and limited basis.

CHAPTER 1

1. Farr, "Going to Buffalo," 3–24.

2. National Park Service Ice Age Floods website, www.nps.gov/iceagefloods. See also Alt, *Glacial Lake Missoula and its Humongous Floods.*

3. Woody, "History of Missoula"; *Missoula Gazette,* Jan. 1, 1890.

4. Audra Browman files, UM.

5. *Missoula Gazette,* Jan. 1, 1890; Missoula Board of Trade, *Missoula Illustrated.*

6. Digital Sanborn Maps, June 1891.

7. *City Directory for Missoula,* 1891; Kenneth Lottich, ed., "My Trip to Montana Territory, 1879," 12–25.

8. *Missoula Gazette,* Jan. 1, 1890; *Helena Weekly Inter-Mountain,* Jan. 1, 1882.

9. *Weekly Missoulian,* Oct. 28, 1891.

10. U.S. Census, 1880, 1890.

11. Leeson, ed., *History of Montana;* Missoula Board of Trade, *Missoula Illustrated,* 22.

12. Fahey, *Flathead Indians,* 161–66.

13. Ibid.; Carrington, "Exodus of the Flathead Indians," chapter 7, folio 8.

14. Carrington, "Exodus of the Flathead Indians," chapter 7, folio 8, quotes from folio 7.

15. Ibid., chapter 7, folio 7–8, chapter 10, folio 12; Carrington and Charlo quotes from ibid., chapter 8, folio 2–3; Ronan to Commissioner of Indian Affairs in Fahey, *Flathead Indians,* 252.

16. Carrington, "Exodus of the Flathead Indians," chapter 11, folio 4.

17. *Missoula Gazette,* June 15, 1891. Although the newspaper reported that the residences would be completed by winter, the Panic of 1893 halted construction, and neither Daly nor Hammond would finish the mansions they began.

18. Carrington, "Exodus of the Flathead Indians," chapter 12, folio 1–3; Fahey, *Flathead Indians,* 254.

19. Farr, "Going to Buffalo, part 2, 28–43.

20. Claude Elder oral history, UM.

21. *Missoulian,* Oct. 21, 1891. Arthur Stone provides personal recollections in *Following Old Trails.*

22. Carrington, "Exodus of the Flathead Indians," chapter 12, folio 1–2.

CHAPTER 2

Epigraph. Thoreau, *Maine Woods,* 121.

1. George McLeod, "Story of the Hammond Lumber Company," 2.

2. Morison, *Oxford History of the American People,* 61–65; Roland Hammond, *History and Genealogy of William Hammond,* 9; Kilburn, *Fathers and Mothers,* 56–57.

3. Weber, *Protestant Ethic and the Spirit of Capitalism;* F. Warne and Co., *Dictionary of Quotations,* 354.

4. Kilburn, *Fathers and Mothers,* 68.

5. Raymond, *River St. John,* 134; Pilon, "Settlement and Early Development," preface, 38; MacNutt, *New Brunswick,* 37, 67; On-line Institute for Advanced Loyalist Studies, www.royalprovincial.com; LaPointe, *Grande-Rivière,* 37.

6. Land Petitions of Archelaus Hammond, 1786 F1031, 1787 F16302, 1790, PNAB; Land Petition of John Coombes, 1786, F1030, PANB; Crown Land Grant Map, no. 125, New Brunswick Dept. of Natural Resources; Kilburn, *Fathers and Mothers,* 68–69; Jean Hammond, "Hammond Family History," MC 1, PANB; Pilon, "Settlement and Early Development," 5.

7. LaPointe, *Grande-Rivière,* 38.

8. Raymond, *River St. John,* 134; MacNutt, *New Brunswick,* 88.

9. Cox, *Lumberman's Frontier,* 13, 40, 76.

10. Ibid., 54; Craig, "Agriculture and the Lumberman's Frontier," 125–37; Wood, *History of Lumbering in Maine,* 68.

11. Moore, "The Modern World-System as Environmental History?"

12. Wynn, *Timber Colony,* 29–30, quote 33.

13. Ibid., 31.

14. Because of the merging of both Acadians and Quebecois or Canadians, I use the generic term "French" to indicate those who share a common language, religion, and culture and "English" to designate Anglo-Saxon Protestants rather than using the terms to refer to actual nationalities.

15. Paradis, *Papers of Prudent C. Mercure,* xxv-xxxi; MacNutt, *New Brunswick,* 198, 80.

16. MacNutt, *New Brunswick,* 61; Paradis, *Papers of Prudent C. Mercure,* xxvii.

17. LaPointe, *Grande-Rivière,* 61; MacNutt, *New Brunswick,* 211.

18. Crown land grant records, Leonard Coombes, 1828, F 16313, PANB; 1833 Census Report of James MacLauchlan, 243; "Deane and Kavanagh Report on the Inhabitants of Madawaska Settlements," July-Aug. 1831 (transcription available at www.upperstjohn.com/aroostook/deane-kavnorth.htm), reported that Coombes "holds commission under the British: zealous in his support of their usurpations."

19. Andrew, *Development of Elites,* 19.

20. Ibid., 6; LaPointe, *Grande-Rivière,* 38.

21. Craig, "Agriculture and the Lumberman's Frontier," 125–37; "Deane and Kavanagh Report"; Andrew, *Development of Elites,* 27–28.

22. "Deane and Kavanugh Report"; Andrew, *Development of Elites,* 12.

23. The standard measurement of lumber, a board foot equals a plank one foot wide by one inch thick. Board feet is often abbreviated "bf "or simply "feet."

24. Wynn, *Timber Colony,* 22.

25. Ibid., 47, 49, 98.

26. Ibid., 79, 130, 149, 124.

27. Ibid., 96, 111, 50.

28. Joslin family history.

29. LaPointe, *Grande-Rivière,* 38; Baird, *Seventy Years of New Brunswick Life,* 94.

30. Wood, *History of Lumbering in Maine,* 65; LaPointe, *Grande-Rivière,* 40, 62; Paradis, *Papers of Prudent C. Mercure,* section 2, 41, 47; The website www.upperstjohn.com provides a thoroughly documented account of the Aroostook War.

31. Wynn, *Timber Colony,* 41–43.

32. Wood, *History of Lumbering in Maine,* 69; Quote from Craig, "Agriculture and the Lumberman's Frontier," 133.

33. Paradis, *Papers of Prudent C. Mercure,* section 2, 47; Wood, *History of Lumbering in Maine,* 69.

34. MacNutt, *New Brunswick,* 269.

35. Paradis, *Papers of Prudent C. Mercure,* xxvii.

36. Ibid., xxix, xxxvii; LaPointe, *Grande-Rivière,* 71; MacNutt, *New Brunswick,* 222, 269, 309; Merk, *Fruits of Propaganda in the Tyler Administration,* 72–73.

37. Wynn, *Timber Colony,* 114–17.

38. Ibid., 77, 114–18.

39. Ibid., 78; Craig, "Agriculture and the Lumberman's Frontier"; Title deeds, Madawaska County vol. 10:482, vol. 4:529, record of attachments #204, #213, #67.

40. Title deeds, Madawaska County, Maine, vol. 10:482 #67, #84; Wynn, *Timber Colony,* 52.

41. Wynn, *Timber Colony,* 131–34.

42. 1851 census, Victoria County, www.upperstjohn.com.

43. LaPointe, *Grande-Rivière,* 307; George McLeod, "Story of the Hammond Lumber Company," 16; Alfred Bell oral history, Nov. 1983, 6, Bill Stoddard file, author's collection.

44. Wynn, *Timber Colony,* 22, 23, 82, quote on 84; Wynn, "Deplorably Dark and Demoralized Lumberers," 168–87.

45. LaPoint, *Grande-Rivière,* 200–205, 232–33.

46. Craig, "Agriculture and the Lumberman's Frontier."

47. Title deeds, Madawaka County, record of attachments vol. 4:539, (3):346 (5):266; 1860 census, Aroostook County, 382; Guy Dubay, "The Richest Men in Van Buren."

48. MacNutt, "Politics of the Timber Trade," 126.

49. Wynn, *Timber Colony,* 63; Cooper and Clay, *History of Logging,* 32; Williams, *Americans and Their Forests,* 203.

50. Cooper and Clay, *History of Logging,* 32.

51. Wynn, *Timber Colony,* 61; Prouty, *More Deadly Than War,* 7.

52. Wynn, *Timber Colony,* 64–67; Marceau, "Old Time Logging Camps," 1396–97; Wood, *History of Lumbering in Maine,* 97; Flanagan, *Skid Trails,* 95.

53. Wynn, *Timber Colony,* 64–67; Wood, *History of Lumbering in Maine,* 97.

54. Thoreau, *Maine Woods,* 128.

55. Abrams, "Eastern White Pine Versatility," 967–78; Lorimer, "Eastern White Pine Abundance," 253–60.

56. Abrams, "Eastern White Pine Versatility," 967–78.

57. Wynn, *Timber Colony,* 43.

58. Cooper and Clay, *History of Logging,* 5–6.

59. Ibid.; Baxter, "Environmental Effects of Dams and Impoundments," 255–83; Ligon et al., "Downstream Ecological Effects of Dams" 183–92.

60. Cooper and Clay, *History of Logging,* 44.

61. Wood, *History of Lumbering in Maine,* 228; George McLeod, "Story of the Hammond Lumber Company," 2.

62. Williams, *Americans and Their Forests,* 162, 186; Cox, *Lumberman's Frontier,* 101.

CHAPTER 3

1. To preserve the historical flavor and accuracy, I generally use the terms (woodhawk, lumberjack, tomahawk, etc.) that were contemporaneous with the particular time period and were used by the historical actors themselves.

2. Maximilian, *Travels in the Interior,* 47.

3. *Progressive Men of Montana,* 1678.

4. Ibid.

5. While some historians consider "fur trade" as primarily focused on beaver pelts and distinct from the "hide trade" that concentrated on bison, others use the term "fur trade" to encompass the hide trade as well. See Barbour, *Fort Union and the Upper Missouri Fur Trade,* and Mattison, "Upper Missouri Fur Trade."

6. Flores, "Wars over Buffalo," 153–72; White, "*It's Your Misfortune and None of My Own,*" 94.

7. Barbour, *Fort Union and the Upper Missouri Fur Trade,* 8.

8. Hewitt, ed., *Journal of Rudolph Freiderich Kurz,* 81.

9. Hoig, *Chouteaus,* 175; Smith, "Fort Peck Agency," 43. An ironic turn of events occurred in 1869, when smallpox-infected bison robes arrived in Philadelphia, spawning an epidemic in that city. See Koch, Diary, 1869–1870, MHS, SC 950, 10.

10. White, "The Winning of the West," 319–43. Smith, "Fort Peck Agency," 43.

11. West, *Contested Plains,* 36, 62; Dodge, *Our Wild Indians,* 503.

12. Sherow, "Workings of the Geodialectic," 61–84; U.S. Congress, *Report of an Expedition Led by Lieutenant Abert,* 9.

13. Sherow, "Workings of the Geodialectic," 89; Montana Historical Society, *Not in Precious Metals Alone,* 13.

14. Morley, Diary, 1862–1865, 5.

15. Lass, *History of Steamboating,* 39; Overholser, *Fort Benton,* 171, 64, 56, 149.

16. Overholser, *Fort Benton,* 45; Lass, *History of Steamboating,* 41.

17. Casler, *Steamboats of the Fort Union Fur Trade,* 26; Lepley, *Birthplace of Montana;* Overholser, *Fort Benton,* 152.

18. Ashby, Reminiscence, 3–5; Chittenden, *History of Early Steamboat Navigation,* 421, 438.

19. Spitzley, Diary, 1867, 3–6; Koch, Diary, 1869–1870. Neither of the Hammonds left a written record of their journey; therefore what follows is based on the observation of other travelers.

20. Petersen, ed., "Log of the Henry M. Shreve," 537–78; Napton, "My Trip on the Imperial in 1867."

21. Spitzley, Diary, 12; Ashby, Reminiscence, 1867; Lass, *A History of Steamboating,* 13.

22. Weston, Diary, 3–6; Wilcox, "Up the Missouri River to Montana," 1; Spitzley, Diary, 10.

23. Raynolds, *Report on the Exploration of the Yellowstone River,* 148.

24. Barbour, *Fort Union,* 212.

25. Wilcox, "Up the Missouri River to Montana," 5.

26. Barbour, *Fort Union,* 216.

27. Ibid., 215; Smith, "Fort Peck Agency," 133; Quote from Phillips, ed. "Upham Letters from the Upper Missouri," 4.

28. Lass, *History of Steamboating,* 48.

29. Denig, *Five Indian Tribe of the Upper Missouri,* 27.

30. Lapenteur, *Forty Years a Fur Trader,* 441; Barbour, *Fort Union,* 231; Athearn, "Fort Buford 'Massacre,'" 675–84.

31. Napton, "My Trip on the Imperial," 5.

32. Bureau of Indian Affairs, "Council of the Indian Peace Commission," in *Papers Relating to Talks and Councils,* 97, 103.

33. Weston, Diary, 13; Napton, "My Trip on the Imperial," 5.

34. Napton, "My Trip on the Imperial," 5.

35. *Helena Herald,* June 12, 1867; *Progressive Men of Montana,* 1678; Leeson, ed., *History of Montana,* 1308.

36. *Progressive Men of Montana,* 1678.

37. Lass, *History of Steamboating,* 42; *Leavenworth Daily Conservative,* May 8, 1868.

38. Weston, Diary, 14. Quote from Morley, Diary, 13; Smith, "Fort Peck Agency," 108; *Helena Weekly Herald,* Sept. 19, 1867.

39. *Helena Weekly Herald,* Sept. 19, 1867; *Redwood Log,* Mar. 1950.

40. McLeod, "Story of the Hammond Lumber Company," 3; Koch, Diary, 14.

41. Koch, ed., "Journal of Peter Koch," 20–21, quote on 14; Smith, Reminiscence.

42. Flores, *Caprock Canyonlands,* 59; Koch, ed., "Journal of Peter Koch," 20–22.

43. Two Bears, Bureau of Indian Affairs, "Council of the Indian Peace Commission with the Various Bands of Sioux Indians at Fort Rice," in *Papers Relating to Talks and Councils,* 102.

44. Bureau of Indian Affairs, *Papers Relating to Talks and Councils,* 110–11.

45. For a colorful if somewhat embellished account, see Thorp and Bunker, *Crow Killer.* Hampton, ed., "Life at the Mouth of the Musselshell"; Koch, Diary, 9–11.

46. Wells, "First Connected Account of the Warfare," 2–4; Koch, Diary, 8.

47. Bureau of Indian Affairs, "Council of the Indian Peace Commissions on Board Steamer Agnes," In *Papers Relating to Talks and Councils,* 106.

48. Lapenteur, *Forty Years a Fur Trader,* 389; Barbour, *Fort Union,* 228; Lass, *History of Steamboating,* 39, 62; Koch, Diary, 24.

49. Lass, "Elias H. Durfee and Campbell K. Peck," 9–19; Overholser, *Fort Benton,* 32.

50. George McLeod, "Story of the Hammond Lumber Company," FHS, 3.

CHAPTER 4

1. Overholser, "Big Muddy Builds Montana."

2. Dickinson, "Diary," 1/4; Lepley, *Packets to Paradise,* 137; Lepley, *Birthplace of Montana,* 48–66.

3. *Progressive Men of Montana,* 1678; George McLeod, "Story of the Hammond Lumber Company," FHS, 2.

4. Leeson, ed., *History of Montana,* 1308; *Progressive Men of Montana,* 1678.

5. Quote from Woody, "History of Missoula"; *Missoula Gazette,* Jan. 1, 1890; *Missoulian,* Dec. 25, 1904; Koelbel, *Missoula,* 41.

6. Quote from Dunbar and Phillips, *Journals and Letters,* 116; Malone et al., *Montana,* 68.

7. Leeson, ed., *History of Montana,* 847–52; *Progressive Men of Montana,* 1678; Ficken, *Forested Land,* 31.

8. Ficken, *Forested Land,* 31.

9. Coman and Gibbs, *Time, Tide and Timber,* 48.

10. Ibid., 70.

11. Ficken, *Forested Land,* 33, quote on 26.

12. *San Francisco Journal of Commerce,* Mar. 1875, quoted in Coman and Gibbs, *Time, Tide and Timber,* 116.

13. Ficken, *Forested Land,* quotes on 42, 44.

14. Coman and Gibbs, *Time, Tide and Timber,* 112; Ficken, *Forested Land,* 41.

15. Ficken, *Forested Land,* 31.

16. *Progressive Men of Montana,* 1678; *Redwood Log* (HLC newsletter) Mar. 1950; Burlingame, *Montana Frontier,* 93.

17. George McLeod, "Story of the Hammond Lumber Company," FHS, 2; *Missoula and Cedar Creek Pioneer,* Jan. 12, 1871, Jan. 26, 1871, Feb. 23, 1871, Aug. 24, 1871; Leeson, ed., *History of Montana,* 1678; Burlingame, *Montana Frontier,* 93; Koelbel, *Missoula,* 49.

18. Quote from Dickinson, "Diary"; *Missoulian,* Dec. 25, 1902.

19. *Missoula Gazette,* Jan. 1, 1892; Johnson, "Andrew B. Hammond," 11.

20. *Misssoulian,* Nov. 17, 1870; Woody, *History of Missoula;* Dunbar and Phillips, *Journals and Letters,* 80–81.

21. *Helena Weekly Herald,* Oct. 3, 1867; *Missoulian* Apr. 27, 1872, Nov. 3, 1870.

22. *Missoula and Cedar Creek Pioneer,* Oct. 5, 1871; Johnson, "Andrew B. Hammond," 12.

23. McLeod, "Story of the Hammond Lumber Company," UM 96/4, 1. This edited version differs slightly from the same manuscript at FHS.

24. *Weekly Missoulian,* Jan. 21, 1873, Jan. 22, 1874, Jan. 20, 1875, quote from Apr. 6, 1872.

25. Ibid., Feb. 17, 1875, Aug. 4, 1875, Sept. 29, 1875.

26. Ibid., Aug. 19, 1874.

27. Ibid., Apr. 25, 1873.

28. Ibid., May 3, 1876.

29. Ibid., Feb. 14, 1877.

30. Ibid., Jan. 18, 1878.

31. Leeson, ed., *History of Montana,* 137–38; Josephy, *Nez Perce Indians,* 562–72.

32. Higgins quote in Phillips, ed., "Battle of the Big Hole;" 5; Leeson, ed., *History of Montana,* 146; *Weekly Missoulian,* July 6, 1877; Josephy, *Nez Perce Indians;* Phillips, ed., "Battle of the Big Hole," 4.

33. Sutherland, *Howard's Campaign against the Nez Perce Indians,* 20.

34. Rawn to adjutant general, Dept. of Dakota, Sept. 12, in Phillips, ed., "Early Days at Fort Missoula," 6, *Sources of Northwest History,* 1929.

35. Chauncey Barbour to Potts July 31, 1877, in Phillips, "Battle of the Big Hole," 6; *Weekly Missoulain,* Aug. 9, 1877; Quote from Toole, *Red Ribbons,* 14.

36. Buck, "Story of the Nez Perce Indian Campaign"; Brown, *Flight of the Nez Perce,* 232.

37. For a full account see Brown, *Flight of the Nez Perce,* and Josephy, *Nez Perce Indians.*

38. Boyer and Morais, *Labor's Untold Story,* 60–62; Zinn, *People's History of the United States,* 245–47.

39. Coon, *Economic Development,* 77; *Weekly Missoulian,* Aug. 8, 1877.

40. *Weekly Missoulian,* Aug. 8, 1873.

41. Ibid., May 31, 1878; Dale Johnson, pers. comm., Mar. 10, 2009, quoting Tommie Lu Worden, descendent of Frank Worden, associate of C. P. Higgins.

42. Johnson, "Andrew B. Hammond," 24–25.

43. Ibid.; quote from *Weekly Missoulian,* Feb. 27, 1880, Jan. 9, 1880.

CHAPTER 5

1. Toole, *Men, Money and Power,* 9.

2. *House Committee on Pacific Railroads Report;* Schwinden, "Northern Pacific Land Grants in Congress," 30; *Report of the Secretary of Interior,* 1885, vol. 1, 41;

Jensen and Draffan, *Railroads and Clearcuts,* 7. Schwantes, in *Pacific Northwest,* 144, estimated the original grant at 60 million acres, but noted that after forfeiture of some lands, the Northern Pacific actually received title to 39 million acres.

3. Smalley, *History of the Northern Pacific Railroad,* 161, 213.

4. For more on Samuel Hauser, see Hakola, "Samuel T. Hauser."

5. Smalley, *History of the Northern Pacific Railroad,* 174. For more on the Panic of 1873, see Lubetkin, *Jay Cooke's Gamble.*

6. Coon, *Economic Development,* 68.

7. *Missoulian,* July 26, 1878, Nov. 27, 1878; Coon, *Economic Development,* 78.

8. Gould to Hauser, Dec. 14, 1878, Hauser papers, 9/25, MHS; Malone, *Battle for Butte,* 23.

9. Coon, *Economic Development,* 95.

10. Smalley, *History of the Northern Pacific Railroad,* 408.

11. O'Bannon to Hauser, July 26, 1881, Hauser papers, 17/7; Smalley, *History of the Northern Pacific Railroad,* 409.

12. Taylor, *Rails to Gold and Silver,* 11.

13. W. J. McCormick to Hauser, May 26, 1884, Hauser papers, 15/18; C. H. McLeod to Herbert Peet, Apr. 9, 1940, McLeod papers, 48/4; Coon, *Economic Development,* 94–100; *Weekly Missoulian,* Aug. 19, 1881; Johnson, "Andrew B. Hammond," 35–36.

14. Johnson, "Andrew B. Hammond," 35–36.

15. *Missoulian,* Aug. 19, Aug. 26, Sept. 2, Sept. 2, Sept. 29, Oct. 21, 1881; *New Northwest* (Portland, Ore.), May 26, 1882.

16. *Missoulian,* May 9, 1879; *Missoula Gazette,* Jan. 1, 1892.

17. *Missoulian,* May 19, 1881, May 14, 1886; Records of Trinity Anglican Church, Perth Andover, Grand Falls Museum; *St. John Daily News,* Nov. 12, 1875; *Moncton Times,* Apr. 25, 1890. John Keith proved the exception, as he and Hammond engaged in a long and bitter dispute over management of the bank in the early twentieth century.

18. Quote from *Helena Weekly Inter-Mountain,* Jan. 1, 1882; *Missoulian,* Mar. 10, 1882.

19. Hammond to C. H. McLeod (unless otherwise noted, all McLeod correspondence is to/from C. H. McLeod and all Hammond correspondence is from McLeod papers MSS. 01, UM), Aug. 8, 1929, 22/4; Northern Pacific Railway Company records, series V, box 102, vol. 89, UM.

20. *Missoulian,* June 9, 1882; *Missoula County Times,* Nov. 21, 1883.

21. Johnson, "Prospectors Turned Tiehacks"; *New Northwest,* May 26, 1882; Smalley, *History of the Northern Pacific Railroad,* 415–16; Swartout, "From Kwangtung to the Big Sky,"42–53; Renz, *History of the Northern Pacific Railroad,* 97.

22. *Missoulian,* June 22, 1883. Quotes from J. B. Brondell recollection in Scott, *Missoula: Trading Post to Metropolis,* 1; Toole, *Red Ribbons,* 23.

23. Koelbel, *Missoula: The Way It Was,* 79; Deverell, *Railroad Crossing,* 4–6, 175; Toole, *Red Ribbons,* 22.

24. Mercer, *Railroads and Land Grant Policy,* 54; Schwantes, *Railroad Signatures across the Pacific Northwest,* 64.

25. Coon, *Economic Development,* 114–16.

26. Hakola, "Samuel T. Hauser," 117–24.

27. Toole, *Red Ribbons,* 27.

28. Leeson, ed., *History of Montana,* 878; *Weekly Missoulian,* Dec. 1, 1882; Hakola, "Samuel T. Hauser," 99.

29. Coon, *Economic Development,* 79; quote from Hammond to Hauser, Jan. 7, 1885, 10/3; Bonner to Hammond, May 15, 1885, 2/2, Hauser papers, MHS.

30. Hauser to Hammond, Jan. 10, 1885, 10/3, Hauser papers, MHS.

31. Higgins to Hauser, Jan. 1, 1885, 11/13, Hauser papers, MHS.

32. Kennett to Hauser, Jan. 13, 1885, 13/13, Hauser papers, MHS.

33. Ibid.

34. Hammond to Hauser, Feb. 12, 1885, 10/3, Hauser papers, MHS.

35. Bonner to Hammond, May 15, 1885, Hauser papers, 2/2, MHS.

36. Kennett to Hauser, Jan. 11, 1887, Aug. 9, 1887, 13/14, Hauser papers, MHS.

37. Taylor, *Rails to Gold and Silver,* 28; Anderson to Hauser, Aug. 8, 1881, Aug. 12, 1881, Sep. 16, 1886.

38. *Missoula County Times,* Oct. 3, 1883; quote from Taylor, *Rails to Gold and Silver,* 28.

39. *Weekly Missoulian,* Mar. 19, 1886, June 4, 1886; Taylor, *Rails to Gold and Silver,* 21–22, 84.

40. Hauser to Hammond, Nov. 23, 1886, 10/3, Hauser papers, MHS; *Weekly Missoulian,* Jan. 21, 1887; Taylor, *Rails to Gold and Silver,* 83, 99, 121; Haloka, "Samuel T. Hauser," 175.

41. Hammond to Hauser, Feb. 7, 1887, 10/3, Hauser papers, MHS.

42. *Weekly Missoulian,* Apr. 29, 1887; quote from Hammond to Hauser, Feb. 7, 1887, 10/3; Hammond to Hauser, Apr. 14, 1887, 10/3, Hauser papers, MHS.

43. Northern Pacific Railway records, series 32, box 324, vol. 450, UM; Johnson, "Andrew B. Hammond," 101.

44. *Weekly Missoulian,* Dec. 23, 1887.

45. Hammond to Hauser, Feb. 7, 1887, 10/3, Hauser papers, MHS; Taylor, *Rails to Gold and Silver,* 104–105.

46. Taylor, *Rails to Gold and Silver,* 101–102; *Weekly Missoulian,* Apr. 29, 1887; "Traffic Agreement between Northern Pacific and Missoula and Bitter Root Valley Railroad," Northern Pacific Railway records, series 40, 128(xl),1 microfilm, UM.

47. L. S. Miller to George Sheldon, Apr. 5, 1896; E. H. McHenry to Edward Adams, Mar. 4, 1896; both in Taylor, *Rails to Gold and Silver,* 140–41.

48. Hakola, "Samuel T. Hauser," 179–83. Oakes to Hauser, July 13, 1891, box 17, folder 13, Hauser papers, MHS.

49. L. S. Miller to George Sheldon, Apr. 5, 1896; branch line reports in Taylor, *Rails to Gold and Silver,* 141, 120. The Bitterroot branch continues to operate,

although the spate of lumber mill closings in the 1990s greatly curtailed its viability.

50. Kennett to Hauser, Oct. 26, 1887, Feb. 6, 1888, 13/14. Hauser papers, MHS; Hammond to Hauser, July 20, 1888, 10/5 Hauser papers, MHS.

51. Kennett to Hauser, Sept. 4, 1888, 13/14, Hauser papers, MHS.

52. Kennett to Hauser, Nov. 4, 1888, 13/14, Hauser papers, MHS.

CHAPTER 6

1. Hibbard, *History of the Public Land Policies,* 170; quote from Vernon Carstensen, ed., *Public Lands,* xxi; Donaldson, *Public Domain,* 519.

2. *Weekly Missoulian,* July 7, Aug. 4, Dec. 29, 1882.

3. Coon, *Economic Development,* 104; *Missoula County Times,* Jan. 7, Feb. 4, Apr. 7, 1885.

4. For a full account of the Cramer Gulch war, see Stone, *Following Old Trails,* 263–69; quotes on 263–64.

5. Stone, *Following Old Trails,* quotes on 264, 269.

6. The National Forest system would eventually encompass 193 million acres. Acreage estimate as of Dec. 1883, Donaldson, *Public Domain,* 531; Cox, *Lumberman's Frontier,* quote on 17, 40.

7. Barrett, "Relationship of Indian-Caused Fires"; Barrett and Arno, "Indian Fires as an Ecological Influence," 647–65.

8. U.S. Senate, *Explorations and Surveys for Pacific Railroad,* vol. I; McCay and Acheson, eds., *Question of the Commons,* 17.

9. Lefevre, *English Commons and Forests,* 1–24; McQuaig, *All You Can Eat,* 161–64.

10. This semantic confusion reached its apogee in "Tragedy of the Commons," an article by ecologist Garrett Hardin published in 1968. Hardin suggested that common resources eventually would be overexploited because each individual seeks to maximize his or her own gain while costs are passed on to society as a whole. Hardin cited examples of pastureland, open oceans, and national parks as commons where access is open to all and thus individuals will overgraze, overfish, or overuse and exploit as much of the resource as possible before someone else does. What Hardin called "commons" are more accurately described as "open-access resources" and applied to concerns over deep-sea fishing and air pollution. Commons, in contrast, are traditionally regulated by social norms, customary rights, and historical claims. See Hardin, "Tragedy of the Commons," 1243–48; Brander and Taylor, "Simple Economics of Easter Island," 119–38.

11. *Appendix to the Congressional Globe,* "Man's Right to the Soil," 425.

12. Letter from Secretary of Interior, S. Exec. Doc. 9, 45th Cong., 2d Sess., Serial Set vol. no. 1780, No. 1, 3; Ise, *United States Forest Policy,* 24.

13. *River and Harbor Bill,* 32d Cong., 1st Sess., *Congressional Globe,* Appendix, 851.

14. Statistics of mines and mining in the state and territories west of the Rocky Mountains, U.S. House Exec. Doc. 207, 41st Cong., 2d sess., 1424.

15. U.S. General Land Office, *Annual report* (hereafter *GLO Report*), 1872, 26, 27.

16. Koelbel, *Missoula: The Way It Was,* 29.

17. Ise, *United States Forest Policy,* 27.

18. White, *Land Use, Environment and Social Change,* 77–79; *GLO Report,* 1875, 11; Donato, "Post-wildfire Logging," 352.

19. *GLO Report,* 1875, 11.

20. Ibid.

21. *Statutes at Large* 19, 73 quoted in Ise, *United States Forest Policy,* 53.

22. *Report of the Secretary of the Interior,* 1877, part 5, 16. For more on the influence of European forestry see Peluso, *Rich Forests, Poor People,* 7–9.

23. *Weekly Missoulian,* Dec. 21, 1877.

24. Quote from *GLO Report,* 1878, 124; *Report of the Secretary of the Interior,* 1877, 16.

25. Robbins, *Colony and Empire,* 66. Cosponsored by Senator Jerome Chaffee (R-Colo.) and Montana Democratic Delegate Martin Maginnis, this bipartisan effort suggests that public land policies hinged more upon sectional differences than upon party lines.

26. Coon, "Economic Development," 99.

27. *Report of the Secretary of the Interior,* 1878, xxiii, xiv.

28. *GLO Report,* 1878, 119.

29. Dunham, *Government Handout,* 103; *Decisions of Department of Interior,* 1881–83, 601, 608–10.

30. *Weekly Missoulian,* Sept. 22, 1882; *GLO Report,* 1885, 82; *River Press* [Fort Benton], Oct. 21, 1885; *Helena Independent,* July 12, 1883.

31. Charter legislation from *Decisions of Department of Interior,* 1885, 13 Stat., 365, 68; Sparks quotes from *Report of the Secretary of the Interior,* 1885, 234–35.

32. *GLO Report,* 1883, 9; McFarland quotes from *Decisions of Department of Interior,* 1881–83, 611.

33. *Leavenworth R. R. Co. v. United States,* 92 U.S. 741; quote from *Decisions of Department of Interior,* 1883–84, 829.

34. *Inter Mountain* [Butte], Nov. 1, 1885.

35. *Missoula Country Times,* Dec. 5, 1883, Dec. 26, 1883; *Weekly Missoulian,* Aug. 31, 1883, Jan. 4, 1884, Mar. 12, 1884; *Helena Independent,* July 12, 1883.

36. *Helena Independent,* July 12, 1883.

37. *Decisions of Department of Interior,* 1884–85, 66, italics in original.

38. Ibid., 66–67.

39. Brewster quote from *Decisions of Department of Interior,* 1884, 66–67; *Report of Secretary of Interior,* 1882, vol. 2, 14, 56.

40. Oakes to Hauser, June 2, 1885, 17/7; Hauser to Oakes, June 1885; Letter books, box 28, both in Hauser papers, MHS.

CHAPTER 7

1. Malone et al., *Montana,* 96–100; *Missoulian,* June 27, 1884, Nov. 21, 1884.

2. Quoted in Ginger, *Age of Excess,* 107.

3. Quote from Nevins, *Grover Cleveland,* 197, 224; Malone, ed., *Dictionary of American Biography,* vol. 17; American National Biography Online, www.anb.org.

4. Nevins, *Grover Cleveland,* 216; Dunham, *Government Handout,* 173.

5. Letter from Secretary of Interior, Senate Ex. Doc. 170, 49th Cong., 1st Sess., 1–3; quotes from *GLO Report,* 1886, 43.

6. *GLO Report,* 1885, 3.

7. *GLO Report,* 1885, 3; 1886, 101, 442; 1887, 83.

8. *Report of Secretary of Interior,* 1886. Both the "Timber and Stone Act" and the "Free Timber Act" were names that were later applied to the act of June 3, 1878. The former provided for the sale of timberlands in 160-acre tracts but was not applied to Montana until 1891. The latter allowed for cutting of timber on "mineral lands" in the public domain.

9. Ibid., 448–49, emphasis in original.

10. *Helena Herald,* Sept. 17, 1885.

11. *Butte Semi-weekly Inter-mountain,* Sept. 16, 1885.

12. *GLO Report,* 1887, 87.

13. *GLO Report,* 1885, 311–14, emphasis in original.

14. Quote from *GLO Report,* 1885, 312–13; similar quote in Sparks to Lamar, Nov. 12, 1886, National Archives, RG 60, file 7308, #7608.

15. *Decisions of Department of Interior,* 66–67.

16. Quote from *Butte Semi-weekly Inter-mountain,* Nov. 1, 1885. It is possible that Bonner and Henry Villard forged an agreement without Hammond's knowledge. Apparently, Villard had drafted the contract, but the new president of the railroad failed to approve it. *Missoulian,* July 31, 1885; *St. Paul Pioneer Press,* Nov. 4, 1885.

17. *St. Paul Pioneer Press,* Nov. 4, 1885.

18. *Butte Semi-Weekly Inter-mountain,* Sept. 16, 1885; *Missoulian,* Sept. 16, 1885.

19. *Butte Semi-Weekly Inter-mountain,* Sept. 16, 1885.

20. See Johnson, "Andrew B. Hammond"; tax reports from *Missoulian,* Nov. 13, 1885, Nov. 11, 1887.

21. *Helena Herald,* Oct. 13, 1885. When filed in federal court, this was revised to thirty-six million board feet of lumber, seventy-five thousand ties, ten thousand posts, one million cords, and fifteen million shingles, although the dollar amount was raised to $1.1 million; S. M. Stockslager to Attorney General, Mar. 8, 1889, Dept. of Justice, file 7308, #2195. In later suits, the figure was dropped to twenty-one million board feet; see *Hammond v. U.S.,* 246 F. 40 (1917). Charges against others from John Noble to Attorney General, Jan. 27, 1890, Dept. of Justice, file 7308, #1202.

22. Hammond to Hauser, Sept. 24, 1885, Oct. 12, 1886, Hauser papers, 10/4, MHS; Bonner quote from *Helena Herald,* Sept. 17, 1885; Hammond to Hauser, Oct. 14, 1885, 10/4, Hauser papers, MHS; *Missoulian,* Jan. 15, 1886, Apr. 30, 1886.

23. Quote from Toole to Hauser, Sept. 24, 1885, 21/19; Hammond to Hauser, Sept. 26, 1885, 7/ 2; Bonner to Hauser, Oct. 27, 1885, 2/13. All in Hauser papers, MHS.

24. Hauser to Daly Oct. 9, 1885; quote from Daly to Hauser, Nov. 13, 1885, 7/4; Vest to Hauser Nov. 16, 1885 22/12. All in Hauser papers, MHS.

25. Hammond to Hauser, Oct. 14, 1885.

26. *Missoulian,* Jan. 29, 1885, Jan. 15, 1885, July 16, 1886.

27. *Helena Independent,* Nov. 26, 1885; quote from Hammond to Hauser, Oct. 12, 1885, 10/4, Hauser papers, MHS.

28. Toole to Hauser, Nov. 17, 1885, 21/9 Hauser papers; *Helena Independent,* Nov. 26, 1885.

29. *Missoula County Times,* Aug. 12, Aug. 26, 1885.

30. *Missoulian,* Dec. 4, 1885, Dec. 25, 1885; *Missoula County Times,* Feb. 3, 1886.

31. *Missoula County Times,* Mar. 17, 1886; *Missoulian,* May 14, 1886; *Report of Secretary of Interior,* 1886, vol. 2, 447.

32. *Missoula County Times,* July 28, 1886, Aug. 11, 1886; *Missoulian,* July 19, 1888, Dec, 3, 1886, Dec. 15, 1886, Aug. 6, 1886.

33. Quote from *Missoulian,* June 11, 1886; *Report of Secretary of Interior,* 1886, vol. 2, 447–48.

34. *Report of Secretary or Interior,* 1886, vol. 2, 447–48, italics in original.

35. *GLO Report,* 1886, 43.

36. *Helena Herald,* Oct. 13, 1885.

37. Sparks to Lamar, Nov. 12, 1886, Dept. of Justice, files 7308, 7608.

38. *Missoulian,* Aug, 21, 1885, Oct. 16, 1885; *Missoula County Times,* Aug. 24, Aug. 24, 1885; Johnson, "Andrew B. Hammond," 85–86.

39. *Missoula County Times,* Dec. 15, 1886; quotes from *U.S. v. NPRR,* 6 Mont. 351 (1887). Second quote from *GLO Report,* 1887, 479; *Leavenworth R. R. Co. v. U.S.,* 92 U.S. 741.

40. *GLO Report,* 1887, 477.

41. Ibid.

42. Quotes from Hobson to Attorney General, June 15, 1888, Dept. of Justice, file 7308, #5030; Hammond to Hauser, Nov. 26, 1887, 10/4, Hauser papers.

43. *GLO Report,* 1887, 84.

44. Missoula County Records, Articles of Incorporation, files 13 and 14, UM; *Missoulian,* Mar. 18, 1887.

45. *Report of Secretary of Interior,* 1886, 448.

46. Ibid, 1885, vol. 1, 41.

47. Ibid, 1886, vol. 1, 28–29, emphasis in original.

48. Quotes from *GLO Report,* 1887, 87; *Report of Secretary of Interior,* 1884, 19, 1885, 41. Sparks advocated a five-year residency clause for homestead entries to ensure the lands would go to settlers rather than timber companies. The Forest Reserve Act of 1891 provided the authority for the president to designate such reserves.

49. Quote from *GLO Report,* 1885, 41, 180; Nevins, *Grover Cleveland,* 223.

50. *GLO Report,* 1888, 17; Nevins, *Grover Cleveland,* 225–27.

51. S. M. Stockslager to Attorney General, Mar. 8, 1889, Dept. of Justice, file 7308, #2195.

CHAPTER 8

1. Lewis Gould, "Party Conflict: Republicans versus Democrats," 265–67, in Calhoun, ed., *Gilded Age.*

2. *Missoula County Times,* June 25, July 30, 1884, Johnson files.

3. City of Missoula Records, vol. 255, 66–127.

4. Herbert Peet to K. Ross Toole, Apr. 5, 1948, 1/4, Herbert M. Peet Collection, MHS; *Butte Inter-mountain,* Apr. 7, 1888; *Missoula Gazette,* May 26, 1888; *Missoulian,* Apr. 25, 1888, Johnson files.

5. Peet to Toole, Apr. 5, 1948.

6. Daly quote from Toole, "Marcus Daly," 20–22; Clark quote from Malone, *Battle for Butte,* 12.

7. Toole, "Marcus Daly," 19; Daly quote from Malone, *Battle for Butte,* 18, 35.

8. Toole, "Marcus Daly," 57; *Missoula Gazette,* Sept. 15, 1888, Johnson files.

9. Power to Daly, Sept. 15, 1888, T. C. Power papers, MHS; *Missoulian,* Sept. 19, 1888, Johnson files.

10. Lent to Power, Oct. 22, 1888, 1/8. T. C. Power papers, MHS; Peet to Toole, Apr. 6, 1949, "Clark-Daly Feud," MSS 89, 1/ 4, MHS; *Evening Item* [Missoula] Aug. 18, 1888; *Missoulian,* Oct. 31, 1888, Johnson files; Waldron and Wilson, *Atlas of Montana Elections,* 7.

11. D. Emmons, "A Reconsideration of the Clark-Daly Feud," in Fritz et al., eds., *Montana Legacy,* 87.

12. In a later interview, Bonner candidly admitted the conspiracy and their reasons for electing Carter. See "Footprints of Time, a True Story of the Montana Career of Thomas H. Carter," *Helena Independent,* Oct. 11, 1900. Subsequent investigations suggest that Daly opposed Clark on orders of his financial backer, James Ben Ali Haggin, whom Clark had insulted. See Peet to C. H. McLeod, Apr. 4, 1940, C. H. McLeod Papers, 48/4.

13. Peet to Toole, Apr. 6, 1949, "Clark-Daly Feud," 6; "Footprints of Time," *Helena Independent,* Oct. 11, 1900.

14. *Butte Miner,* Nov. 14, 1888, 1/4, Peet Collection.

15. Emmons, "Reconsideration of the Clark-Daly Feud," 87–102.

16. Peet to Toole, Apr. 6, 1949, "Clark-Daly Feud," 12; quote from Malone, *Battle for Butte,* 86.

17. *Helena Record,* Nov. 21, 1888, 1/4. Peet Collection.

18. Malone, *Battle for Butte,* 87; quote from *Helena Independent,* Oct. 11, 1900.

19. *Helena Independent,* Oct. 11, 1900.

20. Malone, *Battle for Butte,* 87.

21. Malone, *Battle for Butte,* 90; Daly to Hauser, July 29, 1889, 7/3, Hauser papers.

22. Johnson, "Andrew B. Hammond," 115; quote from *Missoulian,* Aug. 21, 1889.

23. Hammond to Power, Aug. 28, 1889, 1/21, T. C. Power papers.

24. Oakes quote from Oakes to G. W. Dickinson, Sept. 6, 1889, Hauser papers; J. W. Buskett to Hauser, Sept. 2, 1889, Johnson files.

25. *Missoulian,* Sept. 11, 1889; quote from J. W. Buskett to Hauser, Sept. 2, 1889, Johnson files; McLeod to Hammond, Sept. 23, 1910, 18/5; Moser to Power, Sept. 3, 1889, 1/31, T. C. Power papers.

26. Quote from Daly to Hauser, Sept. 23, 1889, 7/3, Hauser papers; W. H. Hammond account in Zimmerman, "Missoula Mercantile Company."

27. John McCormick to Power, Sept. 28, 1889, T. C. Power papers 1/39; Johnson, "Andrew B. Hammond," 132–33; Malone, *Battle for Butte,* 90–91; Waldron and Wilson, *Atlas of Montana Elections,* 11–13.

28. Glasscock, *War of the Copper Kings,* 111; Malone, *Battle for Butte,* 90.

29. Johnson, "Andrew B. Hammond," 137–38; Bonner to Hauser, Nov. 13, 1889, Hauser papers.

30. Quote from McLeod to Hammond, Sept. 23, 1910, 18/5; *Missoula Gazette,* Apr. 4, 1890, May 3, 1890.

31. Johnson, "Andrew B. Hammond," 143–44; *Missoula Gazette,* May 15, 1890, Feb. 26, 1891, July 23, 1890, Johnson files.

32. Johnson, "Andrew B. Hammond," 146; *Missoula Gazette,* Apr. 17, 1890, Oct. 15, 1891, Johnson files.

33. S. G. Murray to T. C. Power, Feb. 3, 1890, 3/1, T. C. Power papers, MHS.

34. Quote from *Butte Inter-Mountain,* Sept. 25, 1888; editorial cartoon from *Missoulian,* 1891, UM archives 72–0701.

35. Johnson, "Andrew B. Hammond," 149.

36. *Weekly Missoulian,* Sept. 23, Sept., 30, 1891.

37. *Anaconda Standard,* Mar. 30, 1892, Johnson files.

38. Berman, *Radicalism in the Mountain West,* 40.

39. *Anaconda Standard,* May 10, 1891, Johnson files.

40. Ibid., May, 11, 1891.

41. *Missoula Gazette,* May 10, 1892, Johnson files.

42. Ibid., May 29, 1892.

43. Malone, *Battle for Butte,* 103; *Missoulian,* Jan. 5, 1893, Johnson files; Hammond to Hauser, Feb. 9, 1893.

44. Hammond to Power, May 28, 1894, T. C. Power papers.

45. Hammond to McLeod, Apr. 30, 1895, 16/7; quote from Hammond to Hauser, Nov. 1, 1894, 10/5, MHS.

46. Quote from Hammond to Hauser, Nov. 4, 1894; Malone, *Battle for Butte,* 99.

CHAPTER 9

1. Howard, *Montana: High, Wide, and Handsome,* 3.
2. Chandler, "Beginnings of 'Big Business,'" 1–31; Hakola, "Samuel T. Hauser," 187–88.
3. Johnson, "Andrew B. Hammond," 24; *Missoulian,* Aug. 10, 1883, Johnson files.
4. *The Bohemian Club of San Francisco,* 96, MSS 2295, HEH.
5. George McLeod, "Story of the Hammond Lumber Company," FHS, 8.
6. Hammond to Hauser, June 10, 1893, 10/6, Hauser papers, MHS.
7. Hammond to Hauser, June 12, 1893, emphasis in original.
8. Hammond to Hauser, June 14, 1893; George McLeod, "Story of the Hammond Lumber Company," UM, 8–9.
9. Hammond to Hauser, Aug. 2, 1893, 10/6, Hauser papers, MHS.
10. Toole, *Red Ribbons,* 33.
11. Hammond to Hauser, July 11, 1893, emphasis in original.
12. Hammond to Hauser, Aug. 8, 1893.
13. Taylor, *Rails to Gold and Silver,* 97; Malone, *Battle for Butte,* 54–55.
14. Hammond to Hauser, Dec. 7, 1893.
15. First National Bank of Missoula Report, Jan. 1, 1894, 10/6, Hauser papers; Hammond to McLeod, Jan. 10, 1895, 16/7; Malone, *Battle for Butte,* 56.
16. White, *Railroaded,* 17.

CHAPTER 10

1. Chandler, "Beginnings of 'Big Business'"; Robbins, *Colony and Empire,* 103.
2. Robbins, *Colony and Empire,* 14, 103; quote from Robbins, *Landscapes of Promise,* 103; see also Cronon, *Nature's Metropolis.*
3. George McLeod, "Story of the Hammond Lumber Company," FHS, 10; Scott, "Yaquina Railroad," 228–45; *Oregonian,* Dec. 15, 1893, 9.
4. *Astoria Daily Budget,* Dec. 1, 1894; *Oregonian,* Dec. 26, 1894; "Transcript of Railroad Commission of Oregon Investigation of the C. and E.R.R," Oct. 14, 1907, Weatherford Papers, box 13, OHS.
5. *Daily Astorian,* July 26, 1895, Dec. 17, 1897; *Oregonian,* July 25, 1895, 8; Scott, "Yaquina Railroad;" quote from Hammond to C. P. Huntington, Aug. 28, 1895, HEH, 2383, box 43.
6. Scott, "History of Astoria Railroad," 221–40.

7. Scott, "History of Astoria Railroad"; "Classified List and Estimated Value of Lots and Acreage in the Astoria Railroad Subsidy Map," vol. 25, HLC, MSS 1716, OHS.

8. *Daily Astorian,* Apr. 21, 1894.

9. Hammond to McLeod, Dec. 5, 1894, 16/7.

10. "History of Subsidy Transfers to A. B. Hammond for Building A. & C. R. R. R," HLC, vol. 25, OHS; quote from *Astoria Daily Budget,* Dec. 1, 1984.

11. Hammond to McLeod, Dec. 5, 1894; quote from Hammond to McLeod, Jan. 22, 1895, 16/7.

12. Hammond to McLeod, Dec. 5, 1894, Oct. 31, 1895, 16/7; *Oregonian,* Dec. 26, 1894, 2, Jan. 3, 1895.

13. Hammond to McLeod, Jan. 22, Feb. 20, Mar. 7, 1895, 16/7.

14. Hammond to McLeod, Sept. 9, Mar. 7, 1895, 16/7; Scott, "History of Astoria Railroad"; Hammond to C. P. Huntington, Aug. 28, 1895, HEH 2383, box 43.

15. Hammond to C. P. Huntington, Aug. 28, 1895.

16. *Daily Astorian,* May 4, 1895.

17. Ibid.; "History of Subsidy Transfers," HLC files; newspaper quotes from *Astoria Daily Budget,* Apr. 11, 1895; Hammond quote from Hammond to McLeod, July 15, 1895, 16/7.

18. *Daily Astorian,* May 21, 1894.

19. Ibid., July 26, 1895, 1; *Oregonian,* July 26, 1895, 3.

20. *Daily Astorian,* July 26, 1895, 1.

21. *Oregonian,* July 24, 1895, 1, 8; George McLeod, "Story of the Hammond Lumber Company," FHS, 24; "Report of the Railroad Commission of Oregon in the matter of the Corvallis and Eastern Railroad Company," Weatherford Papers, box 13; Hammond to McLeod, Mar. 5, 1895, Apr. 5, 1895, 16/7.

22. Hammond to McLeod, Apr. 16, Apr. 30, 1895.

23. *Daily Astorian,* Apr. 25, 1896, Aug. 7, 1895, May 16, 1896; quote from *Oregonian,* July 24, 1895, 1; *Oregonian,* Dec. 19, 1896, This new city included the subdivisions of Warrenton, Flavel, and New Astoria.

24. Quote from Hammond to McLeod, July 15, 1895, 16/7; *Daily Astorian* July 4, 1895.

25. *Oregonian,* Sept. 18, 1896, Dec. 18, 1896, 10.

26. George McLeod, "Story of the Hammond Lumber Company," FHS, 43.

27. *Daily Astorian,* May, 23, 1897.

28. *Oregonian* (Portland), May 22, 1897.

29. H. E. Huntington to Hammond, Nov. 18, 1902, HEH 5502, box 97.

30. *Oregonian,* June 13, 1897.

31. *Oregonian,* May 17, 1898.

32. Astoria and Columbia River Railroad, *Oregon Coast.*

33. *Oregonian,* Sept. 7, 1904: 4

34. Hammond to John Claflin, Apr. 14, 1932, 23/3.

35. *Daily Astorian,* Sept. 22, 1899, 9, May 28, 1901, 1.

36. *Oregonian,* Sept. 13, 1896, 6, Nov. 4, 1900, 10; Hammond to Thomas Hubbard, June 3, 1902, HEH 2478, box 43. Quote in Hammond to Claflin, Apr. 14, 1932, 23/3.

37. *Daily Astorian,* May 28, 1901, 1; Hammond to H. Huntington, Nov. 8, 1901, HEH 2385, box 43; quote from H. Huntington to Hammond, Dec. 30, 1901, HEH 5488, box 97.

38. Hammond to H. Huntington, Aug. 15, 1902, HEH 2404, box 43.

39. *Timberman,* Nov. 1906, 40; Robbins, *Landscapes of Promise,* 220–21; Hammond to Huntington, Nov. 13, 1902, HEH 2410; Hammond to Huntington, Aug. 13, 1904, HEH 2469.

40. *Oregonian,* Sept. 7, 1904, 4.

41. *Oregonian,* Apr. 9, 1906, 12; *Columbia River and Oregon Timberman,* Nov. 1899, 9; Astoria Company, HLC files, box 3; quote from *Oregonian,* Jan. 5, 1901, 1.

42. Donan, *Columbia River Empire,* 39; Meany, "History of the Lumber Industry," 27; *Oregonian,* Dec. 19, 1896, 14.

43. Donan, *Columbia River Empire,* 38; quote from *Oregonian,* Jan. 7, 1907.

44. Astoria Chamber of Commerce, *Astoria and Clatsop County Oregon; Pacific Lumber Trade Journal,* Feb. 1907, 44; Giles, "Notes of Columbia River Salmon," 144.

45. Robbins, *Landscapes of Promise,* 220; quote from *Oregonian,* Sept. 7, 1904, 4; Astoria and Columbia River Railroad, *Oregon Coast.*

46. Edwin Stone to Hammond, Dec. 21, 1901, folder 2, misc. financial papers, HEH 11/4/2, Huntington papers.

47. "Transcript of the Railroad Commission of Oregon investigation of the Corvallis and Eastern Railroad, Oct. 14, 1907," 12–16, box 13, Weatherford papers, OHS.

48. Ibid.; *Oregonian,* Jan. 1, 1907, 20, Oct. 22, 1903, 10; *Timberman,* Mar. 1906, 37; Hammond to Huntington, Nov. 4, 1902, HEH 2406.

49. Quotes from "Report of the Railroad Commission of Oregon," director's minutes, Corvallis and Eastern, box 5, Weatherford papers, OHS; *Daily Astorian,* Apr. 3, 1895; Hammond to Henry Huntington, HEH 2416, box 44.

50. Scott, "History of the Astoria Railroad."

CHAPTER 11

1. *Missoula Daily Democrat-Messenger,* Feb. 26, 1898, June 17, 1898; Hammond to McLeod, July 24, 1905, 16/9; *Los Angeles Times,* Feb. 20, 1904, 6.

2. Quote from Buckley, "Building the Redwood Region," 523–45; George McLeod, "Story of the Hammond Lumber Company," FHS, 32.

3. George McLeod, "Story of the Hammond Lumber Company," FHS, 20.

4. Ficken, *Forested Land;* George McLeod, "Story of the Hammond Lumber Company," FHS.

5. See Wiebe, *Search for Order;* Chandler, *Visible Hand;* Porter, *Rise of Big Business;* and Lamoreaux, *Great Merger Movement in American Business.*

6. Roy, *Socializing Capital,* 4, 14.

7. George McLeod, "Story of the Hammond Lumber Company," FHS, 9–10.

8. Donohue to Daly, Apr. 21, 1898, Anaconda Forest Products Company Records, 1/1, UM.

9. George McLeod, "Story of the Hammond Lumber Company," FHS, 9–10.

10. Ibid., 13.

11. *Pacific Lumber Trade Journal,* Oct. 1899, 18.

12. HLC files, vol. 27; Weatherford papers, box 3, OHS; Evangelyn Fleetwood to Jack Blanchard, Sept. 4, 1992, author's file; *Columbia River and Oregon Timberman,* Mar. 1900, 7, June 1900, 17, Nov. 1903, 20.

13. *Astoria Daily Budget,* May 29, Oct. 17, 1901.

14. Ibid., Nov. 25, 1903.

15. George McLeod, "Story of the Hammond Lumber Company," FHS, 25.

16. Roy, *Socializing Capital,* 57, 151–55; Porter, *Rise of Big Business,* 74; Lamoreaux, *Great Merger Movement,* 2; Nace, *Gangs of America,* 79–80; Hammond to McLeod, Dec. 26, 1905, 17/1.

17. Williams, *Americans and Their Forests,* 5; Prudham, *Knock on Wood,* 91; Buckley, "Building the Redwood Region," 81; Van Tassell, *Mechanization in the Lumber Industry,* xvii.

18. Chandler, *Visible Hand,* 2–12, quote on 12.

19. Chandler, *Visible Hand,* 2.

20. Prudham, *Knock on Wood,* 86–87.

21. Ibid., 102–103.

CHAPTER 12

1. *Columbia River and Oregon Timberman,* May 1900, 14, quote from Nov. 1899, 9. The Nov. 1899 issue reported 150,000 and 300,000 acres, respectively, for Hammond and Willamette Pulp and Paper, but these figures seem highly inflated. Ficken, "Weyerhaeuser," 146–54.

2. Hidy et al., *Timber and Men,* 207, quote on 213.

3. Ficken, *Forested Land,* 94.

4. *Columbia River and Oregon Timberman,* Nov. 1899, 9.

5. *Pacific Lumber Trade Journal,* Apr. 1899, 22; HLC files, box 8, file 562. OHS.

6. Timber and Stone Act, 1878; *GLO Report,* 119; *Report of the Public Lands Commission,* vi.

7. Puter and Stevens, *Looters of the Public Domain;* HLC files, box 3, box 4 title abstracts; Astoria Company to R. B. Montague, Apr. 4, 1901, Weatherford papers, box 12.

8. Donaldson, *Public Domain,* 1082; Ise, *United States Forest Policy;* Libecap and Johnson, "Property Rights," 129–42.

9. Bureau of Corporations, *Lumber Industry,* part 2, 7, 223, 228. Southern Pacific's percentage was greatly reduced by forfeiture. In addition, Hammond, because of his financial relationship with the Southern Pacific, probably enjoyed access to the railroad's timber.

10. Bureau of Corporations, *Lumber Industry,* 57; HLC files, box 8 contracts; *Oregonian,* Jan. 5, 1901, 1; Deed no. 448-E, HLC files, box 7, OHS.

11. *Report of the Secretary of the Interior,* 1903, 58th Cong., 2d Sess., House Document, vol. 18.

12. *Report of the Public Lands Commission,* Senate Doc. 189, 266; Warranty Deed 31828, box 9, HLC files, OHS; box 4, HLC files; bill of sale no. 768, box 8, HLC files; Bureau of Corporations, *Lumber Industry,* part 1, 26, 96; Smith quoted in *American Lumberman,* Nov. 12, 1910, 79, from Bureau of Corporations, *Lumber Industry,* part 1, 38; George McLeod to Weatherford, Nov. 4, and Nov. 21, 1901, Weatherford papers, box 3.

13. Rakestraw, *History of Forest Conservation,* 12, quote on 59.

14. Forest Management Act, June 4, 1897; Ise, *United States Forest Policy,* 176; McLeod to Weatherford, Nov. 4, 1901, quote from Nov. 21, 1901, Weatherford papers.

15. Ise, *United States Forest Policy,* 176.

16. General Land Office Records, land patents search, www.glorecords.blm/gov; Ise, *United States Forest Policy,* 180–82; George McLeod, "Story of the Hammond Lumber Company," UM, 29.

17. Bureau of Corporations, *Lumber Industry,* 226; Ise, *United States Forest Policy,* 184–85; Hidy et al., *Timber and Men,* 250.

18. Ise, *United States Forest Policy,* 184–85; Rakestraw, *History of Forest Conservation,* 159.

19. Robbins, *Hard Times in Paradise,* 31, quote on 32.

20. Puter and Stevens, *Looters of the Public Domain,* 44, 23, 32.

21. Ibid., 35–44.

22. Ibid., 362–64.

23. Ibid. In what would now be considered a conflict of interest, in the nineteenth century it was not unusual for lawyers holding public office to represent private clients.

24. McLeod to Weatherford, Nov. 11, 1911, HLC files, folder 3, OHS; Puter and Stevens, *Looters of the Public Domain,* 148, quote on 50.

25. Puter and Stevens, *Looters of the Public Domain,* 52–65.

26. Ibid., 66, 304.

27. Ibid., 298–99; Mengel, "A History of the Samoa Division, 28–29.

28. *Humboldt Times,* June 19, 1900, 6. Mengel, "A History of the Samoa Division," 32; George McLeod, "Story of the Hammond Lumber Company," FHS, 24.

29. George McLeod, "Story of the Hammond Lumber Company, FHS, 24.

30. Puter and Stevens, *Looters of the Public Domain,* 300.

31. Ibid.; *Humboldt Times,* Sep. 6, 1900, 4.

32. Puter and Stevens, *Looters of the Public Domain,* 68.

33. Ibid., 67; McLeod to Ware, Jan. 25, 1902, HLC files, Deeds, OHS.

34. Puter and Stevens, *Looters of the Public Domain,* 81.

35. Ibid., 80–88.

36. Ibid., 88.

37. John Holland to George McLeod, Jan. 4, 1901, Weatherford papers, box 3, HLC files, OHS, emphasis in original.

38. Puter and Stevens, *Looters of the Public Domain,* 89.

39. See Hays, *Conservation and the Gospel of Efficiency.*

40. *GLO Report,* 1897, 79.

41. Messing, "Public Lands," 35–66; Ise, *United States Forest Policy,* 182; Hermann, "Shadows in Public Life," OHS.

42. *Report of the Secretary of the Interior,* 1903, 13; Messing, "Public Lands."

43. Puter and Stevens, *Looters of the Public Domain,* 94, 96, 137.

44. Messing, "Public Lands"; Puter and Stevens, *Looters of the Public Domain,* 125–26, 138, 146.

45. McLeod to Fulton, Sept. 21, 1903, HLC files 561–131, OHS; Puter and Stevens, *Looters of the Public Domain,* 157.

46. Puter and Stevens, *Looters of the Public Domain,* 7, 172–74, 452.

47. Land fraud investigation figures from James L. Penick, Ethan Allen Hitchcock entry in American National Biography Online, www.anb.org; Puter and Stevens, *Looters of the Public Domain,* 445–48.

48. Quote from Puter and Stevens, *Looters of the Public Domain,* 179; *Oregonian,* Jan. 1, 1905, 4.

49. Puter and Stevens, *Looters of the Public Domain,* 182; quotes from *Oregonian,* Jan. 1, 1905, 4.

50. *Report of the Secretary of the Interior,* 1903, 13.

51. *Oregonian,* Jan. 1, 1905, 1.

52. Messing, "Public Lands;" *Oregonian,* June 29, 1905.

53. Puter and Stevens, *Looters of the Public Domain,* 227, 452; Messing, "Public Lands."

54. Puter and Stevens, *Looters of the Public Domain,* 445–48, 10.

55. Weatherford to W. S. Burnett, Dec. 16, 1910, HLC files, folder 2, OHS.

56. Hibbard, *History of the Public Land Policies,* 469.

57. *Report of the Public Lands Commission,* 266; quote from Bureau of Corporations, *Lumber Industry,* part 1, xviii.

58. Hidy et al., *Timber and Men,* 296.

59. Ibid., quote on 296.

60. Letter addressed to 1907 Public Lands Convention, quoted in Puter and Stevens, *Looters of the Public Domain,* 461.

61. Congressional Record, Feb., 18, 190, quoted in Ise, *United States Forest Policy,* 197; *Columbia River and Oregon Timberman,* May 1904, 1.

62. Ise, *United States Forest Policy,* 197–200.

63. Cornwall quote from *Columbia River and Oregon Timberman,* Oct. 1903, 1; Lane quote from Puter and Stevens, *Looters of the Public Domain,* 456.

64. Messing, "Public Lands," 63; Puter and Stevens, *Looters of the Public Domain,* 463.

65. Bureau of Corporations, *Lumber Industry,* 96; Rakestraw, *History of Forest Conservation,* 277, 319.

66. Rakestraw, *History of Forest Conservation,* 276; Jenson, *Lumber and Labor,* 24.

CHAPTER 13

1. Palais and Roberts, "History of the Lumber Industry," 1–14; Mengel, "History of the Samoa Division," 2–15; Carranco, *Redwood Lumber Industry,* 145.

2. Palias and Roberts, "History of the Lumber Industry."

3. *Humboldt Times,* Jan. 27, 1892.

4. Melendy, "One Hundred Years," 20; Gates, *Falk's Claim,* 38; Carranco, *Redwood Lumber Industry,* 145.

5. Melendy, "One Hundred Years," 281, 322.

6. Cornford, *Workers and Dissent,* 79.

7. Buckley, "Building the Redwood Region," 73.

8. For specific examples see Mengel, "History of the Samoa Division," 16–26.

9. Buckley, "Building the Redwood Region," 91; quote from Sklar, *Corporate Reconstruction of American Capitalism,* 13.

10. Sacks, *Carson Mansion and Ingomar Theater,* 18.

11. Palais and Roberts, "History of the Lumber Industry"; Melendy, "Two Men and a Mill," 61–71.

12. Sacks, *Carson Mansion and Ingomar Theater,* 18; quote from Buckley, "Building the Redwood Region," 137, 127.

13. Melendy, "Two Men and a Mill"; Mengel, "History of the Samoa Division," 19; Cornford, *Workers and Dissent,* 101–12.

14. Cornford, *Workers and Dissent,* 101–102; Melendy, "One Hundred Years," 322; quote from transcript of speech by Wallace Martin, Nov. 28, 1961, Humboldt County Historical Society; Cox, *Lumberman's Frontier,* 204. The original Labor Day as recognized by organized labor was on May 1.

15. McLeod to Hammond, Apr. 7, 1906, 17/3; McLeod to Hammond, May 3, 1907, 17/4; Hammond to McLeod, Aug. 25, 1909, 18/2.

16. Hammond to McLeod, Oct. 7, 1905, 17/1.

17. Hammond to McLeod, Apr. 12, 1907, 17/4.

18. Chandler, *Visible Hand;* Porter, *Rise of Big Business;* Klein, *Genesis of Industrial America.*

19. *GLO Report*, 1888, 55, 1886, 43, 95; Dunham, *Government Handout*, 264–66; Melendy, "One Hundred Years," 85–95; Puter and Stevens, *Looters of the Public Domain*, 18; Palias and Roberts, *History of the Lumber Industry;* Cornford, *Workers and Dissent*, 65–66.

20. Cornford, *Workers and Dissent*, 29–31; quote from *Humboldt Times*, July 29, 1890, 2.

21. Puter and Stevens, *Looters of the Public Domain*, 300; *Arcata Union*, Sept. 1, 1900; quote from *Blue Lake Advocate* (Arcata, Calif.), Sept. 8, 1900, Fountain papers, 47:45; *Columbia River and Oregon Timberman*, Sept. 1900, 5.

22. Melendy, "One Hundred Years," 90–95; quote from *Humboldt Times*, Nov. 1, 1900, 6.

23. In 1908 Merrill and Ring sold their half interest to Hill-Davis, a Weyerhaeuser concern, for $800,000, making Hammond and Weyerhaeuser joint owners.

24. Mengel, "A. B. Hammond Built"; *Los Angeles Times*, May 21, 1902, 3. The value of redwoods was recognized so early that nearly every grove ended up in private hands. Neither the railroads nor the government owned much in the way of redwoods, in contrast to the rest of the Pacific Northwest timberlands.

25. Kolko, *Triumph of Conservatism*, 71–75, quotes on 75, 131, 132–33.

26. Bureau of Corporations, *Lumber Industry*, part 2, 109, quote on 155.

27. Ibid, part 1, 39.

28. Clark, *Mill Town*, 69.

29. Sklar, *Corporate Reconstruction*, 58.

30. *Humboldt Times*, Mar. 20, 1902.

31. Harpster to Hanify, Mar. 22, 1904, Mar. 12, 1904, letterbooks, vol. 21, 76, ERLC.

32. Harpster to Hanify, May 12, 1904, vol. 21, 190.

33. *Timberman*, Feb. 1905, 40.

34. *Humboldt Times*, Apr. 13, 1902, July 15, 1902.

35. Ibid., Apr. 18, 1902, Oct. 28, 1903, 3; Hammond to Thomas Hubbard, July 7, 1903, HEH 5520, box 97.

36. *Columbia River and Oregon Timberman*, Nov. 1900, 3, Oct. 1902, 15.

37. *Humboldt Times*, Dec. 25, 1903, May 12, 1902.

38. *Pacific Lumber Trade Journal*, June 1901, 30, quote from July 1901, 22.

39. Hammond to Huntington, Feb. 3, 1901, HEH 2391, box 43.

40. Melendy, "One Hundred Years," 183–95; Mengel, "History of the Samoa Division," 46–49.

41. Hammond to Huntington, Feb. 14, 1903, HEH 5511, box 97; quote from Hammond to Huntington, Sept. 3, 1903, HEH 2436, box 44.

42. Hammond to Huntington, Nov. 13, 1902, Sept. 3, 1903, box 44.

43. *Humboldt Standard*, Jan. 10, 1903; quote from Hammond to Huntington, Feb. 14, 1903, HEH 5511, box 97.

44. *Humboldt Times,* Apr. 11, 1903:3.

45. Sklar, *Corporate Reconstruction,* 167.

46. Ibid., 56; Palais and Roberts, "History of the Lumber Industry."

47. William Carson to William Mugan, May 13, 1905, vol. 23, 341, quote from July 8, 1910, vol. 26, 121, Feb. 28, 1908, vol. 25, 223, D&C, letterbooks.

48. Hanify to Hapster, Jan. 10, 1905, ERRC, letterbooks, vol. 22, 475.

49. Cox, *Mills and Markets,* 273–77; Melendy, "One Hundred Years," 289; Robbins, *Lumberjacks and Legislators,* 7.

50. Robbins, *Lumberjacks and Legislators;* Sacks, *Carson Mansion and Ingomar Theater,* 16; George McLeod, "Story of the Hammond Lumber Company," FHS, 39; Melendy, "One Hundred Years," 290.

51. Carson to Mugan, Dec. 17, 1904; Dec. 9, 1905, vol. 24, 98; quotes from Dec. 30, 1904, Jan. 5, 1905, letterbooks, vol. 23, 125; and Mar. 9, 1905, vol. 23, 246.

52. Carson to Mugan, quotes from Dec. 30, 1904, vol. 23, 125, Dec. 17, 1904, vol. 23, 125.

53. In 1900 the firm owned 16,000 acres, and their Bay Mill was assessed at $1.3 million. Palais and Roberts, *History of the Lumber Industry.*

54. Carson to Mugan, Mar. 13, 1906, vol. 24, 213. For more on the Sherman Antitrust Act see Sklar, *Corporate Reconstruction.*

55. Harpster to Hanify, quote from Mar. 31, 1904, vol. 21, 90; Mar. 11, 1905, vol. 23, 123.

56. Harpster to Hanify, Nov. 17, 1904, vol. 22, 338.

57. Harpster to Hanify, quotes from Aug. 4, 1904, vol. 22, 91, and Aug. 16, 1904, vol. 22, 112.

58. Carson to Mugan, Feb. 28, 1908, vol. 25, 223, Mar. 31, 1908, vol. 25, 441, quote from Mar. 21, 1906, vol. 24, 226; Hammond to Tyson, May 23, 1912, file 307.

59. Quote from Sklar, *Corporate Reconstruction,* 172; Williams, *Americans and Their Forests,* 221. See Kolko, *Triumph of Conservatism.*

CHAPTER 14

1. Stanley Borden, "The Northwest Pacific Railroad," *Humboldt Historian,* no. 5, 1961; *Humboldt Times,* Apr. 13, 1900, Apr. 26, 1900; Mengel, "History of the Samoa Division," 26.

2. *Columbia River and Oregon Timberman,* Sept. 1900, 5; Borden, "Northwest Pacific Railroad"; Fountain papers, 104:451.

3. *Eureka and Klamath v. California and Northern Railway,* transcript of settlement, Oct. 22, 1901, 188–218, HSU; Borden, "Northwest Pacific Railroad."

4. *Eureka and Klamath v. California and Northern Railway;* Stindt and Dunscomb, *Northwestern Pacific Railroad,* 41.

5. *Blue Lake Advocate* (Arcata, Calif.), Mar.16, 1901, Fountain papers, 50:193, HSU.

6. *Humboldt Times,* May 7, 1903; *Columbia River and Oregon Timberman,* Sept. 1900, 5.

7. Borden, "Northwest Pacific Railroad"; Carranco, *Redwood Lumber Industry,* 133; *Humboldt Times,* May 12, 1903, 1; Hammond to Huntington, Apr. 24, 1903, HEH 2425, box 44.

8. Hammond to Huntington, Apr. 24, 1903, HEH 2425, May 12, 1903, HEH 2428, box 44.

9. Friedricks, *Henry E. Huntington,* 46, 7. While Frank Norris applied the octopus analogy to the title of his 1901 novel about the Southern Pacific (*The Octopus: A Story of California*), no doubt it had been in popular usage previously. As early as 1891, Hammond had been called "the Missoula Octopus" for his economic domination of western Montana.

10. Hammond to McLeod, May 24, 1906.

11. See Mills, *Power Elite;* Friedricks, *Henry E. Huntington,* 135–36; *Los Angeles Times,* Feb. 20, 1904, 6.

12. Melendy, "One Hundred Years," 37.

13. Melendy, "One Hundred Years," 177; Mengel, "History of the Samoa Division," 12; Carranco, *Redwood Lumber Industry,* 60–69.

14. Melendy, "One Hundred Years," 39; Carranco, *Redwood Lumber Industry,* 18.

15. Stindt and Dunscomb, *Northwestern Pacific Railroad,* 48; Hammond to Huntington, May 29, 1903, HEH 2430, box 44.

16. Mengel, "History of the Samoa Division," 26; *Humboldt Times,* Nov. 25, 1905; Fountain papers, 70:534; George Murray to William McGillivary, June 15, 1906, file 17, LRRC files.

17. Dolbeer to McGillivary, June 16, 1906, D&C company files; freighting agreement, file 63, LRRC files.

18. Correspondence between Hammond and Crannell, Nov. 26, 1906, Apr. 6, 1907, May 2, 1907, file 1, LRCC.

19. Humboldt Northern and LRRC agreement, July 5, 1907, file 64, LRRC.

20. Robbins, *Colony and Empire,* 104; Melendy, "One Hundred Years," 289; *Los Angeles Times,* Dec. 15, 1915, part 2, 7. In 1915 the shipping rate from Oregon to Southern California remained at $4.50 per mbf.

21. Cox, "Single Decks and Flat Bottoms," 65–74; Mengel, "History of the Samoa Division," 29; Bill Stoddard, "The Hammond Navy" (manuscript, author's possession); *Columbia River and Oregon Timberman,* Dec. 1901, 15.

22. *Los Angeles Times,* Mar. 24, 1902, 3. George McLeod, "Story of the Hammond Lumber Company," 33, FHS.

23. *Humboldt Times,* Mar. 27, Dec. 25, Dec. 30, 1903; Hammond to C. H. McLeod, Jan. 30, 1906, 17/2.

24. Cox, "Single Decks and Flat Bottoms."

25. "Ships of the Hammond Navy" complied by Jack Blanchard, author's possession; Dan Strite, "45th Annual Report of Board of Harbor Commissioners" (MSS copy, author's possession).

26. *Los Angeles Times,* May 16, 1913, part 2, 9; Hammond to C. H. McLeod, Jan. 22, 1912, 19/1; quote from Hammond to McLeod, Feb. 5, 1916, 20/3.

27. Meany, "History of the Lumber Industry," 253; Adams, "Blue Water Rafting," 16–27.

28. Adams, "Blue Water Rafting;" Stoddard, "The Hammond Navy," 6.

29. *Pacific Lumber Trade Journal,* Dec. 1906, 24; Jack Blanchard log rafts file, author's possession.

30. *Los Angeles Times,* Sept. 11, 1911, part 1, 3; Hammond to C. H. McLeod, Apr. 16, 1912, 19/1; George McLeod, "Story of the Hammond Lumber Company," 8, FHS.

31. Melendy, "One Hundred Years," 310; Buckley, "Building the Redwood Region," 637–38.

32. Hammond to McLeod, Oct. 9, 1906, 17/3.

33. Buckley, "Building the Redwood Region," 422–31; Knight, *Industrial Relations,* 375; *Los Angeles Times,* Oct. 30, 1903, part A, 7.

34. Carson to Mugan, Mar. 27, 1905, D&C records, vol. 23, 266.

35. George McLeod, "Story of the Hammond Lumber Company," 33, UM, 7.

36. George McLeod, "Story of the Hammond Lumber Company," FHS, 25; Whiting, *Autobiography of Perry Whiting,* 166–73, quote on 173.

37. Whiting, *Autobiography of Perry Whiting,* 167.

38. Ibid., 215.

39. McLeod to Hammond, Oct. 17, 1905, Hammond to McLeod, Mar. 22, 1906, 17/2.

40. George McLeod, "Story of the Hammond Lumber Company, 29–32, FHS.

41. Ibid., 31–32; Hammond to McLeod, May 12, 1906, 17/2.

42. Quotes from Hammond to McLeod, May 12, 1906; Hammond to John Claflin, May 24, 1906, McLeod papers, 17/2.

43. Hammond to Thomas Hubbard, May 14, 1906, Hammond to McLeod, May 24, 1906, McLeod to Hammond, May 24, 1906, McLeod papers, 17/2.

44. McLeod to Hammond, Nov. 28, 1905, 17/1; quote from Hammond to McLeod, May 28, 1906, 171.

45. George McLeod, "Story of the Hammond Lumber Company," 27, FHS.

46. Hammond to McLeod, quote from June 20, 1907; Aug. 5, 1907, 17/4.

47. C. H. McLeod to Hammond, Jan. 13, 1908; Hammond to McLeod, Jan. 13, 1907, Aug. 5, 1907, 17/4.

48. Hammond to McLeod, Feb. 20, 1906; quote from Whiting, *Autobiography of Perry Whiting,* 216.

49. Dan Strite to Bill Stoddard, June 14, 1983, author's file.

50. Hammond to C. H. McLeod, quotes from Jan. 4, 1908; Jan. 6, 1908, 17/6.

51. *Astoria Daily Budget,* June 17, 1908; *Timberman,* Nov. 1909, 39; Strite, "Growth Rings" MSS, author's possession; George McLeod, "Story of the Hammond Lumber Company," FHS, 11.

52. Quote from Hammond to J. S. Alexander, Feb. 11, 1909, 18/1; Nearing, *Wages in the United States,* 144.

CHAPTER 15

1. Alaback, "Biodiversity Patterns in Relation to Climate: The Coastal Temperate Rainforests of North America," in Lawford et al., eds., *High-Latitude Rainforests.*

2. Meany, "History of the Lumber Industry," 8–9; Evarts and Popper, eds., *Coast Redwood,* 14; Agee, *Fire Ecology,* 34; Alaback, "Biodiversity Patterns."

3. Hobbs et al., *Forest and Stream Management,* 37–38; Agee, *Fire Ecology,* 128–29; Tennsma et al., *Preliminary Reconstruction and Analysis,* 52. In this discussion, I focus on the forest ecology of the Pacific Northwest. Douglas fir, a widespread and generalist species, exhibits very different characteristics in the Rocky Mountain region. Ecologists Jerry Franklin and Thomas Spies developed a working definition of old-growth as a forest containing "multiple structural characteristics," including large, old trees, snags, down logs and multiple canopies. See Jerry Franklin and Thomas Spies, "Ecological Definitions of Old-Growth Douglas Fir Forests," in *Wildlife and Vegetation of Unmanaged Douglas Fir Forests* (USDA, PNW-GTR-285, May 1991.)

4. Meany, "History of the Lumber Industry," 8–9; Alaback, pers. comm., Nov. 4, 2007; Green and Franklin, "Middle Santiam Research Natural Area;" Poage and Tappeiner, "Long-Term Patterns of Diameter and Basal Area," 1232–43.

5. Meany, "History of the Lumber Industry," 8–9; quote by Fred Lockley in *Pacific Monthly,* 1908, as quoted in Robbins, *Landscapes of Promise,* 234.

6. Meany, "History of the Lumber Industry," 16, 18; Donan, *Columbia River Empire,* 38; Franklin, "Natural Regeneration of Douglas Fir"; Walker, *North American Forests,* 289, 312.

7. Agee, *Fire Ecology,* 205–209; Tennsma et al., *Preliminary Reconstruction and Analysis,* 25–32.

8. Agee, *Fire Ecology,* 205–209; Tennesma et al., *Preliminary Reconstruction and Analysis,* 7–52.

9. Agee, *Fire Ecology,* 324, 189–90. Agee shows that fire-return intervals can vary widely even within forest types. For example, ponderosa pine forests show a fire-return interval ranging from 1.9 to 16 years. Tennsma et al., *Preliminary Reconstruction and Analysis,* 25.

10. Agee, *Fire Ecology,* 59.

11. G. A. McBean, "Factors Controlling the Climate of the West Coast of North America" in Lawford et al., eds., *High-Latitude Rainforests;* Noss, ed., *Redwood Forest,* 8–25; Evarts and Popper, eds., *Coast Redwood,* 14–17.

12. Noss, ed., *Redwood Forest,* 103–105, quote on 105; Dawson, "Fog in the California Redwood Forest," 476–85.

13. Evarts and Popper, eds., *Coast Redwood,* 9; Noss, ed., *Redwood Forest,* 27, 111–15; Brown and Baxter, "Fire History in Coast Redwood Forests," 147–58.

14. Noss, ed., *Redwood Forest,* 86, 112.

15. Evarts and Popper, eds., *Coast Redwood,* 26, 59.

16. Ibid., 24; Melendy, "One Hundred Years" 13, 21; Carranco, *Redwood Lumber Industry,* 39; Noss, ed., *Redwood Forest,* 91.

17. Quoted in Francois Leydet, *Last Redwoods,* 77.

18. Cherry, *Redwood and Lumbering in California Forests,* 34.

19. Noss, ed., *Redwood Forest,* 95–97.

20. Ibid., 35, 56–57, 141, quote on 146.

21. Louma, *Hidden Forest,* 96–123; Barnes et al., *Forest Ecology,* 384.

22. Noss, ed., *Redwood Forest,* 155, 161; Slauson et al., "Distribution and Habitat Associations of the Humboldt Marten," 3.

23. Boyd, *Indians, Fire and the Land,* 8, 167; quote from Noss, ed., *Redwood Forest,* 177.

24. Melendy, "One Hundred Years," 31; Holbrook, *Holy Old Mackinaw,* 164, quote on 173.

25. Melendy, "One Hundred Years," 30–33; Franklin et al., *Ecological Characteristics,* 76.

26. Melendy, "One Hundred Years," 34.

27. White, *Land Use, Environment and Social Change,* 88, 109; Carranco, *Redwood Lumber Industry,* 55.

28. White, *Land Use, Environment and Social Change,* 91; Rajala, *Clearcutting the Pacific Rainforest,* 91; Wright and Isaac, "Decay following Logging Injury."

29. Noss, ed., *Redwood Forest,* 16; Meany, "History of the Lumber Industry," 258; Robbins, *Hard Times in Paradise,* 64.

30. Noss, ed., *Redwood Forest,* 190; Louma, *Hidden Forest,* 186.

31. Noss, ed., *Redwood Forest,* 194–96.

32. Garrison and Rummell, "First-Year Effects of Logging," 708–13.

33. Melendy, "One Hundred Years," 33; White, *Land Use, Environment and Social Change,* 87.

34. Williams, *Americans and Their Forests,* 303.

35. Melendy, "One Hundred Years," 44–47.

36. Cox, *Mills and Markets,* 233.

37. Rajala, *Clearcutting the Pacific Rainforest,* 91; Murphy et al., *Hinton Forest;* Cox, *Mills and Markets,* 232.

38. Melendy, "One Hundred Years," 48–49, 200.

39. Carranco, *Redwood Lumber Industry,* 69.

40. Rajala, *Clearcutting the Pacific Rainforest,* 99.

41. White, *Land Use, Environment and Social Change,* 88, 109.

42. Garrison and Rummell, "First-Year Effects of Logging." Their study of logging in Oregon's ponderosa pine forest calculated total forest disturbance from horse logging at 17 percent, as compared with 30 percent for cable logging. Rajala, *Clearcutting the Pacific Rainforest,* 91.

43. Williams, *Americans and Their Forests,* 315–16, quote on 257.

44. Ibid.; Rajala, *Clearcutting the Pacific Rainforest,* 35; quote from *Columbia River and Oregon Timberman,* Dec. 1904, "The Evolution of Coast Logging," 21–24.

45. Holbrook, *Holy Old Mackinaw,*" 186.

46. Isaac and Hopkins, "Forest Soil of the Douglas Fir Region," 264–79.

47. Ibid.

48. Cissel et al., "Landscape Plan Based on Historical Fire Regimes," 20; Franklin et al., *Ecological Characteristics.*

49. Rajala, *Clearcutting the Pacific Rainforest,* 92–115, quote on 110.

50. Franklin, *Natural Regeneration of Douglas Fir;* Twight, *Ecological Forestry for the Douglas Fir Region,* 5; Rajala, *Clearcutting the Pacific Rainforest,* Munger quotes on 108, 117.

51. Rajala, *Clearcutting the Pacific Rainforest,* 91, 94, 118.

52. Ibid., 3–19.

CHAPTER 16

1. *Humboldt Standard,* Oct. 27, 1903.

2. Ibid.

3. Dubofsky, *Industrialism and the American Worker,* 37–38.

4. Much of my assessment of Humboldt County's early labor history is based on Daniel Cornford's excellent account, *Workers and Dissent in the Redwood Empire,* and his Ph.D. dissertation, "Lumber, Labor, and Community in Humboldt County, CA, 1850–1920," quote on 438.

5. *Humboldt Times,* Mar. 29, 1905; Cornford, *Workers and Dissent,* 135–55.

6. Cornford, *Workers and Dissent,* 150.

7. Cornford, *Workers and Dissent,* 137, 148, 184; Cornford, "Lumber, Labor, and Community," 426–27; quote from Dubofsky, *Industrialism and the American Worker,* 117.

8. Weintraub, *Andrew Furuseth,* 7, 14–15.

9. Ibid., 35.

10. Hammond to Elizabeth Clements, Oct. 26, 1918, C. H. McLeod papers, box 22, file 7.

11. Kazin, *Barons of Labor,* 53; *Labor News,* July 23, 1910, 1.

12. Kazin, *Barons of Labor,* 50–56; Sailor's Union of the Pacific website, www.sailors.org/history.

13. Weintraub, *Andrew Furuseth,* 72–74.

14. *Oakland Tribune,* June 2, 1906, 19

15. Ibid.

16. *Labor News* (Eureka), June 23, 1906, 1.

17. *Los Angeles Times,* Aug. 18, 1907, part 3, 1; *Labor News,* Nov. 10, 1906, 1.

18. Nace, *Gangs of America,* 125; Weintraub, *Andrew Furuseth,* 77, Furuseth quote on 97.

19. Weintraub, *Andrew Furuseth,* 76.

20. Hammond to McLeod, Oct. 9, 1906, 17/6.

21. Weintraub, *Andrew Furuseth,* 76; *Labor News,* Feb. 2, 1907, 8; quote supplied by Judge William Denman in *Symposium on Andrew Furuseth,* 16.

22. *Los Angeles Times,* Aug. 18, 1907, part 3, 1; *Labor News,* Apr. 20, 1907, 4; Carson to Mugan, Nov. 12, 1906, vol. 24, 417, D&C resords; Carson quote from Carson to Mugan, Sept. 4, 1906, vol. 24, 369, D&C records.

23. *Los Angeles Times,* Aug. 18, 1907:III1; *Labor News,* Oct. 2, 1909:2.

24. *Labor News,* Jan. 19, 1907, 2.

25. Ibid., June 6, 1909, 4.

26. Ibid., Feb. 2, 1907, 8.

27. *Los Angeles Times,* Aug. 18, 1907, part 3, 1; *Labor News,* Apr. 13, 1907, 1; Hammond to James Rolph, Mar. 29, 1917, 20/4.

28. *Labor News,* June 12, 1909, 3.

29. Harpster to Hanify, Apr. 29, 1907, vol. 26, 328, ERLC.

30. Hammond to C. H. McLeod, Feb. 28, 1908, 17/6.

31. *Humboldt Times,* June 2, 1909, 4.

32. Ficken, *Forested Land,* 31.

33. *Humboldt Times,* Jan. 27, 1903, May 17, 1903, Oct. 28, 1903.

34. *Humboldt Times,* Oct. 28, 1903.

35. *Humboldt Times,* Jan. 15, 1904.

36. *Humboldt Times,* Oct. 28, 1903, Jan. 1, 1904; *Arcata (Calif.) Union,* Aug. 12, 1905.

37. *Labor News,* July 28, 1906, 2.

38. Howd, *Industrial Relations,* 41–42; Jenson, *Lumber and Labor,* 106; quote from American Red Cross report to Pacific Logging Congress 1917, 37, in Prouty, *More Deadly than War,* 31.

39. Meany, "History of the Lumber Industry," 331; quote from Prouty, *More Deadly than War,* 27; Rader, "Montana Lumber Strike of 1917," 189–207; *Labor News,* June 20, 1908, 1.

40. Cornford, *Workers and Dissent,* 27.

41. Hammond to McLeod, Jan. 30, 1915, 20/2.

42. Quote from "Proceedings from Second Annual Session Pacific Logging Congress," 1910, 28, in Cornford, *Workers and Dissent,* 199.

43. Cornford, *Workers and Dissent,* 153; *Labor News,* May 11, 1907, 1; Hammond to McLeod, July 13, 1908, 17/6. The name of Johnson's firm was no reflection of his labor practices.

44. Cornford, *Workers and Dissent;* 157; Howd, *Industrial Relations,* 42; *Labor News,* Feb. 16, 1907, 1, Mar. 6, 1907, 1; William Carson to J. R. Hanify, Jan. 11, 1905, vol. 23, 141, ERLC; Meany, "History of the Lumber Industry," 281.

45. Carson to Hanify, Jan. 11, 1905, vol. 23, 141–42, ERLC.

46. Harpster to Hanify, Apr. 10, 1905, vol. 23, 169, ERLC.

47. Harpster to Hanify, Apr. 15, 1905, vol. 23, 192, ERLC.

48. *Labor News,* Apr. 14, 1906, 2, May 5, 1906, 2.

49. *Labor News,* June 1, 1907, 2.

50. Payroll records, box 8, HLC files, MSS 1716, OHS.

51. Cornford, *Workers and Dissent,* 158–60.

52. *Blue Lake Advocate,* May 4, 1907; Cornford, *Workers and Dissent,* 164, quote on 163.

53. Cornford, *Workers and Dissent,* 164; *Labor News,* Feb. 16, 1907, 1; quote from Ernest Pape to Little River Lumber Company, Apr. 15, 1907, file 42, LRRC files.

54. Cornford, *Workers and Dissent,* 164.

55. *Labor News,* Feb. 28, 1907:5, Mar. 9, 1907, 2; *Humboldt Times,* May 1, 1907.

56. *Humboldt Times,* Fenwick quote from May 1, 1907, 1; May 2, 1907, 1; *Times* quote from May 3, 1907, 1; May 7, 1907, 1.

57. Ibid., May 1, 1907:1.

58. Ibid., quote from May 1, 1907, 1, May 2, 1907, 1, May 3, 1907, 5.

59. Ibid., May 4, 1907, 1; *Labor News,* May 4, 1907, 1, May 11, 1907, 4.

60. Carson to William Mugan, Apr. 27, 1907, letterbooks vol. 25, 19, D&C records.

61. Carson to San Francisco office, Apr. 30, 1907, vol. 25, 22, D&C records.

62. Carson to Mugan, May 7, 1907, vol. 25, 22, D&C records.

63. *Humboldt Times,* May 5, 1907, 1.

64. Carson to Mugan, May 7, 1907, vol. 25, 22, D&C records.

65. Harpster to Hanify, May 2, 1907, vol. 26, 323, ERLC records.

66. Harpster to Hanify, Apr. 9, 1907, vol. 26, 267, ERLC records.

67. *Labor News,* May 11, 1907, 1, 6; quote from *Humboldt Times,* May 11, 1907, 4.

68. *Labor News,* May 11, 1907, 1.

69. Carson to Mugan, May 13, 1907, vol. 25, 29, D&C records.

70. *Humboldt Times,* May 3, 1907, 1; *Labor News,* Apr. 27, 1907, 2.

71. *Labor News,* May 18, 1907, 1; Carson to Mugan, June 1, 1907, vol. 25, 40, D&C records.

72. Carson to Mugan, May 28, 1907, vol. 25, 35, D&C records; *Humboldt Times,* May 21, 1907, 1, May 22, 1907, 4, May 24, 1907, 1, June 4, 1907, 1.

73. *Labor News,* June 8, 1907, 12.

74. *Labor News,* June 12, 1909, 7.

CHAPTER 17

Epigraph. Abraham Lincoln, "Annual Address before the Wisconsin State Agricultural Society, 1859," in *The Complete Works of Abraham Lincoln,* vol. 5, eds. John Nicolay and John Hay (Ithaca, N.Y.: Cornell University Library, 2009), 249.

1. Kolko, *Triumph of Conservatism.* Progressivism and Kolko's thesis are the subject of seemingly endless debate among historians. Compelling alternatives to Kolko include Wiebe, *Search for Order,* Link and McCormick, *Progressivism,* Rodgers, *Atlantic Crossings,* and McGerr, *Fierce Discontent;* Dubofsky, *Industrialism and the American Worker,* 113.

2. Hammond to Elizabeth Clements, Oct. 26, 1918, 20/7.

3. *Labor News,* Feb. 15, 1908, 1. For more on the Union Labor Hospital, see Claasen, *A Card for All Seasons.*

4. Classen, *Card for All Seasons,* 20; *Labor News,* Jan. 16, 1909, 3; quote from Prouty, *More Deadly than War,* xxvii.

5. See Brandes, *American Welfare Capitalism.* Many otherwise conservative lumbermen also endorsed state workers' compensation plans.

6. *Labor News,* July 4, 1908, 1, Oct. 10, 1908, 4, Dec. 5, 1908, 1. While the *Labor News* certainly displayed an editorial bias, it published the transcripts verbatim from the hearing of June 22, 1908, held in the Superior Court of Humboldt County. Much of the evidence I use on the Union Labor Hospital comes from the testimony of George Fenwick and others. In 1907, Pacific Lumber had built their own hospital and imposed the $1 deduction beginning in January 1908. Classen, *A Card for All Seasons,* 14.

7. HLC payroll records, Jan. 1914, 1921, box 8, MSS 1716, OHS.

8. *Labor News,* June 20, 1908, 1; HLC payroll, Jan. 1914, OHS.

9. *Labor News,* July 4, Oct. 10, 1908, 4, Dec. 5, 1908, 1; July 25, Oct. 10, 1; jury quote Nov. 14, 1908, 1, Bayside quote Nov. 28, 1908, 1, emphasis in original.

10. Quote from *Labor News,* June 12, 1909, 6; July 17, 1909, 1; file 963, Sequoia Hospital, LRRC files.

11. *Humboldt Times,* June 1, 1909, 4; quotes from June 2, 1908, 4; May 30, 1908, 1; union label league quote from June 5, 1908, 4–6.

12. Ibid., May 5, 1908, 1.

13. *Humboldt Times,* June 1, 1909, 4.

14. *Labor News,* June 5, 4–6.

15. *Humboldt Times,* June 4, 1909, 4.

16. *Labor News,* June 12, 1909, 7.

17. Ibid., June 12, 1909, 7.
18. Ibid., May 16, 1908, 1.
19. Ibid., May 15, 1909, 1
20. Ibid., May 15, 1909. 1.
21. Ibid., June 12, 1909, 1.
22. Ibid., June 12, 1909, 1.
23. *Humboldt Times,* July 4, 1909, 1.
24. *Labor News,* Feb. 5, 1910, 8, quote from Mar. 12, 1910, 2.
25. Kazin, *Barons of Labor,* 204–207; Starr, *Inventing the Dream,* 269.
26. Hammond to McLeod, Oct. 29, 1909, 18/2.
27. McLeod to Harry Cole, Aug. 20, 1934, 24/2, McLeod papers, UM.
28. *Humboldt Times,* May 26, 1950; *Los Angeles Times,* Feb. 20, 1904, 6; McLeod to Hammond, Sept. 22, 1904, 16/8; Hammond to McLeod, Oct. 7, 1905 17/1, Apr. 23, 1910, 18/4; Hammond to McLeod, Jan. 31, 1911; quote from Feb. 6, 1911, 18/7.
29. Hammond to McLeod, Mar. 16, 1909, 18/1.
30. T. C. Marshall to Hammond, Jan. 15, 1910, 18/4; Charles Bonaparte to Secretary of Interior, Dec. 30, 1907; Sharp quote from J. E. Wilsen to Attorney General, July 6, 1907, National Archives, RG 48, file 2–7, GLO Timber Trespasses, Montana, Johnson files.
31. Frank Pierce to Attorney General, Apr. 27, 1910, RG 48, National Archives; T. C. Marshall to Hammond, Jan. 15, 1910; quote from McLeod to Hammond, Jan. 20, 1910, 18/4.
32. Hammond to McLeod, Jan. 24, 1910, 18/4.
33. Frank Pierce, Assistant Secretary of Interior to Attorney General, Apr. 27, 1910, RG 48, National Archives.
34. Marshall to Hammond, Jan. 15, 1910.
35. Hammond to McLeod, Jan. 24, 1910, 18/4.
36. McLeod to Hammond, May 10, 1910, 18/5.
37. Hammond to McLeod, June 30, 1910.
38. Hammond to McLeod June 28, 1910; Hammond quote from July 7, 1910; W. S. Burnett to Robert Bevlin, U.S. Attorney, Aug. 14, 1911, 18/5, McLeod papers, UM; witness quote from McLeod to Hammond, July 8, 1912, 19/3.
39. Quotes from Hammond to McLeod, Sept. 27, 1910; McLeod to Hammond, Sept. 23, 1910, 18/5.
40. Hammond to McLeod, Sept. 27, 1910. 18/5.
41. U.S. House Report 1301, "Instructing the Attorney General to Institute Certain Suits," Serial Set no. 5225, vol. no. 1, 60th Cong., 1st Sess.; W. S. Burnett to McLeod, Mar. 1, 1912, 19/1, McLeod papers, UM; *Los Angeles Times,* Feb. 21, 1913, part 1, 4.
42. W. S. Burnett to McLeod, Mar. 1, 1912, 19/1; *Hammond et al. v. Oregon & C. R. Co.,* 117 Or. 244, 243, p. 767.

43. W. S. Burnett to McLeod, Mar. 1, 1912, 19/1; McLeod to Hammond, Aug. 22, 1912, 19/4. The Oregon and California Railroad grant lands not privatized ultimately fell under the Bureau of Land Management and were heavily logged precipitating the spotted owl crisis of the 1980s. Nearly all of the spotted owl management units are on the old Oregon and California Railroad grant lands.

44. Hammond to McLeod, Feb. 2, Feb. 1, 1913; *Labor News*, Jan. 18, 1913, 1; McLeod to Hammond, Feb. 11, 1913, 19/5.

45. Butcher, "Analysis of Timber Depredations."

46. Hammond to McLeod, Feb. 25, 1913, 19/5.

47. Hammond to McLeod, Feb. 10, 1913, 19/5.

CHAPTER 18

1. Hammond to McLeod, Aug. 12, Oct. 17, Nov. 11, 1914, 20/1; McLeod to Hammond, Dec. 4, 1914, 20/1; Knight, *Industrial Relations*, 285; quote from Hammond to Ronan, Jan. 30, 1915, 20/1.

2. Quotes from Hammond to McLeod, Dec. 12, 1914, and Feb. 9, 1916, 20/3; Levi, *Committee of Vigilance*, 17.

3. Knight *Industrial Relations*, 302; *Los Angeles Times*, June 2, 1916, part 1, 4, June 7, 1916, part 1, I4; Hammond to McLeod, Jan. 2, 1917, 20/4.

4. *Los Angeles Times*, June 22, 1916, part 2, 1, June 27, 1916, part 2, 1; Hammond to McLeod, Jan. 2, 1917.

5. *Los Angeles Times*, July 11, 1916, part 2, 1, July 31, 1916, part 2, 3.

6. Knight, *Industrial Relations*, 305.

7. San Francisco Chamber of Commerce, *Law and Order in San Francisco*, 13–17, quote on 13; Hammond to McLeod, July 26, 1916.

8. Knight, *Industrial Relations*, 124.

9. Ibid., 136, 140–41, 202, 291; *Labor News*, July 23, 1910, 1.

10. Levi, *Committee of Vigilance*, 30.

11. Ibid., 21, 32, quote from Rolph to Koster, July 10, 1916, on 129.

12. Knight, *Industrial Relations*, 309–22.

13. Ibid., 311; Levi, *Committee of Vigilance*, 45, quote on 112.

14. Levi, *Committee of Vigilance*, 46; San Francisco Chamber of Commerce, *Law and Order in San Francisco*, 17, 36.

15. Levi, *Committee of Vigilance*, 81–83; Knight, *Industrial Relations*, 320–21.

16. Levi, *Committee of Vigilance*, 65, 67, 90, 133; Knight, *Industrial Relations*, 313; Hammond to McLeod, Aug 15, 1916.

17. San Francisco Chamber of Commerce, *Law and Order in San Francisco*, 36.

18. Knight, *Industrial Relations*, 317. Although Mooney and Billings were convicted and sentenced to hang, California governor William Stephens commuted their sentences to life imprisonment. Subsequent investigations revealed their

innocence as well as serious irregularities in the trial. Finally, in 1939, Governor Culbert Olson pardoned both men. Levi, *Committee of Vigilance,* 126.

19. Levi, *Committee of Vigilance,* 103; quote from Webb, "United States Wooden Steamship Program," 278; *Timberman,* Apr. 1917, 37; Neil Price, "The Hammond Shipyard," *Humboldt Historian* (May-June, 1976), 9–10; Hammond to McLeod, June 21, 1917, 20/4.

20. Hammond to McLeod, July 2, 1917, 20/4.

21. Hammond to Edgar Piper, Aug. 9, 1918, 20/6.

22. Hammond to McLeod, Aug. 6, 1917, 20/6.

23. Hammond to McLeod, July 2, 1917; G. W. Fenwick to Hammond, Mar. 12, 1917, 20/4.

24. *Labor News,* Aug. 11, 1917, 1; Hammond to Rolph, Apr. 5, 1917, 20/4.

25. Rolph to Hammond, Apr. 2, 1918, 20/6.

26. Hammond to Rolph, Mar. 22, quotes from Mar. 29, 1917, 20/4.

27. Ibid., Mar. 29, 1917.

28. Ibid., Apr. 11, 1918, 20/5.

29. Dubofsky and Dulles, *Labor in America,* 212; Howd, *Industrial Relations,* 70; quote from Jenson, *Lumber and Labor,* 3.

30. Jenson, *Lumber and Labor,* 3–21.

31. Howd, *Industrial Relations,* 22.

32. Dubofsky, *We Shall Be All,* 25; Meany, "History of the Lumber Industry," 295.

33. Tyler, *Rebels of the Woods,* 25–27.

34. Howd, *Industrial Relations,* 41.

35. Ibid., 84; LRRC files 581, 774, 779, 804.

36. Dubofsky, *We Shall Be All,* 349; Howd, *Industrial Relations,* 2.

37. Ficken, *Forested Land,* 143; Tyler, *Rebels of the Woods,* 92; Dubofsky, *We Shall Be All,* 362

38. Hammond to McLeod, July 3, 1917.

39. Ibid., Aug. 2, 1917.

40. *Los Angeles Times,* Sept. 20, 1917, part 1, 2; *Astoria Daily Budget,* July 9, 1914; Hammond to William Wilson, Oct. 11, 1917, 20/5; quote from McLeod to Hammond, Sept. 19, 1917, 20/5.

41. Hammond to McLeod, July 26, 1917, 20/4.

42. Ficken, *Forested Land,* 143; Dubofsky, *We Shall Be All,* 363.

43. Hammond to F. W. Leadbetter, Oct. 2, 1917, 20/5.

44. Hammond to McLeod, Aug. 2, 1917.

45. Ibid., June 11, 1918, 20/7.

46. Alaback, "Biodiversity Patterns," in Lawford et al., eds., *High-Latitude Rainforests;* Agee, *Fire Ecology of the Pacific Northwest Forests,* 189–93; quote from F. W. Leadbetter to Hammond, Oct. 1, 1917. Leadbetter's statement is a bit of an exaggeration. While Hammond held extensive stands of Sitka spruce on the Oregon

Coast, both the U.S. and Canadian governments owned far more than any individual as Sitka spruce grows along the Pacific coast from Oregon to Alaska.

47. Hammond to Leadbetter, Oct. 2, 1917.

48. Tyler, *Rebels of the Woods,* 97; Howd, *Industrial Relations,* 75; quote from Ficken, *Forested Land,* 141.

49. Hyman, *Soldiers and Spruce,* 190.

50. Ibid., 77–79.

51. Ibid., 91–96.

52. Ibid., 109, 112.

53. Ibid., 2.

54. Ibid., 50.

55. Ibid., 84.

56. Ibid., 110, 305.

57. Ibid., quote on 174, 306–308.

58. Hyman, *Soldiers and Spruce,* 131, 148, 156, 159; Leadbetter to Hammond, Oct. 3, 1917, 20/5.

59. E. A. Selfridge to R. H. Downman, Dec. 17, 1917, 20/5, McLeod papers, UM.

60. Hyman, *Soldiers and Spruce,* 223; Crumpacker files, MS 4490, folder 2, OHS.

61. Hyman, *Soldiers and Spruce,* 282, 337–38.

62. Ibid., 108, 190; Hammond to McLeod, May 27, 1918, 20/6; quote from Hammond to McLeod, Sept. 24, 1918, 20/7.

63. *Timberman,* Jan. 1918, 48, Oct. 1918, 38.

64. Ficken, *Forested Land,* 152.

65. Hammond to McLeod, Oct. 30, 1917; Hammond to Louis Post, Oct. 11, 1917, 20/5; Kazin, *Barons of Labor,* 225, 229; quote from *Final Report of the Commission on Industrial Relations,* S. Doc. 415. 165.

66. Tyler, *Rebels of the Woods,* 99; quotes from Weinstock to Hammond, Oct. 31, 1917, 20/5.

67. Quote from Hammond to Weinstock, Nov. 1, 1917; Weinstock to Hammond, Nov. 5, 1917, 20/5.

68. Quote from Hammond to H. W. McLeod, Aug. 27, 1917; Hammond to Weinstock, Nov. 1, 1917, 20/5.

69. Quotes from Hammond to Weinstock, Nov. 1 (Weinstock quotes embedded in his Oct. 5 letter to Planing Mills Assn.; Hammond to Weinstock, Nov. 8, 1917, 20/5.

70. Hammond to McLeod, Jan. 9, 1918, 20/6; quote from H. W. McLeod to Hammond, Oct. 1917, 20/5.

71. Quote from Hammond to Capt. A. F. Pillsbury, Nov. 12, 1917, 20/5; Hammond to G. W. Fenwick, Nov. 28, 1917; "intolerable" quote from California Metal Trades Association to William Wilson, Nov. 17, 1917, 20/5, McLeod papers, UM.

Like Union Lumber Company, the name of Union Iron Works bore no relation to its labor practices.

72. Wolff, "They're Building Ships Out There," 36.

73. Hammond to Disque, Jun. 4, 1918, 20/6.

74. Quote from Hammond to McLeod, Oct. 10, 1918, 20/6; Hammond to Edgar Piper, Aug. 9, 1918, 20/6.

75. Dubofsy and Dulles, *Labor in America,* 191–216.

76. Hammond to G. W. Cousins, Nov. 25, 1919, 21/5.

77. Dumenil, *Modern Temper,* 220; quotes from Hammond to McLeod, Dec. 26, 1919, 21/1, Jan 7, 1922, 21/5. See also Cohen, *Making a New Deal.*

78. Allen, *Only Yesterday,* 38–62; Goldberg, *Discontented America,* 168.

79. Dumenil, *Modern Temper,* 226.

CHAPTER 19

1. Goldberg, *Discontented America,* 41–42, quote on 64.

2. Hammond to Charles Donnelly, Apr. 5, 1925, 21/10. For more on the 1920s see Dumenil, *Modern Temper.*

3. Robbins, *Lumberjacks and Legislators,* 109–10.

4. Hammond to Charles Donnelly, Apr. 5, 1925, 21/10.

5. Goldberg, *Discontented America,* 168.

6. Ibid., 87. For more on the latent depression during the 1920s see Kennedy, *Freedom from Fear.*

7. Quotes from Hammond to McLeod, Jan. 28, 1920, and Dec. 23, 1919; Sept. 1, 1919, Apr. 19, 1920; "lumber prices" quote from Apr. 4, 1920, all in 21/2.

8. Hammond to McLeod, Oct. 4, 1920, 21/3; quotes from Jan. 19, 1921, 21/4, and Feb. 26, 1921, 21/4; Mar. 1, 1921, 21/4.

9. Goldberg, *Discontented America,* 57.

10. McLeod to Hammond, Dec. 23, 1921, 21/4.

11. Quote from McLeod to Hammond, Jan. 9, 1922; Jan. 25, Feb. 4, 1922; Hammond to McLeod, Jan. 28, 1922, 21/5.

12. McLeod to Hammond, Feb. 4, 1922; Hammond to McLeod, Jan. 30, 1922, 21/5.

13. *Oregonian,* Jan. 4, 1923.

14. Quote from *Los Angeles Times,* Nov. 28, 1920, part 5, 7; Hammond to McLeod, Nov. 7, 1922, 21/7; *Oregonian,* Jan. 4, 1923; *Los Angeles Times,* Nov. 28, 1920, part 5, 7.

15. Hammond to McLeod, Apr. 19, 1920, 21/2, Apr. 1, 1924, 21/9, Apr. 26, 1923, 21/8; Mengel, "History of the Samoa Division," 77; *Los Angeles Times,* July 20, 1931, A8, June 4, 1933, 17.

16. Hammond to McLeod, Aug. 10, 1923, 21/8, June 22, 1920, 21/2.

17. J. A. Rankin to Hammond, Feb. 22, 1929, 22/5.

18. Hammond to McLeod, Jan. 31, 1922, 21/5, Mar. 21, 1929, 22/5.

19. "Shaws" quote from Hammond to McLeod, Mar. 3, 1920; "field marshal" quote from Feb. 26, 1927, 22/3.

20. Dan Strite to Bill Stoddard, June 14, 1983, author's file.

21. Hammond to McLeod, Feb. 28, 1922, 21/5.

22. Dumenil, *Modern Temper.* For more on consumer culture see Leach, *Land of Desire* and Fox and Lears, eds., *Culture of Consumption.* For welfare capitalism see Brandes, *American Welfare Capitalism.*

23. *Arcata Union,* Feb. 1, 1923; Mengel, "History of the Samoa Division," 85; *Timberman,* Mar. 1920, 53.

24. *Morning Astorian,* Sept. 15, 1918.

25. McLeod to Hammond, Mar. 2, 1922, 21/ 6.

26. Hammond to Huntington, Jan. 6, 1927, HEH 12228, box 43; Robert Schad to Max Farrand, Jan. 1, 1934, HEH 19/3, folder 2.

27. Hammond quotes from Hammond to McLeod, Jan. 6, 1920, 21/2; McLeod quotes from McLeod to Hammond, Aug. 24, 1923, 21/8.

28. Hammond to George L. Hammond, Jan. 6, 1930, 22/6.

29. Hammond to Joseph Stack, July 17, 1922, 21/6.

30. Hammond to McLeod, Aug, 8, 1929, 22/5.

31. Hammond to McLeod, Apr. 27, 1926, 22/1.

32. Hammond to Thomas Hubbard, Feb. 14, 1914, 19/8.

33. Ibid.; quotes from Hammond to McLeod, Nov. 17, 1916.

34. Mengel, "History of the Samoa Division," 81–83; Hammond to McLeod, Dec. 23, 1921, 21/4; Apr. 26, 1920, June 22, 1920, 21/2; *Pacific Lumber Trade Journal,* Apr. 1908, 42, Aug. 1908, 24, Blanchard files, author's possession.

35. Quote from *Astoria Daily Budget,* Sept. 12, 1922; *West Coast Lumberman,* Sept. 15, 1922, 48; *Timberman,* Sept. 1922, 31, Blanchard files, author's possession; Hammond to McLeod, Aug. 20, 1922, 21/6.

36. *Astoria Daily Budget,* Sept. 12, Dec. 10, 1922.

37. Hammond to McLeod, June 20, 1925, 21/10.

38. McLeod to Hammond, Jan. 3, 1924; Malone et al., *Montana: A History of Two Centuries,* 283; McLeod quote from McLeod to Hammond, Feb. 8, 1924; Hammond to McLeod, Feb. 19, 1923, Jan. 24, 1924; "Soviet Russia" quote from July, 17, 1924, 21/9.

39. Hammond to McLeod, June, 6, 1925; Hammond to James Woodworth, Oct. 30, Nov. 18, 1925, 21/10, Mar. 19, 1926, 22/1.

40. Hammond to McLeod, Jan. 4, 1926, 22/1.

41. Hammond to McLeod, Sept. 19, 1927, 22/3.

42. Hammond to Rose Hammond, Jan. 30, 1922, McLeod to Hammond, Jan. 25, Jan. 28, 1922, 21/5; Hammond to McLeod, May 29, June 22, 1922, folder 6.

43. Ann Hammond to A. B. Hammond, Oct. 12, 1927; quotes from A. B. to Ann Hammond, Oct. 17, 1927, 22/3; Hammond to Leonard Hammond, Dec. 6, 1926, 22/3; Hammond to McLeod, May 22, 1930, 23/1.

44. Hammond to Mayannah Seeley, May 6, 1924, 22/1.

45. Hammond to George L. Hammond, Jan. 6, 1930, 22/6.

46. Merger file, 1926, Harry Cole papers, Bancroft Library.

47. Ibid.; Harry Cole to LRRC, Sept. 28, 1928; quote from L. S. DeGraff to Harry Cole, Oct. 9, 1928; N. P. Wheeler, Jr., to Harry Cole, Oct. 9, 1928; E. C. Cronwell to W. A. Dusenbury Oct. 16, 1928, redwood merger file 1928–29, Harry Cole papers; Hammond to McLeod, June 16, 1927, 22/3.

48. Quote from N. P. Wheeler, Jr., to Cole, Oct, 9, 1928, Cole papers; Dusenbury to Cole, Oct. 27, 1928, Cole papers. The Prairie Creek and Del Norte tracts eventually formed the core of Redwood National Park.

49. Cronwell to Dusenbury, Oct. 20, 1928, Cole papers.

50. "Dominated" quote from N. P. Wheeler, Jr. to Cole, Oct, 9, 1928, Cole papers; Mengel, "History of the Samoa Division," 132; "best property" quote from Dusenbury to Cole, Oct. 27, 1928; "pointless" quote from Dusenbury to Cronwall, Dec. 13, 1928, Cole papers.

51. McLeod to Hammond, Mar. 31, 1927; Hammond to McLeod, Apr. 4, June 16, Sept. 27, 1927, Mar. 27, 1928, Mar. 8, 1929, quote from Dec. 30, 1929, Feb. 25, 1930, all box 22; HLC Statement of Assets and Liabilities, Dec. 31, 1927, folder 4, Dec. 31, 1928, 22/5.

52. McLeod to Hammond, Mar. 31, 1927; Hammond to McLeod, Apr. 4, June 16, Sept. 27, 1927, 22/3. For more on the Whitney Lumber Company see Strite's four-part series, "Up the Kilchis."

53. Strite, "Up the Kilchis"; Stoddard, "Hammond in the Tillamook," 6–13; Hammond-Tillamook Lumber Company tax records, OHS.

54. Hammond to McLeod, Oct. 3, Oct. 9, 1929, 22/5.

55. Adams, *Logging Railroads of the West,* 47; Stoddard, "Hammond in theTillamook."

56. Cole to LRRC, Sept. 28, 1928; Mengel, "History of the Samoa Division," 100; quote from Hammond to McLeod, Jan. 6, 1930, 22/6.

57. Hammond to Claflin, Apr. 14, 1932, 23/3; quote from Hammond to McLeod, Dec. 30, 1929, 22/5.

58. J. A. Rankin to Hammond, May 1, 1930; Arthur Jones to Hammond, Dec. 26, 1930, 23/1; Hammond to Claflin, Jan. 3, 1931; McLeod to Hammond, Jan. 29, 1931, 23/2; quote from Hammond to Claflin, Apr. 14, 1932; Hammond to Claflin, Jan. 24, 1931, 23/2.

59. Quote from Hammond to McLeod, Feb. 2, 1931, 23/2; Hammond to Claflin, Dec. 24, 1930, 23/1.

60. Hammond to Claflin, Dec. 24, 1930, 23/1; *Wall Street Journal,* Jan. 29, 1931; *Blue Lake Advocate,* Feb. 21, 1931.

61. Hammond to McLeod, Feb. 26, 1931, 23/2; agreement between HLC and MMC, Oct. 1, 1931, 23/2.

62. Hammond to Claflin, Mar. 5, 1932, 24/1.

63. Ibid.

64. Hammond to Claflin, Feb. 23, 1932.

65. "Disturbed" quote from Hammond to McLeod, Apr. 8, 1932; H. W. McLeod to Hammond, Mar. 25, 1932, 24/1; Claflin quote from Claflin to Hammond, 24/1.

66. Hammond to Claflin, Mar. 18, 1932.

67. Hammond to Claflin, Apr. 14, 1932, 23/3

68. Quote from Hammond to Claflin, Apr. 14, 1932, 23/3; Hammond to McLeod, Apr. 8, 1932, 24/1.

69. *New York Herald Tribune,* Jan. 16, 1934. 7

70. Kemp, *Epitaph for the Giants,* 1–4; Wells, *Tillamook,* 7.

71. Photos 87–2B-5, 87–2B-27, Tillamook Pioneer Museum, Tillamook, Ore.

72. Holbrook, *Holy Old Mackinaw,* 184.

73. Wells, *Tillamook,* 7–11; U.S. Forest Service, "Summary of the Tillamook Burn Study"; Case, "Big Timber Gets Religion,"14–16; Tennsma et al., *Preliminary Reconstruction andAnalysis;* Kemp, *Epitaph for the Giants,* 32–33.

74. Case, "Big Timber Gets Religion"; Kemp, *Epitaph for the Giants* 79; G. B. McLeod to Dan Strite, May 7, 1954, author's collection; Strite, "Hurrah for Garibaldi," 341–68; State of Oregon, recapitulation of 1933 fire, copy in author's possession.

75. Tennsma et al., *Preliminary Reconstruction and Analysis;* quote from Isaac and Meagher, *Natural Reproduction on the Tillamook Burn,* 11.

76. Tennsma et al., *Preliminary Reconstruction and Analysis;* Wells, *Tillamook,* 11.

77. Wells, *Tillamook,* 11; quote from H. C. Patton to Tillamook County, July 22, 1940; quitclaim deed, Nov. 7, 1941, misc. correspondence, HLC files, OHS.

EPITAPH

1. L. C. Hammond to McLeod, Feb. 20, 1940, 23/12. All L. C. Hammond correspondence from C. H. McLeod papers, UM.

2. Gus Kirby to L. C. Hammond, Jan. 24, Jan. 23, 1934, 23/12.

3. Report to R. W. Lea, May 24, 1935, 71, misc. papers, Cole papers.

4. Ibid., 45.

5. Ibid., 46.

6. Ibid., 67, Mengel, "History of the Samoa Division," 120.

7. Mengel, "History of the Samoa Division," 120–24; *Timberman,* Mar. 1936: 109; quote from L. C. Hammond to McLeod, Dec. 29, 1937, 23/12.

8. Mengel, "History of the Samoa Division," 130; Melendy, "One Hundred Years,"113, "sales" quote on 300; "financial condition" quote from L. C. Hammond to McLeod, Dec. 29, 1937, Dec. 12, 1940, Mar. 14, 1941.

9. L. C. Hammond o McLeod, Feb. 16, 1942.

10. Onstine, *Great Lumber Strike,* 8–9.

11. Ibid.; *Humboldt Standard,* June 11, 1935.

12. Onstine, *Great Lumber Strike,* 14–16; *Humboldt Standard,* June 21, 1935.

13. L. C. Hammond to McLeod, Apr. 10, 1942.

14. Dubofsky and Dulles, *Labor in America,* 326; historical sketch, Redwood District Council of Lumber and Sawmill Workers Collection, HSU.

15. Ibid.; Mengel, "History of the Samoa Division," 147.

16. Mengel, "History of the Samoa Division," 143.

17. HLC Memoriam of Leonard Coombes Hammond, copy in author's possession.

18. Mengel, "History of the Samoa Division," 148–50; Redwood District Council.

19. George McLeod interview with Elwood Maunder and Clodaugh Neiderhauser, Feb. 19, 1957, FHS, 22–24.

20. *Humboldt Standard,* Aug. 26, 1950, 1.

21. Etulain and Malone, *American West,* 29; *Redwood Log,* Aug. 1950; Mengel, "History of the Samoa Division," 158, 162.

22. Mengel, "History of the Samoa Division," 170–72.

23. Ibid., 172–74; Robbins, *Hard Times in Paradise,* 247; *Arcata Union,* Nov. 2, 1956. The original offer was closer to $80 million. Mengel, "History of the Samoa Division," 166; quote from *San Francisco Chronicle,* Oct. 23, 1956.

24. Mengel, "History of the Samoa Division," 175–87; *Arcata Union,* Oct. 11, 1963; Fountain papers, vol. 96, 129, 131.

25. Leydet, *Last Redwoods,* 19, 11; Schrepfer, *Fight to Save the Redwoods,* 6, 12.

26. Schrepfer, *Fight to Save the Redwoods,* 21; Mengel, "History of the Samoa Division," 122; Save the Redwood League 1922 report, HSU; *Timberman,* Apr. 1922, 118.

27. Schrepfer, *Fight to Save the Redwoods,* 109.

28. Ibid., 111, 120; Metsker Maps, Humboldt County, HSU; *New York Times,* Sept. 9, 1966, 32, Dec. 8, 1967, quote from editorial on Dec. 29, 1967.

29. Schrepfer, *Fight to Save the Redwoods,* 158.

30. Ibid., 186, quote on 188.

31. Ibid., 224–26.

32. Ibid., 200–204.

33. Quotes from *Business Week,* Dec. 22, 1986; Mengel, "History of the Samoa Division," 206.

34. St. Clair and Cockburn, "Decadent, Old-Growth Timber Baron"; Louisiana-Pacific Corporation History, www.fundinguniverse.com/company-histories/LouisianaPacific-Corporation-Company-History. See also Durbin, *Tongass.*

35. Evarts and Popper, eds., *Coast Redwood,* 150.

36. See Harris, *Last Stand,* and Widick, *Trouble in the Forest,* 13–20.

37. Harris, *Last Stand;* Widick, *Trouble in the Forest;* Evarts and Popper, eds., *Coast Redwood,* 152–56.

38. *National Forest Redwoods Act of 1991,* 12; marbled murrelet species profile, U.S. Fish and Wildlife Service: www.ecos.fws.gov.

39. Widick, *Trouble in the Forest,* 5, 18, 49.

40. Ibid., 14.

41. Paradis, ed., *Papers of Prudent C. Mercure,* lxxvi. See also Craig, *Backwoods Consumers and Homespun Capitalists.*

Bibliography

ARCHIVAL COLLECTIONS AND MANUSCRIPTS

Albert, Thomas. "History of Madawaska, according to the Historical Researches of Patrick Therriault and the Handwritten Notes of Prudent Mercure." Manuscript, 1920. Provincial Archives of New Brunswick, Fredericton.

Anaconda Forest Products. Papers. University of Montana.

Ashby, Shirley Carter. Reminiscence. SC 283. Montana Historical Society, Helena.

Big Blackfoot Milling Company. Records, 1882–1899. University of Montana.

Bohemian Club of San Francisco, annual report, 1907. Manuscript 2295, Huntington Library, Pasadena, CA.

Browman, Audra. Files. University of Montana.

Browman, Audra. "Chronicle of Western Montana before 1864." Manuscript, 1964. Missoula Public Library.

Buck, Henry. "The Story of the Nez Perce Indian Campaign during the Summer of 1877." Manuscript. Montana Historical Society.

Carrington, Henry B. "The Exodus of the Flathead Indians from Their Ancestral Home in the Garden Valley, Montana, to the Jocko Reservation, Montana." Manuscript. University of Montana.

Cole, Harry. Papers. Bancroft Library, University of California.

Cook, Mary, E. "Diary of Mrs. Mary E. Cook, Written while Coming up the Mo. River in 1868." Montana Historical Society, Helena.

Crumpacker, Maurice. Files. Oregon Historical Society, Portland.

Dickinson, Emma. "The Diary of Emma Slack Dickenson, Written before and during Her Journey up the Missouri to Fort Benton and Overland to Missoula in 1869." Montana Historical Society, Helena.

Dickinson, William H., and Emma Dickinson. Papers. University of Montana.

Dolbeer and Carson Lumber Company. Records. Bancroft Library, University of California.

Elder, Claude. Oral history. University of Montana.

Elk River Mill and Lumber Company Records. Bancroft Library, University of California.

Fenwick, Georgina. Papers. University of Montana.

Fountain, Susan Baker. Papers. Humboldt State University.

Hammond, Jean. "Hammond Family History." MC 1, Provincial Archives of New Brunswick, Fredericton.

Hammond family. Files. Provincial Archives of New Brunswick, Fredericton.

Hammond Lumber Company. Files. Humboldt State University.

Hammond Lumber Company. Files. Oregon Historical Society, Portland.

Hampton, H. D., ed. "Life at the Mouth of the Musselshell: Journal of Cornelius M. Lee, 1868–1874." Unpublished manuscript, 2009, author's possession. (Currently in book form as *Life and Death at the Mouth of the Musselshell: Montana Territory 1868–1872*. Stevensville, Mont.: Stoneydale Press, 2011.)

Hauser, Samuel T., Papers. Montana Historical Society, Helena.

Hermann, Binger. "Shadows in Public Life." Manuscript. Oregon Historical Society, Portland.

Huntington, Henry. Collection. Huntington Library, San Marino, Calif.

Johnson, Dale. Files. Author's collection.

Joslin family. History, manuscript. King's Landing, Prince William, New Brunswick.

Koch, Hans Peter. Diary, 1869–1870. SC 950. Montana Historical Society, Helena.

Little River Redwood Lumber Company. Files. Timber Heritage Association, Eureka, Calif.

Marceau, Margaret. "Old Time Logging Camps." In *Grand Falls Past and Present* series, Apr. 18, 1984, 1393–95. Grand Falls Museum, New Brunswick.

McLeod, Charles Herbert. Papers, 1865–1953. K. Ross Toole Archives, University of Montana.

McLeod, Evelyn. "Early Missoula." Manuscript. Missoula Public Library, Missoula, Mont.

McLeod, George B. "The Story of the Hammond Lumber Company." Oral history manuscript. Forest History Society, Durham, N.C. (A slightly different version is at the University of Montana archives.)

Mengel, Lowell S., II. "A History of the Samoa Division of Louisiana-Pacific Corporation and its Predecessors, 1853–1973." Manuscript. Humboldt State University, 1974.

Missoula County Records. Collection. University of Montana.

Morley, James. Diary, 1862–1865. SC533. Montana Historical Society, Helena.

Napton, John. "My Trip on the Imperial in 1867." Manuscript. Montana Historical Society, Helena.

National Archives. RG 48, RG 60.

Northern Pacific Railway. Records. University of Montana.

Peet, Herbert M., Collection. Montana Historical Society, Helena.

Power, Thomas C., Papers. Montana Historical Society, Helena.
Smith, Francis Marion. Reminiscence. SC 2254. Montana Historical Society, Helena.
Spitzley, Stephen. Diary, 1867. SC 771. Montana Historical Society, Helena.
Strong, Clarence. Papers. K. Ross Toole Archives, University of Montana.
Trinity Anglican Church, Perth Andover. Records. Grand Falls Museum, Grand Falls, New Brunswick.
Weatherford, John Knox, Papers. Oregon Historical Society, Portland.
Wells, J. A. "First Connected Account of the Warfare Waged by the Sioux Indians against Wood Choppers, Missouri River between Ft. Buford and Ft. Benton 1866–1870." Manuscript. Montana Historical Society, Helena.
Weston, Daniel. Diary, 1866. SC 282. Montana Historical Society, Helena.
Wilcox, A. H. "Up the Missouri River to Montana in the Spring of 1862." SC 981. Montana Historical Society, Helena.
Woody, Francis. "History of Missoula." Manuscript, 1897. Missoula Public Library, Missoula, Mont.
Zimmerman, Lloyd, Jr. "Missoula Mercantile Company." Manuscript, 1962. University of Montana.

LEGAL AND PUBLIC DOCUMENTS

Annals of Congress, 9th Cong., 2d sess.
Appendix to the Congressional Globe, "Man's Right to the Soil," 32d Cong., 1st Sess.
City of Missoula Records.
Decisions of Department of Interior and General Land Office Relating to Lands and Land Claims 1881 through 1884.
Department of Justice file #7308, National Archives, RG 60.
Final Report of the Commission on Industrial Relations, 64th Cong., 1st Sess., S. Doc. 415. Washington, D.C.: U.S. Government Printing Office, 1916.
Hammond et al. v. Oregon & C. R. Co., 117 Or. 244, 243, 1920.
House Committee on Pacific Railroads Report, 46th Cong., 2d Sess., report #691.
Leavenworth R. R. Co. v. United States, 92 U.S. 741, 1875.
Letter from the Secretary of Interior, S. Exec. Doc. 9, 45th Cong., 2d Sess., Serial Set Vol. No. 1780, Sess. Vol. no. 1, 3.
Letter from the Secretary of the Interior, in response to Senate resolution of June 5, 1886, relative to a certain circular issued from the General Land Office. June 22, 1886. Serial Set Vol. No. 2341 Session Vol. No. 8, 49th Congress, 1st Session, S. Exec. Doc. 170
Long v. Hammond, 40 Me. 204, 1855.
National Forest Redwoods Act of 1991, Hearing before the Subcommittee on National Parks and Public Lands, June 15, 1991, serial 102–51. Washington, D.C.: U.S. Government Printing Office, 1992.

Provincial Census of New Brunswick, 1851, 1861: County of Victoria.

Raynolds, William. *Report on the Exploration of the Yellowstone River,* 40th Cong., 2d Sess. S. Exec. Doc. 77.

Report of the Commissioner of the General Land Office, House Documents, 49th Cong., 2d Sess., no. 1 Part 5.

Report of the Public Lands Commission, Senate Doc. 189, 58th Cong., 3d Sess. 1905.

Report of the Secretary of the Interior. Washington, D.C: U.S. Government Printing Office, 1875. Serial Set No. 1680 H. Exec. Doc. 1 pt. 5, vol. 1–2.

Report of the Secretary of the Interior. Washington, D.C.: U.S. Government Printing Office, 1877, Serial Set No. 1749 H. Exec. Doc. 1 pt. 5, vol. 1–2.

Report of the Secretary of the Interior. Washington, D.C.: U.S. Government Printing Office, 1878, Serial Set No. 1800 H. Exec. Doc. 1 pt. 5, vol. 1–2.

Report of the Secretary of the Interior. Washington, D.C.: U.S. Government Printing Office, 1880, Serial Set No. 1912 H. Exec. Doc. 1 pt. 5, vol. 3.

Report of the Secretary of the Interior, 49th Cong., 2d Sess., H. Exec. Doc. 1 pt. 5, vol. 1.
Serial Set Vol. No. 2467 Session Vol. No. 8.

River and Harbor Bill, 32nd Cong., 1st Sess., *Congressional Globe,* Appendix, 851.

Statistics of mines and mining in the state and territories west of the Rocky Mountains. U.S. House Exec. Doc. 207, 41st Cong., 2d Sess.

Timber and Stone Act, 1878.

U.S. Census, 1880, 1890.

U.S. Congress, *Report of an Expedition Led by Lieutenant Abert, on the Upper Arkansas and through the Country of the Camanche Indians, in the Fall of the Year 1845,* 29th Cong., 1st Sess., S. Doc. 438.

U.S. Department of Interior. *Report of the Secretary of the Interior.* Washington, D.C.: U.S. Government Printing Office, 1882–1903.

U.S. General Land Office. *Annual Report of Commissioner of General Land Office.* Washington, D.C.: U.S. Government Printing Office, 1872–97.

U.S. House of Representatives, vol. 1 Executive Documents, no. 1 part 5, "Report of the Secretary of the Interior," 45th Cong., 2d Sess., p. 16, 1877.

U.S. House Report 1301, "Instructing the Attorney General to Institute Certain Suits," Serial Set No. 5225, Vol. No. 1. 60th Cong., 1st Sess.

U.S. Senate Doc. 415. *Final Report of the Commission on Industrial Relations,* 64th Cong., 1st Sess. Washington, D.C.: U.S. Government Printing Office, 1916.

U.S. Senate. *Explorations and Surveys for Pacific Railroad,* vol. I, 33d Cong., 2d Sess., S. Ex. doc. 78, serial #758.

NEWSPAPERS AND JOURNALS

Astoria (Ore.) Daily Budget

Blue Lake Advocate (Arcata, Calif.)

Butte (Mont.) Semi-weekly Inter-Mountain

Collier's Weekly
Columbia River and Oregon Timberman; Timberman beginning 1905
Daily (Ore.) Astorian
Helena (Mont.) Independent
Helena (Mont.) Weekly Herald
Helena (Mont.) Weekly Inter-Mountain
Humboldt (Calif.) Times
Humboldt (Calif.) Historian
Labor News (Eureka, Calif.)
Los Angeles Times
Missoula (Mont.) and Cedar Creek Pioneer
Missoula (Mont.) Daily Democrat-Messenger
Missoula (Mont.) Gazette
Missoulian (Mont.)
Moncton (Wash.) Times
New Brunswick Courier
New Northwest (Portland, Ore.)
New York Herald Tribune
New York Times
Oakland (Calif.) Tribune
Oregonian (Portland)
Pacific Lumber Trade Journal
Redwood Log
River Press (Fort Benton, Mont.)
San Francisco Call-Bulletin
St. John Valley (Maine) Times
St. John (Maine) Daily News

INTERVIEWS AND PERSONAL COMMUNICATIONS

Alaback, Paul. Nov. 4, 2007.
Johnson, Dale. Oct. 20, 2007
Mengel, Lowell, II. May 2, 2007
Robinson, Ty. Apr. 10, 2007.

PUBLISHED SOURCES

Abrams, Marc D. "Eastern White Pine Versatility in the Presettlement Forest." *Bioscience* 51, no. 11 (Nov. 2001).

Adams, Kramer. "Blue Water Rafting: The Evolution of Ocean Going Log Rafts." *Forest History* 15, no. 2 (July 1971).

———. *Logging Railroads of the West.* Seattle, Wash.: Superior, 1961.

Agee, James K. *Fire Ecology of the Pacific Northwest Forests.* Washington, D.C.: Island Press, 1993.

Allen, Frederick Lewis. *Only Yesterday, an Informal History of the 1920s.* New York: Harper and Row, 1931.

Alt, David. *Glacial Lake Missoula and its Humongous Floods.* Missoula, Mont.: Mountain Press, 2001.

Andrew, Shelia. *The Development of Elites in Acadian New Brunswick, 1861–1881.* Montreal, Canada: McGill-Queen's University Press, 1997.

Andrews, H. D. "Summary of the Tillamook Burn Study, 1942–43." Washington, D.C.: U.S. Government Printing Office, Apr. 1944.

Andrews, Ralph W. *Glory Days of Logging.* New York: Bonanza Books, 1966.

Astoria and Columbia River Railroad. *The Oregon Coast: From Portland to a Summer Paradise in Four Hours.* Astoria, Ore.: Astoria and Columbia River Railroad, 1904.

Astoria Chamber of Commerce. *Astoria and Clatsop County Oregon.* Astoria, Ore.: Astoria Chamber of Commerce, ca. 1911.

Athearn, Robert. "The Fort Buford 'Massacre.'" *Mississippi Valley Historical Review* 41, no. 4 (Mar. 1955).

Axtell, Silas Blake, ed. *A Symposium on Andrew Furuseth.* New Bedford, Mass.: Darwin Press, 1948.

Baird, William. *Seventy Years of New Brunswick Life.* St. John, Canada: George Day, 1890.

Barbour, Barton. *Fort Union and the Upper Missouri Fur Trade.* Norman: University of Oklahoma Press, 2001.

Barnes, Burton, Donald Zak, Shirley Denton, and Stephen Spurr. *Forest Ecology.* 4th ed. New York: John Wiley & Sons, 1998.

Barrett, Stephen, and Stephen Arno. "Indian Fires as an Ecological Influence in the Northern Rockies." *Journal of Forestry* 80, no. 10 (Oct. 1982).

Baxter, R. M. "Environmental Effects of Dams and Impoundments." *Annual Review of Ecology and Systematics,* 8 (1988).

Berman, David. *Radicalism in the Mountain West, 1890–1920.* Boulder: University Press of Colorado, 2007.

Boyd, Robert, ed. *Indians, Fire and the Land in the Pacific Northwest.* Corvallis: Oregon State University Press, 1999.

Boyer, Richard, and Herbert Morais. *Labor's Untold Story.* New York: United Electrical, Radio & Machine Workers of America, 1955.

Brander, James A. and M. Scott Taylor. "The Simple Economics of Easter Island: A Ricardo-Malthus Model of Renewable Resource Use." *American Economic Review* 88, no. 1 (Mar. 1998), 119–38.

Brandes, Stuart. *American Welfare Capitalism: 1880–1940.* Chicago: University of Chicago Press, 1970.

Brown, Mark. *Flight of the Nez Perce.* New York: Putnam, 1967.

Brown, Peter, and William Baxter. "Fire History in Coast Redwood Forests of the Mendocino Coast, California." *Northwest Science* 77, no. 2 (2003).

Bureau of Corporations. *The Lumber Industry*. Washington, D.C.: U.S. Government Printing Office, 1913.

Bureau of Indian Affairs. *Papers Relating to Talks and Councils Held with the Indians in Dakota and Montana Territories in the Years 1866–1869*. Washington, D.C.: U.S. Government Printing Office, 1910.

Burlingame, M. *The Montana Frontier*. Helena, Mont.: State Publishing, 1942.

Calhoun, Charles, ed. *The Gilded Age: Perspectives on the Origins of Modern America*. New York: Roman and Littlefield, 2007.

Carranco, Lynwood. *Redwood Lumber Industry*. San Marino, Calif.: Golden West Books, 1982.

Carstensen, Vernon, ed. *The Public Lands: Studies in the History of the Public Domain*. Madison: University of Wisconsin Press, 1963.

Case, Robert Ormond. "Big Timber Gets Religion." *Saturday Evening Post*, Dec. 30, 1944.

Casler, Michael M. *Steamboats of the Fort Union Fur Trade*. Fort Union, Mont.: Fort Union Association, 1999.

Chandler, Alfred D., Jr. "The Beginnings of 'Big Business' in American Industry." *Business History Review* 37, no 1 (1959).

———. *Strategy and Structure: Chapter in the History of the American Industrial Enterprise*. Cambridge, Mass.: MIT Press, 1962.

———. *The Visible Hand: The Managerial Revolution in American Business*. Cambridge, Mass.: Harvard University Press, 1977.

Cherry, Edgar. *Redwood and Lumbering in California Forests*. San Francisco: Edgar Cherry, 1884.

Chittenden, Hiram. *History of Early Steamboat Navigation on the Missouri River*. New York: Francis P. Harper, 1903.

Cissel, John, Frederick J. Swanson, Gordon E. Grant, Deanna H. Olson, Stanley V. Gregory, Steven L. Garman, et al. *A Landscape Plan Based on Historical Fire Regimes for a Managed Forest Ecosystem: The Augusta Creek Study* (USDA PNW-GTR-422). Portland, Ore.: U.S. Department of Agriculture, Forest Service, Pacific Northwest Research Station. 1998.

Claasen, Elizabeth. *A Card for All Seasons: A History of General Hospital, Eureka, CA 1906–2000*. Bayside, CA: Union Labor Health Foundation, 2002.

Clark, Norman. *Milltown: A Social History of Everett, Washington*. University of Washington Press, 1970.

Cohen, Lizabeth. *Making a New Deal: Industrial Workers in Chicago: 1919–1939*. New York: Cambridge University Press, 1990.

Cohen, Stan. *Missoula County Images*. 2 vols. Missoula, Mont.: Pictorial Histories Publishing, 1982, 1993.

Coman, Edwin T., and Helen M. Gibbs, *Time, Tide and Timber: A Century of Pope and Talbot*. Stanford, Calif.: Stanford University Press, 1949.

Cooper, Laurie Armstrong, and Douglas Clay. *History of Logging and River Driving in Fundy National Park: Implication for Ecological Integrity of Aquatic Ecosystems.* Alma, New Brunswick: Parks Canada, 1997.

Cornford, Daniel. *Workers and Dissent in the Redwood Empire.* Philadelphia: Temple University Press, 1987.

Cotroneo, Ross. *The History of the Northern Pacific Land Grant: 1900–1952.* New York: Arno Press, 1979.

Cox, Thomas R. *The Lumberman's Frontier: Three Centuries of Land Use, Society, and Change in America's Forests.* Corvallis: Oregon State University Press, 2010.

———. *Mills and Markets: A History of the Pacific Coast Lumber Industry to 1900.* Seattle: University of Washington Press, 1974.

———. "Single Decks and Flat Bottoms: Building the West Coast's Lumber Fleet, 1850–1929. *Journal of the West* 20 (July 1981).

Craig, Béatrice. "Agriculture and the Lumberman's Frontier in the Upper St. John Valley, 1800–70." *Journal of Forest History* 32, no. 3(July 1988).

———. *Backwoods Consumers and Homespun Capitalists: The Rise of a Market Culture in Eastern Canada.* Toronto: University of Toronto Press, 2009.

Cronon, William. *Nature's Metropolis: Chicago and the Great West.* New York: W. W. Norton, 1991.

Dallison, Robert. *Hope Restored: The American Revolution and the Founding of New Brunswick.* Fredericton, New Brunswick: Goose Lane, 2003.

Dawson, T. E. "Fog in the California Redwood Forest: Ecosystem Inputs and Use by Plants." *Oecologia* 117, no. 4 (1998).

Defebaugh, James. *History of the Lumber Industry in America.* Chicago: American Lumberman, 1906.

Denig, Edwin. *Five Indian Tribe of the Upper Missouri: Sioux, Arickaras, Assiniboines, Crees, Crows.* Norman: University of Oklahoma Press, 1961.

Deverell, William. *Railroad Crossing: Californians and the Railroad, 1850–1910.* Berkeley: University of California Press, 1994.

Dodge, Richard Irving. *Our Wild Indians: Thirty-Three Years Personal Experience among the Red Men of the Great West.* Hartford, Conn.: A. D. Worthington, 1884.

Donaldson, Thomas. *The Public Domain.* Washington, D.C.: U.S. Government Printing Office, 1884.

Donan, P. *The Columbia River Empire.* Portland, Ore.: Oregon Railroad and Navigation, ca. 1902.

Donato, D. C. "Post-wildfire Logging Hinders Regeneration and Increases Fire Risk." *Science* 311, no. 5759 (Jan. 20, 2006).

Dubay, Guy. "The Richest Men in Van Buren," *St. John Valley Times,* Feb. 4, 1976, 18.

Dubofsky, Melvyn. *Industrialism and the American Worker, 1865–1920.* Wheeling, IL: Harlan Davidson, 1996.

———. *We Shall Be All: A History of the Industrial Workers of the World.* Chicago: Quadrangle Books, 1969.

Dubofsky, Melvyn, and Foster Rhea Dulles. *Labor in America.* Wheeling, Ill.: Harlan Davidson, 2004.

Dumenil, Lynn. *The Modern Temper: American Culture and Society in the 1920s.* New York: Hill and Wang, 1995.

Dunbar, Seymour, and Paul Phillips. *The Journals and Letters of Major John Owen.* New York: Edward Eberstadt, 1927.

Dunham, H. *Government Handout: A Study in the Administration of Public Lands, 1875–1891.* New York: Da Capro Press, 1970. Originally published 1941.

Durbin, Kathie. *Tongass: Pulp Policies and the Fight for the Alaska Rain Forest.* Corvallis: Oregon State University Press, 2005.

Elster, Jon, ed. *Karl Marx: A Reader.* Cambridge: Cambridge University Press, 1986.

Etulain, Richard and Michael Malone. *The American West: A Modern History, 1900 to Present.* Lincoln: University of Nebraska Press, 1989.

Evarts, Jon, and Marjorie Popper, eds. *Coast Redwood: A Natural and Cultural History.* Los Olivos, Calif.: Cachuma Press, 2001.

Fahey, John. *The Flathead Indians.* Norman: University of Oklahoma Press, 1974.

Farr, William. "Going to Buffalo." *Montana, the Magazine of Western History* 53, no. 4 (Winter 2003) and 54, no. 1 (Spring 2004).

Fernow, B. E. "The Forest Reservation Policy." *Science* 26, no. 5 (Mar. 26, 1897).

Ficken, Robert E. *The Forested Land: A History of Lumbering in Western Washington.* Seattle: University of Washington Press, 1987.

———. "Weyerhaeuser and the Pacific Northwest Timber Industry 1899–1903" *Pacific Northwest Quarterly* 70, no. 4 (Oct. 1979).

Fisher, Peter. *History of New Brunswick.* St. John, New Brunswick: Chubb and Sears, 1825.

Flanagan, Darris. *Skid Trails: Glory Days of Montana Logging.* Stevensville, Mont.: Stoneydale Press, 2003.

Flores, Dan. *Caprock Canyonlands: Journeys into the Heart of the Southern Plains.* Austin: University of Texas Press, 1990.

———. "Wars over Buffalo: Stories vs. Stories on the Northern Plains," in *Native Americans and the Environment: Perspectives on the Ecological Indian,* edited by Michael Harkin and David Rich Lewis. Lincoln: University of Nebraska Press, 2007.

Fox, Richard Wightman, and T. J. Jackson Lears, eds. *The Culture of Consumption: Critical Essays in American History, 1880–1980.* New York: Pantheon, 1983.

Franklin, Jerry. *Natural Regeneration of Douglas Fir and Associated Species Using Modified Clear Cutting Systems in the Oregon Cascades* (USDA PNW-3). Portland, Ore.: Pacific Northwest Forest and Range Experiment Station, Sept. 1963.

Franklin, Jerry, Kermit Cromack, Jr., William Denison, Arthur McKee, Chris Maser, James Sedell, Fred Swanson, and Glen Juday. *Ecological Characteristics of Old-Growth Douglas-Fir Forests* (USDA PNW-118). Portland, Ore.: Pacific Northwest Forest and Range Experiment Station, 1981.

Frey, Robert L., ed. *Encyclopedia of American Business History and Biography: Railroads in the Nineteenth Century.* New York: Facts on File, 1998.

Friedricks, William B. *Henry E. Huntington and the Creation of Southern California.* Columbus: Ohio State University Press, 1992.

Fritz, Harry, Mary Murphy, and Robert Swartout, Jr., eds. *Montana Legacy.* Helena: Montana Historical Press, 2002.

Galamos, Louis. "The Emerging Organizational Synthesis in Modern American History" *Business History Review,* 44, no. 3 (Autumn 1970).

Garrison, George, and Robert Rummell. 1951. "First-Year Effects of Logging on Ponderosa Pine Forest Range Lands of Oregon and Washington." *Journal of Forestry* 49, no. 10 (Oct. 1951).

Gates, Jon Humboldt. *Falk's Claim: The Life and Death of a Redwood Lumber Town.* Eureka, Calif.: Pioneer Graphics, 1983.

Gates, Paul W. "The Homestead Law in an Incongruous Land System." *American Historical Review* 41, no. 4 (July 1936).

Gaton, Joseph. "The Genesis of the Oregon Railroad System." *Oregon Historical Quarterly* 7, no. 2 (June 1906).

Genzoli, A. *Golden Adventures from the Humboldt Historian.* Eureka, Calif: Humboldt County Historical Society, 1983.

Gerber, M. "The Steamboat and Indians of the Upper Missouri." *South Dakota History* 4, no. 1 (Winter 1973).

Giles, Albion. "Notes of Columbia River Salmon." *Oregon Historical Quarterly* 35, no. 2 (June 1934).

Ginger, Ray. *Age of Excess: American Life from the End of Reconstruction to World War I.* New York: Macmillan, 1965.

Glasscock, C. B. *The War of the Copper Kings.* New York: Grossett and Dunlap, 1935.

Goldberg, David J. *Discontented America, the United States in the 1920s.* Baltimore: Johns Hopkins University Press, 1999.

Green, Sarah E., and Jerry F. Franklin. "Middle Santiam Research Natural Area." In *Federal Research National Areas in Oregon and Washington: A Guidebook for Scientists and Educators,* edited by Jerry Franklin, Fredrick C. Hall, C. T. Dyrness, and Chris Maser (suppl. 24). Portland, Ore.: U.S. Department of Agriculture, 1987.

Hammond, Roland. *A History and Genealogy of William Hammond of London, England, and his wife Elizabeth Penn; through their Son Benjamin of Sandwich and Rochester, Mass. 1600–1894.* Boston: D. Clapp and Sons, 1894.

Hardin, Garrett. "The Tragedy of the Commons." *Science* 162, no. 3969 (1968).

Harris, David. *The Last Stand: The War between Wall Street and Main Street over California's Ancient Redwoods.* San Francisco: Sierra Club Books, 1996.

Hays, Samuel P. *Conservation and the Gospel of Efficiency: The Progressive Conservation Movement, 1890–1920.* Cambridge, Mass.: Harvard University Press, 1959.

Helman, Christopher. "Timber!" *Forbes,* Dec. 12, 2005.

Hewitt, J. N. B., ed. *Journal of Rudolph Freiderich Kurz: An Account of His Experiences among Fur Traders and American Indians on the Mississippi and the Upper Missouri Rivers during the Years 1846 to 1852.* Washington, D.C.: U.S. Government Printing Office, 1937.

Hibbard, B. H. *A History of the Public Land Policies.* Madison: University of Wisconsin Press, 1965.

Hidy, Ralph, Frank Hill, and Allan Nevins. *Timber and Men: The Weyerhaeuser Story.* New York: Macmillian, 1963.

Hobbs, Stephen D., John P. Hayes, Rebecca L. Johnson, Gordon H. Reeves, Thomas A. Spies, John C. Tappeiner II, and Gail E. Wells, eds. *Forest and Stream Management in the Oregon Coast Range.* Corvallis: Oregon State University Press, 2002.

Hoig, Stan. *The Chouteaus: First Family of the Fur Trade.* Albuquerque: University of New Mexico Press, 2008.

Holbrook, Stewart. *Holy Old Mackinaw: A Natural History of the American Lumberjack.* New York: Macmillan, 1938.

Holmes, Oliver Wendell. "James A. Garfield's Diary of a Trip to Montana in 1872." In *Sources of Northwest History,* 21. Missoula: State University of Montana, 1934.

Hosmer, Paul. *Now We're Loggin.'* Portland, Ore.: Binford and Mort, 1930.

Howard, Joseph Kinsey. *Montana: High, Wide, and Handsome.* Lincoln: University of Nebraska Press, 1983.

Howd, Cloice R. *Industrial Relations in the West Coast Lumber Industry* (U.S. Department of Labor, bulletin 349). Washington, D.C.: U.S. Government Printing Office, 1924.

Hyman, Harold M. *Soldiers and Spruce: Origins of the Loyal Legion of Loggers and Lumbermen.* Los Angeles: Institute of Industrial Relations, University of California Press, 1963.

Ingham, John. *Biographical Dictionary of American Business Leaders.* Westport, Conn.: Greenwood Press, 1983.

Isaac, Leo A., and G. S. Meagher. *Natural Reproduction on the Tillamook Burn Two Years after the Fire.* Portland, Ore.: Pacific Northwest Forest Experiment Station, 1936.

Isaac, Leo A., and Howard G. Hopkins. "The Forest Soil of the Douglas Fir Region, and Changes Wrought upon It by Logging and Slash Burning." *Ecology* 18, no. 2 (1937).

Ise, John. *The United States Forest Policy.* New Haven, Conn.: Yale University Press, 1920.

Isenberg, Andrew. *The Destruction of the Bison.* Cambridge: Cambridge University Press, 2000.

Jensen, Derrick, and George Draffan. *Railroads and Clearcuts: Legacy of Congress's 1864 Northern Pacific Railroad Land Grant.* Spokane, Wash.: Inland Empire Public Lands Council, 1995.

Jenson, Vernon. *Lumber and Labor.* New York: Farrar and Rinehart, 1945.

John, Richard R. "Elaborations, Revisions, Dissents: Alfred D. Chandler, Jr.'s, *The Visible Hand* after Twenty Years." *Business History Review* 71, no. 2 (Summer 1997).

Johnson, Olga. "Prospectors Turned Tiehacks." *Western News* (Libby, Mont.), Apr. 20, 1967.

Josephson, M. *The Robber Barons: The Great American Capitalists 1861–1901.* New York: Harcourt, Brace, 1934.

Josephy, Alvin, Jr. *The Nez Perce Indians and the Opening of the Northwest.* New Haven, Conn.: Yale University Press, 1965.

Kappler, Charles. *Indian Affairs: Laws and Treaties,* vol. 2. Washington, D.C.: U.S. Government Printing Office, 1904.

Kazin, Michael. *Barons of Labor: The San Francisco Building Trades and Union Power in the Progressive Era.* Chicago: University of Illinois Press, 1987.

Kemp, J. Larry. *Epitaph for the Giants: The Story of the Tillamook Burn.* Portland, Ore.: Touchstone Press, 1967.

Kennedy, David. *Freedom from Fear: The American People in Depression and War 1929–1945.* New York: Oxford University Press, 2005.

Kennedy, Rebecca S. H., and Thomas A. Spies. "Forest Cover Changes in the Oregon Coast Range from 1939 to 1993." *Forest Ecology and Management* 200 (2004).

Kilburn, Paul Dayton. *Fathers and Mothers: History and Genealogy of Kilburn and Graham.* Arvada, Colo.: Kilburn Press, 2002.

Klein, Maury. *The Change Makers.* New York: Henry Holt, 2003.

———. *The Genesis of Industrial America, 1870–1920.* New York: Cambridge University Press, 2007.

Knight, Robert. *Industrial Relations in the San Francisco Bay Area, 1900–1918.* Berkeley: University of California Press, 1960.

Koch, E., ed. "Journal of Peter Koch—1869 and 1870." In *Sources of Northwest History,* no. 5, edited by Paul Phillips. Missoula: State University of Montana, 1929.

Koelbel, Lenora. *Missoula: The Way It Was.* Rev. ed. Missoula, Mont.: Pictorial Histories, 2004.

Kolko, Gabriel. *The Triumph of Conservatism: A Reinterpretation of American History, 1900–1916.* London: Free Press of Glencoe, 1963.

Kyne, Peter. *Valley of the Giants.* New York: Grosset and Dunlap, 1918.

Lamoreaux, Naomi. *The Great Merger Movement in American Business: 1895–1904.* Cambridge: Cambridge University Press, 1985.

Langille, H. D., Fred Plummer, Arthur Dodwell, Theordore Rixon, and John Leiberg. *Forest Conditions in the Cascade Range Forest Reserve, Oregon,* edited by U.S. Geological Survey. Washington, D.C: U.S. Government Printing Office, 1903.

Lapenteur, Charles. *Forty Years a Fur Trader on the Upper* Missouri. Edited by Elliot Coues. Minneapolis, Minn.: Ross and Haines, 1962.

LaPointe, Jaques. *Grande-Rivière: une page d'histoire Acadienne.* Moncton, New Brunswick: Les Éditions D'Acadie, 1989.

Lass, William E. "Elias H. Durfee and Campbell K. Peck: Indian Traders on the Upper Missouri Frontier." *Journal of the West* 43, no 2. (Spring 2004).

———. *A History of Steamboating on the Upper Missouri River.* Lincoln: University of Nebraska Press, 1962.

Lawford, Richard G., Paul B. Alaback, and Eduardo Fuentes, eds. *High-Latitude Rainforests and Associated Ecosystems of the West Coast of the Americas.* New York: Springer, 1996.

Leach, William. *Land of Desire Merchants, Power, and the Rise of a New American Culture.* New York: Pantheon Books, 1993.

Leeson, Michael, ed. *History of Montana 1739–1885.* Chicago: Warner, Beers, 1885.

Lefevre, G. Shaw. *English Commons and Forests.* London: Cassell, 1894.

Lepley, John. *Birthplace of Montana, a History of Ft. Benton.* Missoula, Mont.: Pictorial Histories, 1999.

———. *Packets to Paradise: Steamboating to Fort Benton.* Fort Benton, Mont.: River and Plains Society, 2001.

Levi, Steven. *Committee of Vigilance, the San Francisco Chamber of Commerce Law and Order Committee, 1916–1919: A Case Study in Official Hysteria.* Jefferson, N.C.: McFarland, 1983.

Leydet, Francois. *The Last Redwoods and the Parkland of Redwood Creek.* San Francisco: Sierra Club Books, 1969.

Libecap, Gary D., and Ronald N. Johnson, "Property Rights, Nineteenth-Century Federal Timber Policy, and the Conservation Movement." *Journal of Economic History* 39, no. 1 (Mar. 1979).

Ligon, Franklin, William E. Dietrich, and William J. Trush. "Downstream Ecological Effects of Dams: A Geomorphic Perspective." *Bioscience* 45, no. 3 (1995).

Link, Arthur, and Richard McCormick. *Progressivism.* Wheeling, Ill.: Harland Davidson, 1983.

Lorimer, Craig G. "Eastern White Pine Abundance in 19th Century Forests: A Reexamination of Evidence from Land Survey and Lumber Statistics" *Journal of Forestry* 106, no. 5 (July/Aug. 2008).

Lottich, Kenneth, ed. "My Trip to Montana Territory, 1879." *Montana the Magazine of Western History* 15, no. 1 (1965).

Louma, Jon R. *The Hidden Forest.* New York: Henry Holt, 1999.

Lubetkin, John M. *Jay Cooke's Gamble: The Northern Pacific, the Sioux, and the Panic of 1873.* Norman: University of Oklahoma Press, 2006.

MacNutt, W. S. *New Brunswick: A History 1784–1867.* Toronto, Ontario: Macmillan, 1963.

———. "The Politics of the Timber Trade in Colonial New Brunswick, 1825–40." *Canadian Historical Review,* 30 (1949), 47–65.

Malone, Dumas, and Harris Elwood Starr, eds. *Dictionary of American Biography,* vol. 17. New York: Charles Scribner's Sons, 1944.

Malone, Michael P. *The Battle for Butte: Mining and Politics on the Northern Frontier, 1864–1906.* Seattle: University of Washington Press, 1981.

Malone, Michael P., Richard Roeder, and William Lang. *Montana: A History of Two Centuries.* Seattle: University of Washington Press, 1976.

Marsh, George Perkins. *Man and Nature; or, Physical Geography as Modified by Human Action.* New York: Scribner, 1869.

Mathews, Allan, James. *A Guide to Historic Missoula.* Helena: Montana Historical Society Press, 2002.

Mattison, Ray H. "The Upper Missouri Fur Trade: Its Methods of Operation." *Nebraska History* 42, no. 1 (Mar. 1961).

Maximilian, Prince of Wied. *Travels in the Interior of North America 1832–1834,* vol. 23. Edited by Reuben Thwaites. Cleveland, Ohio: A. H. Clark, 1906.

Maxwell, Lilan. *The History of Central New Brunswick.* Fredericton, New Brunswick: York-Sunbury Historical Society, 1984.

McCay, Bonnie, and James Acheson, eds. *The Question of the Commons: The Culture and Ecology of Communal Resources.* Tucson: University of Arizona Press, 1987.

McDonald, Duncan. "The Nez Perces War of 1877." *New Northwest,* Dec. 27, 1878.

McGerr, Michael. *A Fierce Discontent: The Rise and Fall of the Progressive Movement.* New York: Free Press, 2003.

McKinney, Gage. "A. B. Hammond, West Coast Lumberman." *Journal of Forest History* 28 (Oct. 1984).

———. "A Redwood Giant of the Past." *Merchant* 56 (Mar. 1978).

McQuaig, Linda. *All You Can Eat: Greed, Lust and the New Capitalism.* Toronto, Ontario: Penguin, 2001.

Melendy, H. Brett. "Two Men and a Mill: John Dolbeer, William Carson, and the Redwood Lumber Industry in California." *California Historical Society Quarterly* 38, no. 1 (Mar. 1959).

Mengel, Lowell S., II. "A. B. Hammond Built a Vast Timber Empire." *Humboldt Historian* 36 (Nov./Dec. 1985).

Mercer, L. J. *Railroads and Land Grant Policy.* New York: Academic Press, 1982.

Merk, Frederick. *Fruits of Propaganda in the Tyler Administration.* Cambridge, Mass.: Harvard University Press, 1971.

Messing, John. "Public Lands, Politics, and Progressives: The Oregon Land Fraud Trials, 1903–1910." *Pacific Historical Review* 35, no. 1 (Feb. 1966).

Mills, C. Wright. *The Power Elite.* New York: Oxford University Press, 1959.

Missoula Board of Trade. *Missoula Illustrated.* Missoula, Mont.: Missoula Board of Trade, 1890.

Moore, Jason. "The Modern World-System as Environmental History?" *Theory and Society* 35, no. 3 (June 2003), 307–77.

Monahan, G. *Montana's Wild and Scenic Upper Missouri River.* Anaconda, Mont.: Northern Rocky Mountain Books, 1997.

Morison, Samuel Eliot. *The Oxford History of the American People.* New York: Oxford University Press, 1965.

Montana Historical Society. *Not in Precious Metals Alone: A Manuscript History of Montana.* Helena: Montana Historical Society Press, 1976.

Murphy, Peter, Roger Udell, and Robert Stevenson. *The Hinton Forest 1955–2000: A Case Study in Adaptive Forest Management.* Hinton, Alberta: Foothills Research Institute, 2002.

Nace, Ted. *Gangs of America: The Rise of Corporate Power and the Disabling of Democracy.* San Francisco: Berrett-Koehler, 2003.

Nearing, Scott. *Wages in the United States: 1908–1910.* New York: Macmillan, 1911.

Nelson, William H. *The American Tory.* London: Oxford University Press, 1961.

Nevins, Allan. *Grover Cleveland: A Study in Courage.* New York: Dodd, Mead, 1962.

———. *Study in Power: John D. Rockefeller, Industrialist and Philanthropist,* vol. 1. New York: Charles Scribner's Sons, 1953.

Noble, David. *America by Design: Science, Technology, and the Rise of Corporate Capitalism.* New York: Alfred Knopf, 1979.

Noss, Reed, ed. *The Redwood Forest: History, Ecology, and Conservation of the Coast Redwoods.* Washington, D.C.: Island Press, 2000.

Onstine, Frank. *The Great Lumber Strike of Humboldt County 1935.* Arcata, CA: Mercurial Enterprises, 1980.

Ostler, Jeffrey. *The Plains Sioux and U.S. Colonialism from Lewis and Clark to Wounded Knee.* New York: Cambridge University Press, 2004.

Overholser, Joel. "The Big Muddy Builds Montana." In *Our Fort Benton: The Birthplace of Montana,* edited by Nora Harber. Fort Benton, Mont.: n.d.

———. *Fort Benton: World's Innermost Port.* Fort Benton, Mont.: J. Overholser, 1987.

Palais, Hyman, and Earl Roberts. "The History of the Lumber Industry in Humboldt County." *Pacific Historical Review* 19, no. 1 (Feb. 1950).

Palmer, L. *Steam toward the Sunsets.* Newport, Ore.: Lincoln County Historical Society, 1982.

Paradis, Roger, ed. *Papers of Prudent C. Mercure: Histoire du Madawaska.* Madawaska, Maine: Madawaska Historical Society, 1998.

Peluso, Nancy Lee. *Rich Forests, Poor People: Resource Control and Resistance in Java.* Berkeley: University of California Press, 1992.

Petersen, William J., ed. "The Log of the Henry M. Shreve to Fort Benton in 1869." *Mississippi Valley Historical Review* 31, no 4. (Mar. 1945).

Petts, G. E. *Impounded Rivers: Perspectives for Ecological Management.* Hoboken, N.J.: John Wiley and Sons. 1984.

Phillips, P. C. "The Battle of the Big Hole: An Episode in the Nez Perce War." In *Sources of Northwest History.* Missoula: State University of Montana, 1929.

Phillips, P. C., ed. "Upham Letters from the Upper Missouri, 1865." In *Sources of Northwest History*. Missoula: State University of Montana, 1933.

Pisani, Donald. "Forests and Conservation 1865–1890." *Journal of American History* 72, no. 2 (Sept. 1985).

Poage, Nathan and John Tappeiner, "Long-Term Patterns of Diameter and Basal Area Growth of Old-Growth Douglas-Fir Trees in Western Oregon." *Canada Journal of Forest Research* 32 (2002).

Porter, Glen. *The Rise of Big Business: 1860–1920.* Wheeling, Ill: Harlan Davidson, 1973.

Progressive Men of Montana. Chicago: A. W. Bowen, 1901.

Prouty, Andrew Mason. *More Deadly than War: Pacific Coast Logging 1827–1981.* New York: Garland, 1985.

Prudham, W. Scott. *Knock on Wood: Nature as Commodity in Douglas-Fir Country.* New York: Routledge, 2005.

Puter, Stephen, and Horace Stevens. *Looters of the Public Domain.* Portland, Ore.: Portland Printing House, 1908.

Rader, Benjamin. "The Montana Lumber Strike of 1917." *Pacific Historical Review* 36, no. 2 (May 1967).

Rajala, Richard A. *Clearcutting the Pacific Rainforest.* Vancouver: University of British Columbia Press, 1998.

Rakestraw, Lawrence. *A History of Forest Conservation in the Pacific Northwest, 1891–1913.* New York: Arno Press, 1979.

Raymer, Robert G. *Montana: The Land and the People.* Chicago: Lewis Publishing, 1930.

Raymond, William O. *The River St. John.* Sackville, New Brunswick: Tribune Press, 1950.

Renz, L. T. *The History of the Northern Pacific Railroad.* Fairfield, Wash.: Ye Galleon Press, 1980.

Richardson, Elmo. *The Politics of Conservation: Crusades and Controversies 1897–1913.* Berkeley: University of California Press, 1962.

Robbins, R. *Our Landed Heritage: The Public Domain 1776–1936.* Princeton, N.J.: Princeton University Press, 1942.

Robbins, William G. *Colony and Empire: The Capitalist Transformation of the American West.* Lawrence: University Press of Kansas, 1994.

———. *Hard Times in Paradise: Coos Bay, Oregon, 1850–1986.* Seattle: University of Washington Press, 1988.

———. *Landscapes of Promise: The Oregon Story 1800–1940.* Seattle: University of Washington Press, 1997.

———. *Lumberjacks and Legislators: Political Economy of the U.S. Lumber Industry, 1890- 1941.* College Station: Texas A & M University Press, 1982.

Rodgers, Daniel. *Atlantic Crossings: Social Politics in a Progressive Age.* Cambridge, Mass.: Harvard University Press, 1998.

Roy, William G. *Socializing Capital: The Rise of the Large Industrial Corporation in America.* Princeton, N.J.: Princeton University Press, 1997.

Sacks, Benjamin. *The Carson Mansion and Ingomar Theater.* Fresno, Calif: Fresno Valley Publishers, 1979.

San Francisco Chamber of Commerce. *Law and Order in San Francisco, a Beginning.* San Francisco: San Francisco Chamber of Commerce, 1916.

Schrepfer, Susan. *The Fight to Save the Redwoods: A History of Environmental Reform, 1917–1978.* Madison: University of Wisconsin Press, 1983.

Schwantes, Carlos. *The Pacific Northwest: An Interpretive History.* Lincoln: University of Nebraska Press, 1989.

———. *Railroad Signatures across the Pacific Northwest.* Seattle: University of Washington Press, 1993.

Scott, Kim Allen, ed. *Splendid on a Large Scale: The Writings of Hand Peter Gyllembourg Koch, Montana Territory, 1869–1874.* Helena, Mont.: Bedrock Editions and Drumlummon Institute, 2010.

Scott, Leslie. "History of Astoria Railroad." *Oregon Historical Quarterly* 15, no. 4 (Dec. 1914).

———. "The Yaquina Railroad." *Oregon Historical Quarterly* 16, no. 3 (Sept. 1915).

Scott, Ruth Boydston. *Missoula: Trading Post to Metropolis.* Missoula, Mont.: R. Boydston, 1997.

Senieur, Matilda. "Bismarck to Fort Benton by Steamboat in the Year 1869." *Montana The Magazine of Western History* 2, no. 2 (1952).

Sherow, James. "Workings of the Geodialectic: High Plains Indians and Their Horse in the Region of the Arkansas River Valley, 1800–1870." *Environmental History Review* 16, no. 2 (Summer 1992).

Sklar, Martin, J. *The Corporate Reconstruction of American Capitalism, 1890–1916.* Cambridge: Cambridge University Press, 1988.

Slauson, Keith M., William J. Zielinski, and Gregory W. Holm. *Distribution and Habitat Associations of the Humboldt Marten (*Martes americana humboldtensis*), and Pacific Fisher (*Martes pennanti pacifica*) in Redwood National and State Parks: A Report to Save-the-Redwoods League.* Arcata, CA: U.S. Department of Agriculture Pacific Southwest Research Station, Redwood Sciences Laboratory, 2003.

Smalley, Eugene. *History of the Northern Pacific Railroad.* New York: Putnam, 1883.

Starr, Kevin. *Inventing the Dream: California through the Progressive Era.* New York: Oxford University Press, 1985.

St. Clair, Jeffrey, and Alexander Cockburn. "A Decadent, Old-Growth Timber Baron Is Chopped Down." *High Country News* (Paonia, Colo.), Aug. 21, 1995.

Steen, Harold, ed. *Origins of the National Forests: A Centennial Symposium.* Durham, N.C.: Forest History Society, 1992.

Stindt, Fred, and Guy Dunscomb. *The Northwestern Pacific Railroad.* Redwood City, Calif.: Fred Stindt, 1964.

Stoddard, Bill. "Hammond in the Tillamook." *Columbia River & Pacific Northwest Timberperson* 2, no. 1 (Winter 1983).

Stone, Arthur L. *Following Old Trails*. Missoula, Mont.: M. J. Elrod, 1913.

Strite, Dan. "Hurrah for Garibaldi." *Oregon Historical Quarterly* 77, no. 3 (Dec. 1976).

———. "Up the Kilchis." *Oregon Historical Quarterly* 72, no. 4 (Dec. 1971–Sept. 1972).

Stuart, Granville. *Diary and Sketchbook of a Journey to "America" in 1866 and Return Trip up the Missouri River to Fort Benton, Montana*. Los Angeles: Dawson's Book Shop, 1963.

Sutherland, Thomas. *Howard's Campaign against the Nez Perce Indians*. Portland, Ore.: A. G. Walling, 1878 (facsimile reproduction, 1933).

Swartout, Robert, Jr. "From Kwangtung to the Big Sky: The Chinese Experience in Frontier Montana" *Montana the Magazine of Western History* 38 (1988).

Taylor, Bill. *Rails to Gold and Silver: Lines to Montana's Mining Camps*, vol. 1. Missoula, Mont.: Pictorial Histories, 1999.

Tennsma, Peter D. A., John T. Rienstra, and Mark A. Yeiter. *Preliminary Reconstruction and Analysis of Change in Forest Stand Age Classes of the Oregon Coast Range from 1850 to 1940* (T/N OR-9). Portland, Ore.: Bureau of Land Management, 1991.

Thompson, Dennis Blake. "Hammond's 17: Surviving the Gap." *Tall Timber Short Lines* 80 (Fall 2005).

Thoreau, Henry David. *The Maine Woods*. Princeton, N.J.: Princeton University Press, 1972.

Thorp, Raymond, and Robert Bunker. *Crow Killer: The Saga of Liver-Eating Johnson*. Bloomington: Indiana University Press, 1958.

Toole, John. *The Baron, the Logger, the Miner and Me*. Missoula, Mont.: Mountain Press, 1984.

———. *Men, Money and Power: The Story of the First Interstate Bank of Missoula*. Missoula, Mont.: First Interstate Bank, 1986.

———. *Red Ribbons: A Story of Missoula and Its Newspaper*. Helena, Mont.: Falcon Press, 1989.

Toole, K. Ross. "The Genesis of the Clark-Daly Feud." *Montana Magazine of Western History* 1 (1951).

———. *Montana: An Uncommon Land*. Norman: University of Oklahoma Press, 1959.

Toole, K. Ross, and Edward Butcher. "Timber Depredations on the Montana Public Domain, 1885–1918." *Journal of the West* 7 (July 1968).

Trachtenberg, Allan. *The Incorporation of America*. New York: Hill and Wang, 1982.

Twight, Peter A. *Ecological Forestry for the Douglas Fir Region*. Washington, D.C.: National Parks and Conservation Association, 1973.

Tyler, Robert L. *Rebels of the Woods: The I.W.W. in the Pacific Northwest*. Eugene: University of Oregon, 1967.

Vandergeest, Peter, and Nancy Lee Peluso. "Territorialization and State Power in Thailand." *Theory and Society* 24 (1995).

Van Tassell, Alfred. *Mechanization in the Lumber Industry* (Report M-5). Philadelphia: Works Projects Administration, 1940.

Waldron, Ellis, and Paul Wilson. *Atlas of Montana Elections: 1889–1976.* Missoula: University of Montana, 1978.

Walker, Laurence. *The North American Forests: Geography, Ecology and Silviculture.* Boca Raton, Fla.: CRC Press, 1999.

Warne, F., and Co. *Dictionary of Quotations from Ancient and Modern English and Foreign Sources.* London: F. Warne, 1899.

Weber, Max. *The Protestant Ethic and the Spirit of Capitalism.* New York: Charles Scribner's Sons, 1930.

Webb, William Joe. "The United States Wooden Steamship Program during World War I." *American Neptune* 35 (Oct. 1975).

Weintraub, Hyman. *Andrew Furuseth, Emancipator of the Seamen.* Berkeley: University of California Press, 1959.

Wells, Gail. *The Tillamook: A Created Forest Comes of Age.* Corvallis: Oregon State University Press, 1999.

West, Elliott. *The Contested Plains: Indians, Goldseekers, and the Rush to Colorado.* Lawrence: University Press of Kansas, 1998.

White, Richard. *"It's Your Misfortune and None of My Own": A New History of the American West.* Norman: University of Oklahoma Press, 1991.

———. *Land Use, Environment and Social Change: The Shaping of Island County, Washington.* Seattle: University of Washington Press, 1980.

———. *Railroaded: The Transcontinentals and the Making of Modern America.* New York: W. W. Norton, 2011.

———. "The Winning of the West: The Expansion of the Western Sioux in the Eighteenth and Nineteenth Centuries." *Journal of American History* 65, no. 2 (Sep. 1978).

Whiting, Perry. *The Autobiography of Perry Whiting, Pioneer Building Material Merchant of Los Angeles.* Los Angles: Perry Whiting, 1930.

Widick, Richard. *Trouble in the Forest: California's Redwood Timber Wars.* Minneapolis: University of Minnesota Press, 2009.

Wiebe, Robert. *The Search for Order: 1877–1920.* New York: Hill and Wang, 1967.

Williams, Michael. *Americans and Their Forests: A Historical Geography.* New York: Cambridge University Press, 1989.

Williams, Richard L. *The Loggers.* New York: Time-Life Books, 1976.

Wolff, William Almon. "They're Building Ships Out There." *Collier's Weekly,* May 25, 1918.

Wood, C. *The Northern Pacific: Main Street of the Northwest.* Seattle, Wash.: Superior Publishing, 1968.

Wood, Richard G. *A History of Lumbering in Maine 1820–1861,* Issue 93, University of Maine Studies. Orono: University of Maine, 1971.

Wright, Ernest, and Leo Isaac. *Decay Following Logging Injury to Western Hemlock, Sitka Spruce, and True Firs* (Technical bulletin 1148). Washington, D.C.: U.S. Department of Agriculture, 1956.

Wright, Esther Clark. *Planter and Pioneers: Nova Scotia 1749–1775.* Hantsport, Nova Scotia: E. C. Wright, 1982.

Wright and Woodward's Missoula City Directory, 1890. Missoula, Mont.: Wright and Woodward, 1891.

Wynn, Graeme. "Deplorably Dark and Demoralized Lumberers? Rhetoric and Reality in Early Nineteenth-Century New Brunswick." *Journal of Forest History* 24, no. 4 (Oct. 1980).

———. *Timber Colony: A Historical Geography of Early Nineteenth Century New Brunswick.* Toronto, Ontario: University of Toronto Press, 1981.

Zinn, Howard. *A People's History of the United States.* New York: HarperCollins, 1999.

UNPUBLISHED THESES AND DISSERTATIONS

Barrett, Stephen. "Relationship of Indian-Caused Fires to the Ecology of Western Montana Forests." M.A. thesis, University of Montana, 1981.

Buckley, James Michael. "Building the Redwood Region: The Redwood Lumber Industry and the Landscape of Northern California, 1850–1929." Ph.D. diss., University of California, Berkeley, 2000.

Butcher, Edward. "An Analysis of Timber Depredations in Montana to 1900." Master's thesis, University of Montana, 1967.

Coon, Shirley Jay. "The Economic Development of Missoula, Montana." Ph.D. diss., University of Chicago, 1926.

Cornford, Daniel. "Lumber, Labor, and Community in Humboldt County, CA, 1850–1920." Ph.D. diss., University of California, Santa Barbara, 1983.

Hakola, John. "Samuel T. Hauser and the Economic Development of Montana: A Case Study in Nineteenth-Century Frontier Capitalism." Ph.D. diss., Indiana University, 1961.

Johnson, Dale. "Andrew B. Hammond: Education of a Capitalist on the Montana Frontier." Ph.D. diss., University of Montana, 1976.

Meany, Edmond, Jr. "The History of the Lumber Industry in the Pacific Northwest to 1917." Ph.D. diss., Harvard University, 1935.

Melendy, Howard Brent. "One Hundred Years of the Redwood Lumber Industry 1850–1950." Ph.D. diss., Stanford University, 1952.

Pilon, Beatrice D. "Settlement and Early Development of the Parish of Kingsclear, York County, New Brunswick, 1784–1840" Master's thesis, University of New Brunswick, 1966.

Schwinden, Theodore. "The Northern Pacific Land Grants in Congress." Master's thesis, Montana State University, 1950.

Smith, Dennis. "Fort Peck Agency Assiniboines, Upper Yanktonais, Hunkpapas, Sissetons, and Wahpetons: A Cultural History to 1888." Ph.D. diss., University of Nebraska, 2001.

Toole, Kenneth Ross. "Marcus Daly, a Study of Business in Politics." M.A. thesis, Montana State University, 1948.

WEBSITES

American National Biography Online, www.anb.org

Canadian-American Center, University of Maine, www.umaine.edu/canam/ham/acadiansettlement.

Digital Sanborn Maps, 1867–1970, http://sanborn.umi.com/mt/5054/dateid-000004.htm?CCSI=2790n, accessed Aug. 10, 2013.

General Land Office Records, www.glorecords.blm/gov.

National Park Service Ice Age Floods, www.nps.gov/iceagefloods.

On-Line Institute for Advanced Loyalist Studies, www.royalprovincial.com.

Sailor's Union of the Pacific, www.sailors.org/history.

The Upper St. John River Valley, historical reports, census records, lands grants, etc., www.Upperstjohn.com.

Index

References to illustrations are in italic type.

WHEN MONEY GREW ON TREES